SETS, MATRICES,
AND
LINEAR PROGRAMMING

SETS, MATRICES,
AND
LINEAR PROGRAMMING

Robert L. Childress

School of Business Administration
University of Southern California

Prentice-Hall, Inc., Englewood Cliffs, New Jersey

Library of Congress Cataloging in Publication Data

CHILDRESS, ROBERT L
  Sets, matrices, and linear programming.

  Includes bibliographies.
  1. Set theory.   2. Matrices.   3. Linear
programming.   I. Title.
QA248.C43   519.7′2   73-17313
ISBN   0-13-806737-6

Printed in the United States of America

10   9   8   7   6   5   4   3   2   1

PRENTICE-HALL INTERNATIONAL, INC., *London*
PRENTICE-HALL OF AUSTRALIA, PTY. LTD., *Sydney*
PRENTICE-HALL OF CANADA, LTD., *Toronto*
PRENTICE-HALL OF INDIA PRIVATE LIMITED, *New Delhi*
PRENTICE-HALL OF JAPAN, INC., *Tokyo*

To Carolyn

Contents

Preface

The importance of modern quantitative techniques in the business, economics, and social science curricula is well established. The primary reason for this importance comes from the fact that the techniques are extremely useful in the administration of a business or governmental enterprise. Quantitative methods are used not only in the traditional areas of production and engineering but are also widely employed in the functional areas of finance, marketing, and accounting. For this reason, the competent administrator must understand certain quantitative tools and the associated language. This text provides an introduction to several of the more important of these tools —sets, matrices, and linear programming.

The objectives of this text are twofold. The first is to provide an introduction to the important quantitative tools of sets, matrices, and linear programming in a manner that permits the nonmathematically inclined student to easily grasp the basic mathematical concepts. This introduction should prove quite useful to both the student pursuing a general course of study as well as the student who is interested in studying additional quantitative techniques. A second, and equally important objective, is to provide numerous examples of the applicability of quantitative techniques in the administration of an enterprise. Based on these examples, the reader will hopefully be able to recognize problems in his job that can be solved using quantitative techniques. He will then be in the enviable position of being able to apply the techniques or, alternatively, of directing the work of the specialist assigned to the problem.

Sets, Matrices, and Linear Programming is designed for a one-semester course. By omitting certain chapters, it can also be used as a supplemental text in the introductory quantitative methods course. Certain instructors may wish to omit Chapter 4, entitled "Additional Elements of Matrix Algebra." If the time available for the study of linear programming is severely limited, it may be desirable to concentrate on the formulation of linear programming problems, Chapter 5, and return to the solution techniques in a later course. Similarly, Chapter 7, "Duality and Sensitivity," and Chapter 9, "Integer Programming," can be deferred to a later course.

The book can also be used as a supplementary text in an "operations research" course. Most students will find that the level of presentation of the more difficult mathematical concepts and the number of examples makes the text a welcome supplement in the traditional operations research course.

Numerous students at the University of Southern California and the University of Missouri-St. Louis have contributed helpful suggestions on ways of improving the text. Mr. Robert Hochhalter, a M.B.A. candidate at the University of Southern California, was especially helpful in reviewing the manuscript. I am also indebted to Dr. Warren Erikson of the University of Southern California and Dr. Robert Markland of the University of Missouri-St. Louis for their helpful suggestions and critique. I am especially grateful to Mrs. Glenell Smoot, Miss Chris Kutschinski, and Mrs. Toni Graham, who worked tirelessly on the difficult task of typing and retyping the manuscript.

ROBERT L. CHILDRESS
Los Angeles, California

Chapter 1

Sets

The concept of sets and the algebra of sets is enjoying increasing popularity and usage in business and the sciences. One reason for this popularity is that an understanding of the basic concepts of sets and set algebra provides a form of language through which the business specialist or practitioner can communicate important concepts and ideas to his associates. Specialists in electrical engineering, for example, use set algebra in the design of circuits. At the opposite extreme, specialists in written communications use sets in the analysis of statements and preparation of reports. Students of business administration use sets in the study of probability, statistics, programming. optimization, etc. This chapter provides an introduction to sets and set algebra along with selected applications. The reader can expect additional applications later in this text as well as in his studies of marketing, finance, economics, production, and statistics.

1.1 Sets Defined

A *set* is defined as a collection or aggregate of objects. To illustrate this definition, the members of the reader's quantitative business methods course are a set of students. Similarly, the members of the American Economic Association are a set of individuals, the faculty of the school of business is a set of instructors, and the *Encyclopedia Britannica* is a set of books.

From these examples the student can see that the concept of a set is relatively straightforward. There are, however, certain requirements for the collection or aggregate of objects that constitute the set. These requirements are:

1. The collection or aggregate of objects must be well defined, i.e., we must be able to determine unequivocally whether or not any object belongs to the set.
2. The objects of a set must be distinct; i.e., we must be able to distinguish between the objects and no object may appear twice.
3. The order of the objects within the set must be immaterial, i.e., *a*, *b*, *c* is the same set as *c*, *b*, *a*.

On the basis of these requirements for a set, suppose we are asked to verify that the students in a quantitative analysis class are a set. To determine if the students are a set, we ask three questions. First, can we determine if a student is registered for the course? Second, is it possible to distinguish between the students? Third, is the order of individuals in the class immaterial, i.e., is the class the same set whether arranged alphabetically by student or by social security number? If the answer to all three questions is yes, we conclude that the class is a set.

Example: Verify that the letters in the word Mississippi satisfy the requirement for a set. The letters in this word are m, i, s, and p. These letters are well-defined, distinct, and the order of the letters is immaterial. Therefore, the letters are a set.

Example: A coin is tossed three times. Denoting a head by H and a tail by T, determine the set that represents all possible outcomes of the three tosses.

The possible outcomes are {HHH, HHT, HTH, THH, HTT, THT, TTH, TTT}, where HHH represents a head on the first toss, a head on the second toss, and a head on the third toss. HHT represents a head on the first toss, a head on the second toss, and a tail on the third toss, etc.

Example: The set of digits is defined as the numbers {0, 1, 2, 3, 4, 5, 6, 7, 8, 9}. Determine which of the following numbers are members of this set: 3, 7.2, IV, 137, $\frac{4}{3}$.

The digit 3 is the only member of the set. The reader should be able to verify that the remaining numbers do not belong to the set of digits.

1.2 Specifying Sets and Membership in Sets

It is customary to designate a set by a capital letter. For instance, the set of digits defined in the above example could be designated as D.

The objects that belong to the set are termed the *elements* of the set or *members* of the set. The elements or the members of the set are designated by one of two methods: (1) the roster method, or (2) the descriptive method. The roster method involves listing within braces all members of the set. The descriptive method (also called the defining property method) involves describing the membership in a manner such that one can determine if an object belongs in the set.

To illustrate the specification of sets and membership, consider again the set of digits. The set of digits is designated by the capital letter D. The elements in the set (or alternatively the members of the set) are shown either by listing all the elements in the set within braces or by describing within braces the membership. If the roster method is used, the set of digits would appear as

$$D = \{0, 1, 2, 3, 4, 5, 6, 7, 8, 9\}$$

This is read as "D is equal to that set of elements 0, 1, 2, 3, 4, 5, 6, 7, 8, 9." If the descriptive method of specifying the set is used, the set would be

$$D = \{x \mid x = 0, 1, 2, 3, \ldots, 9\}$$

This is read as "D is equal to that set of elements x such that x equals 0, 1, 2, 3, . . . , 9." In interpreting the symbolism used in set notation, it is useful to think of the left brace as shorthand for "that set of elements" and the vertical line as shorthand for "such that." Commas are used to separate the elements, and the raised periods mean "continuing in the established pattern." The 9 in the above set is interpreted as the final number of the set. The right brace designates set completion.

Example: The positive integers or "natural numbers" are the numbers 1, 2, 3, 4, 5, ... Show the set of natural numbers.

Assume that the set of natural numbers is represented by N. We cannot, of course, list all of the members of the set. We therefore use the descriptive method of specifying set membership and write

$$N = \{x \mid x = 1, 2, 3, 4, 5, \ldots\}$$

Example: Develop the set notation for the English alphabet.

We can use either the roster method or the descriptive method of specifying set membership. Representing the set of letters in the English alphabet by A, the set is

$$A = \{a, b, c, d, e, f, g, h, i, j, k, l, m, n, o, p, q, r, s, t, u, v, w, x, y, z\}$$

The descriptive method would conserve some space,

$$A = \{a, b, c, \ldots, y, z\}$$

Either method is acceptable. When one is using the descriptive method, however, it is important to remember that the description must be sufficient for one to determine the membership of the set. The elements must be well-defined, distinct, and order must be immaterial.

Example: The possible convention sites for the Western Farm Equipment Association are Los Angeles, San Francisco, Phoenix and Las Vegas. The set of convention sites is

$$S = \{\text{Los Angeles, San Francisco, Phoenix, Las Vegas}\}$$

1.2.1 SET MEMBERSHIP

The Greek letter ϵ (epsilon) is customarily used to indicate that an object belongs to a set. If A again represents the set of letters in the English alphabet, then $a \epsilon A$ means that a is an element of the alphabet. The symbol ϵ (epsilon with a slashed line) represents nonmembership. We could thus write $\alpha \epsilon A$, meaning that alpha is not a member of the English alphabet. Similarly, referring to the convention site set S, we can write Los Angeles ϵS and San Diego ϵS.

1.2.2 FINITE AND INFINITE SETS

A set is termed *finite* or *infinite*, depending upon the number of elements in the set. The set A defined above is finite since it has 26 members, the letters in the English alphabet. The set D is also finite, since it has only the ten digits. The set N of positive integers or natural numbers is infinite, since the process of counting continues infinitely.

Example: Rational numbers are defined as that set of numbers a/b, where a represents all integers, both positive and negative including 0, and b represents all positive and negative integers, excluding 0. Develop the set R of rational numbers and specify whether the set is finite or infinite.

Letting R represent the set of rational numbers, we have

$$R = \{a/b \mid a = \text{all integers including } 0, b = \text{all integers excluding } 0\}$$

The set of rational numbers is infinite. Examples of rational numbers include any number that can be expressed as the ratio of two positive or negative whole numbers, such as $\frac{3}{2}$, $-\frac{6}{2}$, 7, etc.

Example: The individuals who are members of the American Economic

Association comprise a set. Assuming that a list of the membership is available, we could write the set as

$$S = \{\text{all members of the American Economic Association}\}$$

This set is finite, although quite large. It is a set because the elements are well-defined, distinct, and of inconsequential order.

1.3 Set Equality and Subsets

Two sets P and Q are said to be equal, written $P = Q$, if every element in P is in Q and every element in Q is in P. Set equality thus requires all elements of the first set to be in a second set and all elements in the second set to be in the first set. As an example, consider the set

$$P = \{0, 1, 2, 3, 4, 5\} \quad \text{and} \quad Q = \{2, 0, 1, 3, 5, 4\}$$

These sets are equal, since every element in P is in Q and every element in Q is in P.

The student will often have occasion to consider only certain elements of a set. These elements form a *subset* of the original set. As an example, assume that S represents the stockholders of company XYZ. Those stockholders who are employees of the company represent a subset of S. Thus, if Mr. Jones is a stockholder of XYZ and is employed by XYZ, he is a member of the subset of stockholders who are employed by the company.

Subset can be defined as follows. A set R is a subset of another set S if every element in R is in S. For example, if $S = \{0, 1, 2, 3, 4\}$ and $R = \{0, 1, 2\}$, then every element in R is in S, and R is a subset of S. The symbol for subset is $\subseteq$. R is a subset of S is written $R \subseteq S$.

Example: Let A represent the letters in the English alphabet and C represent the letters in the word "corporation." Verify that C is a subset of A.

Since $A = \{a, b, c, d, e, f, g, h, i, j, k, l, m, n, o, p, q, r, s, t, u, v, w, x, y, z\}$ and $C = \{c, o, r, p, a, t, i, n\}$, we see that $C \subseteq A$.

Example: Let D represent the set of digits and I represent the set of all integers including 0. Since $D = \{0, 1, 2, 3, 4, 5, 6, 7, 8, 9\}$ and $I = \{\ldots, -5, -4, -3, -2, -1, 0, 1, 2, 3, 4, 5, \ldots\}$ the finite set D is a subset of the infinite set I, i.e., $D \subseteq I$.

In the preceding example D was defined as the set of all digits including 0. We previously defined N as the set of all positive integers excluding 0. Assume that one is interested in determining if D is a subset of N. By inspection of the sets, D includes the element 0, whereas N does not. Consequently, D is not a subset of N. This is written as $D \nsubseteq N$. The notation for subset $\subseteq$ and not subset $\nsubseteq$ parallels the notation for member ϵ and not member $\notin$.

1.3.1 PROPER SUBSETS

The term *subset* is often differentiated from that of *proper subset.* A proper subset is designated by the symbol $\subset$. A proper subset P is a subset of another set U, written $P \subset U$, if all elements in P are in U but all elements in U are not in P. This simply means that for P to be a proper subset of U, then U must have all elements that are in P plus at least one element that is not in P. As an example, if

$$S = \{0, 1, 2, 3, 4\}$$

and

$$R = \{0, 1, 2\}$$

then R is a proper subset of S, i.e., $R \subset S$.

Example: Verify that the set C, the letters in corporation, is a proper subset of A, the letters in the English alphabet. Since every letter in C is in A, but every letter in A is not in C, we conclude that $C \subset A$.

1.3.2 UNIVERSAL SET

In discussing sets and subsets, the term *universal set* is often encountered. The term "universal set" is applied to the set that contains all the elements the analyst will wish to consider. If, for example, we are interested in categorizing the stockholders of the XYZ Company, the universal set would be all stockholders of the XYZ Company. The various categories of stockholders would then be subsets of the universal set. Similarly, if the analyst were interested in certain combinations of letters, the universal set would be defined as A, the letters of the English alphabet. It would then be possible to specify various subsets of the univeral set A, such as C, the letters in the word corporation.

The universal set contains all elements under consideration. In contrast, the *null* set is defined as a set that has no elements or members. To illustrate, assume that three students are to take an exam. We define the universal set as consisting of the students who score an A on the exam. If we refer to the students as S_1, S_2, and S_3, the results of the test could be $\{S_1, S_2, S_3\}$, $\{S_1, S_2\}$, $\{S_1, S_3\}$, $\{S_2, S_3\}$, $\{S_1\}$, $\{S_2\}$, $\{S_3\}$, $\{\ \}$. Note that there are eight possible outcomes and these are shown as subsets. One of the subsets is the null set $\{\ \}$. The null set, also referred to as the *empty* set, is designated by the Greek letter ϕ (phi). In this example of eight possible subsets, one of the possibilities is the universal set $\{S_1, S_2, S_3\}$ and another is the null set ϕ. This illustrates that both the universal set and the null sets are included as subsets of the universal set. This concept is also illustrated by the following examples.

Example: A stock market analyst is concerned with the price movement of the stock of the three large automobile manufacturers. Using the ticker symbols of Chrysler, Ford, and General Motors, determine the universal set representing upward movements in price of the stocks and specify all possible subsets.

The universal set representing upward price movement in all three stocks is $U = \{C, F, GM\}$. The possible subsets are $\{C, F, GM\}$, $\{C, F\}$, $\{C, GM\}$, $\{F, GM\}$, $\{C\}$, $\{F\}$, $\{GM\}$, $\{\ \}$.

We again note that there are eight possible subsets. One of the subsets is the universal set $U = \{C, F, GM\}$ and another is the null set $\phi = \{\ \}$.

Example: A businessman, vacationing at Santa Anita Racetrack, is considering betting on the first three races. If we denote betting on race 1 as R_1, betting on race 2 as R_2, etc., the universal set is $U = \{R_1, R_2, R_3\}$. The possible subsets are $\{R_1, R_2, R_3\}$, $\{R_1, R_2\}$, $\{R_1, R_3\}$, $\{R_2, R_3\}$, $\{R_1\}$, $\{R_2\}$, $\{R_3\}$, $\{\ \}$.

1.3.3 COUNTING SUBSETS

The number of possible subsets of the universal set can be calculated through the use of a straightforward formula. The number of possible subsets is given by the formula

$$N = 2^n \tag{1.1}$$

where n represents the number of elements in the universal set and N is the number of possible subsets. If the universal set contains three members, there are eight possible subsets. Similarly, for a universal set with four members there are $N = 2^4 = 16$ possible subsets.

1.4 Set Algebra

Set algebra consists of certain operations on sets whereby the sets are combined to produce other sets. As an example, consider a group of students who are enrolled in introductory quantitative methods and another group who are enrolled in introductory finance. From these two sets we can specify, using the algebra of sets, a third set containing as members those individuals who are enrolled in both courses. These operations are most easily illustrated through use of the Venn diagram.

1.4.1 VENN DIAGRAM

The *Venn diagram*, named after the English logician John Venn (1834–83), consists of a rectangle that conceptually represents the universal set. Subsets of the universal set are represented by circles drawn within the rectangle or universal set. In Fig. 1.1 the universal set U is represented by the rectangle

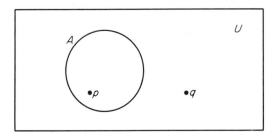

Figure 1.1

and the subset A by the circle. The Venn diagram shows that p is a member of A and that q is not a member of A. Both p and q are members of the universal set.

1.4.2 COMPLEMENTATION

The first set operation we consider is that of *complementation*. Let P be any subset of a universal set U. The complement of P, denoted by P' (read "P complement"), is the subset of elements of U that are not members of P. The complement of P is indicated by the shaded portion of the Venn diagram in Fig. 1.2.

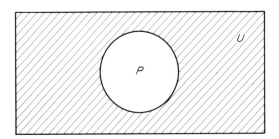

Figure 1.2

Example: For the universal set $D = \{0, 1, 2, 3, 4, 5, 6, 7, 8, 9\}$ with the subset $P = \{0, 1, 3, 5, 7, 9\}$, determine P'.

The complement of P contains all elements in D that are not members of P. Thus, $P' = \{2, 4, 6, 8\}$.

1.4.3 INTERSECTION

A second set operation is *intersection*. Again, let P and Q be any subsets of a universal set U. The intersection of P and Q, denoted by $P \cap Q$ (read "P intersect Q"), is the subset of elements of U that are members of both P and Q. $P \cap Q$ is shown by the shaded area of Fig. 1.3.

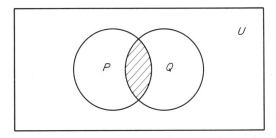

Figure 1.3

Example: For the universal set $D = \{0, 1, 2, 3, 4, 5, 6, 7, 8, 9\}$ with sub-sets $P = \{0, 1, 3, 5, 7, 9\}$ and $Q = \{0, 2, 3, 5, 9\}$, determine the intersection of P and Q.

The intersection of P and Q is the subset that contains the elements in D that are simultaneously in P and Q. Thus,

$$P \cap Q = \{0, 3, 5, 9\}$$

With a little thought the reader can also recognize that

$$P \cap U = P$$

and

$$P \cap P' = \phi$$

Sets such as $P \cap P' = \phi$ which have no common members are termed *disjoint* or *mutually exclusive*.

1.4.4 UNION

A third set operation is *union*. If we again let P and Q be any subsets of a universal set U, then the union of P and Q, denoted by $P \cup Q$ (read "P union Q"), is the set of elements of U that are members of either P or Q. $P \cup Q$ is shown by the shaded portion of the Venn diagram in Fig. 1.4.

Example: For the universal set $D = \{0, 1, 2, 3, 4, 5, 6, 7, 8, 9\}$ with sub-sets $P = \{0, 1, 3, 5, 7, 9\}$ and $Q = \{0, 2, 3, 5, 9\}$, determine the union of P and Q.

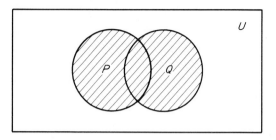

Figure 1.4

The union of P and Q is the subset that contains the elements in D that are in P or in Q. Thus,

$$P \cup Q = \{0, 1, 2, 3, 5, 7, 9\}$$

From the definitions given it also follows that

$$P \cup U = U$$
$$P \cup P' = U$$

1.4.5 OTHER SET OPERATIONS

Two additional set operations are sometimes included in the algebra of sets. The two operations are *difference* and *exclusive union*. Since subsets formed by either of these set operations can also be formed by use of complementation, intersection, and union, these operations are often excluded in discussing set algebra.

Let P and Q be any subsets of the universal set U. The difference of P and Q, denoted by $P - Q$ (read "P minus Q"), is the subset that consists of those elements that are members of P but are not members of Q. This subset is shown in the Venn diagram in Fig. 1.5. The difference of $P - Q$ can also be expressed as $P \cap Q'$.

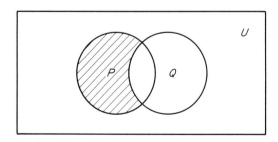

Figure 1.5

The exclusive union of P and Q, denoted by $P \underline{\cup} Q$ (read "P exclusive union Q"), is the set of elements of U that are members of P or of Q but not of both. The subset is shown in the Venn diagram. The subset can also be expressed as $(P \cap Q)' \cap (P \cup Q)$ or $(Q' \cap P) \cup (P' \cap Q)$.

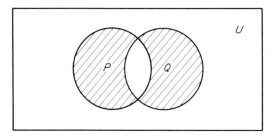

Figure 1.6

1.4.6 COMBINING SET OPERATIONS

Part of the utility of the algebra of sets occurs because of the ability to combine two or more sets into new sets through the use of the set operations. As an example, consider a national firm with district offices. The accounting system of this firm is computerized, and the various accounts are stored in a data bank accessible to the computer. The collection department manager at the national office is concerned with accounts that are sixty days past due. Accounts over sixty days past due are subsets of the set of accounts receivable. In analyzing the accounts due in both the western and southwestern districts, the manager would request the union of the two subsets of past due accounts. If he were interested in customers who simultaneously had accounts past due in both districts, the manager would request the intersection of the subsets.

The use of set algebra to form sets can be illustrated by Fig. 1.7. This

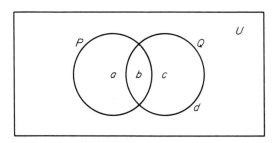

Figure 1.7

figure consists of a Venn diagram, representing the universal set, and two subsets P and Q. Areas in the Venn diagram are shown by the letters a, b, c, and d. The set that consists of areas a, b, c, and d is the universal set U, while the set P consists of a and b, the set Q consists of b and c, etc. Using this notation, we can use the algebra of sets to construct all possible subsets. The student should carefully study Fig. 1.7 and the construction of subsets using the set operations describing the areas a, b, and c presented in Table 1.1

Table 1.1

Area	Set
a, b, c, d	U
a, b	P
b, c	Q
a, d	Q'
c, d	P'
b	$P \cap Q$
a, b, c	$P \cup Q$
d	$(P \cup Q)'$
a	$P \cap Q'$ or $P - Q$
c	$P' \cap Q$ or $Q - P$
a, c	$(P \cap Q') \cup (P' \cap Q)$ or $(P \cap Q)' \cap (P \cup Q)$, or $(P \cup Q)$, or $(P - Q) \cup (Q - P)$
a, c, d	$(P \cap Q)'$

before proceeding.

The construction of subsets using the algebra of sets and Venn diagrams is further illustrated by the following examples.

Example: Business Month is developing a profile of its subscribers. Of the information requests returned by the subscribers to the business magazine, the following data were obtained: 30 percent were in a service industry, 40 percent were self-employed, 20 percent sold through retail channels. Of those who were self-employed, 40 percent were in the service industry, 20 percent were in a retail business, and 10 percent were in a retail service business. The response also indicated that 50 percent of the retail businesses were service-oriented. From these data, develop a Venn diagram that shows the reader profile as subsets of the response set.

The circles shown in the Venn diagram in Fig. 1.8 represent the service industry S, self-employed E, and retail business R. From the data we know that $S = 30$, $E = 40$, and $R = 20$. Since 40 percent of the self-employed are in the service industry, it follows that $E \cap S = 16$. Similarly, $E \cap R = 8$ and $E \cap R \cap S = 4$. We also know from the data that $R \cap S = 10$.

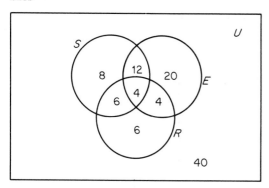

Figure 1.8

The example can be expanded to give the reader additional practice in the use of set operations to specify subsets. This is done by developing set notation for statements describing subsets of the *Business Month* Venn diagram. The statements and subsets are given in Table 1.2 and refer to Fig. 1.8. This table and figure also deserve careful study.

Table 1.2

Statement	*Set*	*Number*
1. Self-employed in retail service	$E \cap R \cap S$	4
2. Self-employed in retail nonservice	$E \cap R \cap S'$	4
3. Self-employed in nonretail nonservice	$E \cap R' \cap S'$	20
4. Self-employed in nonretail service	$E \cap R' \cap S$	12
5. Non-self-employed in nonretail service	$E' \cap R' \cap S$	8
6. Non-self-employed in retail service	$E' \cap R \cap S$	6
7. Non-self-employed in retail nonservice	$E' \cap R \cap S'$	6
8. Non-self-employed, nonservice, nonretail	$E' \cap R' \cap S'$	40
9. Service or self-employed	$S \cup E$	54
10. Nonretail service or nonretail self-employed	$(R' \cap S) \cup (R' \cap E)$	40
11. Non-self-employed in service	$E' \cap S$	14

Example: The Venn diagram shown in Fig. 1.9 is subdivided into areas $a_1, a_2, a_3, a_4, a_5,$ and a_6. The areas are described by the subsets given as follows.

Area	*Set*
a_1, a_2, a_3, a_4	$P \cup Q$
a_5	$R \cap (P \cup Q)'$
a_3, a_4	$(P \cap R) \cup (Q \cap R)$
a_1, a_2, a_5	$(R' \cap P) \cup (R' \cap Q) \cup ((P \cup Q)' \cap R)$

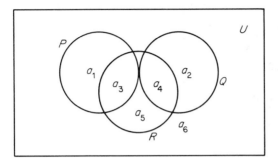

Figure 1.9

1.4.7 LAWS OF THE ALGEBRA OF SETS

The algebra of sets is composed of certain laws. In some cases these laws are similar to the algebra of numbers. We shall state these laws in the form of postulates, thus relieving the reader of the burden of studying mathematical proofs of the laws. The less obvious postulates are illustrated with Venn diagrams.

Postulate 1: $P \cup Q = Q \cup P$
Postulate 2: $P \cap Q = Q \cap P$

Postulates 1 and 2 are termed the *commutative law*. These postulates state that the order in which we combine the union or intersection of two sets is immaterial. This corresponds to the commutative law in ordinary algebra, which states that $p + q = q + p$ and that $p \cdot q = q \cdot p$.

Postulate 3: $P \cup (Q \cup R) = (P \cup Q) \cup R$
Postulate 4: $P \cap (Q \cap R) = (P \cap Q) \cap R$

Postulates 3 and 4 are termed the *associative law*. These postulates state that the selection of two of three sets for grouping in a union or intersection is immaterial. Thus, the order in which the sets are combined is immaterial. These postulates correspond to the associative law in ordinary algebra, which enables us to state that $p + (q + r) = (p + q) + r$ and that $p \cdot (q \cdot r) = (p \cdot q) \cdot r$.

Postulate 5: $P \cup (Q \cap R) = (P \cup Q) \cap (P \cup R)$
Postulate 6: $P \cap (Q \cup R) = (P \cap Q) \cup (P \cap R)$

Postulates 5 and 6 are called the *distributive law*. Postulate 5 has no analogous postulate in ordinary algebra. Postulate 6 corresponds to the distributive law in ordinary algebra that enables us to state that $p(q + r) = p \cdot q + p \cdot r$.

Since Postulates 5 and 6 are not as obvious as Postulates 1 through 4, we shall verify one of them through Venn diagrams. To verify Postulate 5, we must show that the area represented by $P \cup (Q \cap R)$ is the same as that represented by $(P \cup Q) \cap (P \cup R)$. The area representing the set $P \cup (Q \cap R)$ is shown in Fig. 1.10(a). In Fig. 1.10(a) the intersection of Q

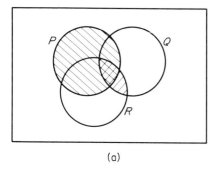
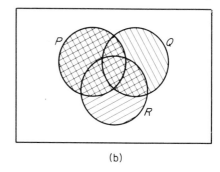

(a)   (b)

Figure 1.10

and R is shown by the shading from lower left to upper right, and P is shown by shading from upper left to lower right. The union of P with $Q \cap R$, represented by both the shaded and crosshatched area in the figure, gives $P \cup (Q \cap R)$.

The area representing the set $(P \cup Q) \cap (P \cup R)$ is shown in Fig. 1.10(b). In Fig. 1.10(b) the union of P and Q is shown by shading from upper left to lower right, and the union of P and R is shown by shading from lower left to upper right. The intersection of the two shaded areas gives the crosshatched area described as the set $(P \cup Q) \cap (P \cup R)$. Since the shaded and crosshatched area in Fig. 1.10(a) equals the crosshatched area in Fig. 1.10(b), we conclude that Postulate 5 is true. Postulate 6 is verified in the same manner as Postulate 5.

Postulate 7: $P \cap P = P$
Postulate 8: $P \cup P = P$
Postulate 9: $P \cup \phi = P$

Postulates 7 through 9 follow directly from the operations of union and intersection. They can be verified by the Venn diagram in Fig. 1.11.

Postulate 10: $P \cap U = P$
Postulate 11: $P \cup P' = U$
Postulate 12: $P \cap P' = \phi$

Postulates 10 through 12 are based upon the definition of the null set, the universal set, and the complement. These postulates are obvious and can also easily be verified by the Venn diagram shown in Fig. 1.11.

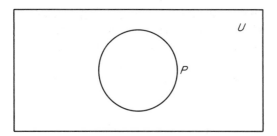

Figure 1.11

Postulate 13: $(P \cup Q)' = P' \cap Q'$
Postulate 14: $(P \cap Q)' = P' \cup Q'$

Postulates 13 and 14 are termed De Morgan's law. Postulate 13 states that the complement of the union of two sets is equal to the intersection of the complement of each set. Postulate 14 states that the complement of the intersection of two sets is equal to the union of the complement of each set. We shall verify Postulate 13 through the use of Venn diagrams. The student is asked to verify Postulate 14.

Figure 1.12(a) shows the set $(P \cup Q)'$ as the area shaded from upper left

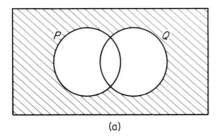 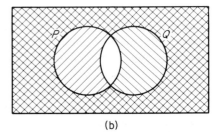

(a)   (b)

Figure 1.12

to lower right. In Fig. 1.12(b), P' is shown by shading from upper left to lower right and Q' is shown by shading from lower left to upper right. The intersection of P' with Q' is shown by the crosshatched area. Since the shaded area in Fig. 1.12(a) equals the crosshatched area in Fig. 1.12(b), it follows that $(P \cup Q)' = P' \cap Q'$.

The laws of set algebra are used analagously with the laws of ordinary algebra. Just as the laws of ordinary algebra can be used to simplify algebraic expressions, the laws of set algebra can be used to simplify sets. This is illustrated by the following examples.

Example: Simplify the set $(A \cup B) \cup (A \cap B)$.

1. Let $P = (A \cup B)$: $P \cup (A \cap B)$
2. Postulate 5: $(P \cup A) \cap (P \cup B)$
3. Substitute for P: $((A \cup B) \cup A) \cap ((A \cup B) \cup B)$
4. Postulate 1: $(A \cup A \cup B) \cap (A \cup B \cup B)$
5. Postulate 7: $(A \cup B) \cap (A \cup B)$
6. Postulate 7: $A \cup B$

In this example, it is important for the student to use parentheses to enclose $P = (A \cup B)$. This is further illustrated by the following example.

Example: On page 11 and again on page 12 we stated that $(P \cap Q)' \cap (P \cup Q)$ was equal to $(Q' \cap P) \cup (P' \cap Q)$. Using the laws of set algebra, verify this relationship.

Given: $(P \cap Q)' \cap (P \cup Q)$
1. Postulate 14: $(P' \cup Q') \cap (P \cup Q)$
2. Let $R = (P' \cup Q')$ $R \cap (P \cup Q)$
3. Postulate 6: $(R \cap P) \cup (R \cap Q)$
4. $(P' \cup Q') = R$ $((P' \cup Q') \cap P) \cup ((P' \cup Q') \cap Q)$
5. Postulate 6: $(P \cap P') \cup (P \cap Q') \cup (P' \cap Q) \cup$
           $(Q \cap Q')$
6. Postulate 12: $(P \cap Q') \cup (P' \cap Q)$

Thus

$$(P \cap Q)' \cap (P \cup Q) = (P \cap Q') \cup (P' \cap Q)$$

Example: Simplify the expression $P \cup (P' \cap Q)$.

Given: $P \cup (P' \cap Q)$
1. Postulate 5: $(P \cup P') \cap (P \cup Q)$
2. Postulate 11: $U \cap (P \cup Q)$
3. Postulate 10: $P \cup Q$

The expression simplifies to $P \cup Q$.

Example: Show that the set $(P \cup Q') \cap (P \cap Q)'$ equals Q'.

Given: $(P \cup Q') \cap (P \cap Q)'$
1. Postulate 14: $(P \cup Q') \cap (P' \cup Q')$
2. Postulate 5: $Q' \cup (P \cap P')$
3. Postulate 12: Q'

The primary difficulty in understanding this example is in converting $(P \cup Q') \cap (P' \cup Q')$ to $Q' \cup (P \cap P')$. This step involves rewriting (Postulate 1) the expression as $(Q' \cup P) \cap (Q' \cup P')$. Postulate 5, the distributive law, is applied to write the expression as $Q' \cup (P \cap P')$. Since $P \cap P'$ is equal to ϕ, the expression reduces to Q'.

1.5 Cartesian Product

Suppose that one is asked to list the possible outcomes of two tosses of a coin. Since either a head or a tail occurs on a single toss, the possible outcomes are described by the set $O = \{(H, H), (H, T), (T, H), (T, T)\}$. There are four elements in this set and, corresponding to the requirements given for a set, the order of the elements is immaterial. Are each of the elements, however, distinct? To answer this question we must ask if the element (H, T) differs from the element (T, H). The answer is that these elements do differ, since the order of the occurrence is important. If, for instance, an individual bet \$1 that a head would occur on the first toss of the coin and \$1 that a tail would occur on the second toss, he would be \$2 richer if element (H, T) occurred and \$2 poorer if (T, H) occurred.

The elements (H, H), (H, T), (T, H) and (T, T) are examples of *ordered pairs*. One of the two components of each ordered pair is designated as the first element of the pair, and the other, which need not be different from the first, is designated as the second element. If the first element is designated as a and the second element is designated as b, we have the ordered pair (a, b). This ordered pair differs from the ordered pair (b, a), and both ordered pairs differ from the set $\{a, b\}$.

Ordered pairs are formed by the *Cartesian product* of two sets. If A and B are two sets, the Cartesian product of the sets, designated by $A \times B$, is the set containing all possible ordered pairs (a, b) such that $a \in A$ and $b \in B$. If the set A contains the elements a_1, a_2, a_3 and the set B contains the elements b_1 and b_2, the Cartesian product $A \times B$ is the set $A \times B = \{(a_1, b_1), (a_1, b_2), (a_2, b_1), (a_2, b_2), (a_3, b_1), (a_3, b_2)\}$. All possible ordered pairs are included in the set.

The concept of the Cartesian product is quite useful in many decision problems. The student of probability and statistics will often be asked to consider the possible outcomes of an experiment. As an example, consider again the problem of determining all possible outcomes of two tosses of a coin. If we define the outcome of the first toss as the set $O_1 = \{H, T\}$ and the outcome of the second toss as the set $O_2 = \{H, T\}$, then the Cartesian product $O_1 \times O_2$ gives all possible outcomes of the two tosses. As we have seen, these outcomes are $O_1 \times O_2 = \{(H, H), (H, T), (T, H), (T, T)\}$.

The Cartesian product of two sets can be determined quite easily with the aid of a box diagram. Figure 1.13 shows the box diagram for the Cartesian

$$O_2$$

		H	T
	H	H, H	H, T
O_1	T	T, H	T, T

Figure 1.13

product of O_1 and O_2. The method of constructing the diagram involves listing the elements of O_1 to the left of the box and O_2 above the box. The blanks in the box are then filled in with the ordered pairs. The Cartesian product $O_1 \times O_2$ consists of the elements in the box, i.e., (H, H), (H, T), (T, H), (T, T).

The Cartesian product can be expanded to combine more than two sets. This means that the concepts discussed for ordered pairs can, for example, be applied to *ordered triplets*. To illustrate, assume that we are asked to list all possible outcomes of three tosses of a coin. Denoting $O_i = \{H, T\}$, where O_i represents the possible outcomes on the *i*th toss (i.e., $i = 1$ represents the first toss, $i = 2$ represents the second toss, etc.), the possible outcomes would be given by the Cartesian product $O_1 \times O_2 \times O_3$. This Cartesian product is determined by finding $O_1 \times O_2$ and then $(O_1 \times O_2) \times O_3$. From Fig. 1.13, we know that $O_1 \times O_2 = \{(H, H), (H, T), (T, H), (T, T)\}$. $(O_1 \times O_2) \times O_3$ is shown in Fig. 1.14. The Cartesian product of $O_1 \times O_2 \times$

$$O_1 \times O_2$$

		H, H	H, T	T, H	T, T
	H	H, H, H	H, H, T	H, T, H	H, T, T
O_3	T	T, H, H	T, H, T	T, T, H	T, T, T

Figure 1.14

O_3 is $\{(H, H, H), (H, T, H), (T, H, H), (T, T, H), (H, H, T), (H, T, T), (T, H, T), (T, T, T)\}$.

The box diagrams of Figs. 1.13 and 1.14 provide a straightforward method of determining of the Cartesian products of sets. With some practice, the student can apply this concept without difficulty. Before illustrating the concept with examples, however, let us carry the Cartesian product one additional step. Assume that we are asked to list all possible outcomes of four tosses of a coin. There are 16 such outcomes. Most individuals would be extremely hard pressed to think of all sixteen. If we use the concept of the Cartesian product, the task becomes routine. The possible outcomes are given by the set $O_1 \times O_2 \times O_3 \times O_4$. The box diagram can be expressed in terms of $(O_1 \times O_2) \times (O_3 \times O_4)$ or by any other grouping of the parentheses. Figure 1.15 shows the Cartesian product of $(O_1 \times O_2) \times (O_3 \times O_4)$.

		$O_3 \times O_4$			
		H, H	H, T	T, H	T, T
$O_1 \times O_2$	H, H	H, H, H, H	H, H, H, T	H, H, T, H	H, H, T, T
	H, T	H, T, H, H	H, T, H, T	H, T, T, H	H, T, T, T
	T, H	T, H, H, H	T, H, H, T	T, H, T, H	T, H, T, T
	T, T	T, T, H, H	T, T, H, T	T, T, T, H	T, T, T, T

Figure 1.15

The elements in the box diagram such as (H, H, H, T) contain four members. Mathematicians call such an element an ordered "4-tuple." Using this term, an ordered pair could be referred to as an ordered 2-tuple, an ordered triplet as an ordered 3-tuple, etc. In general, then, an element containing n members is referred to as an ordered n-tuple.

Example: A retailer specializes in three products: color television, black and white television, and stereos. He offers a service contract with the sale of each of the products, which the customer may or may not elect to purchase. Determine the possible combination of sales options.

Let the products be represented by the set $P = \{C, B, S\}$ and the sales contract by $R = \{E, E'\}$. The Cartesian product of $P \times R$ gives the combination of sales options. The box diagram shows the elements of $P \times R$.

		P		
		C	B	S
R	E	E, C	E, B	E, S
	E'	E', C	E', B	E', S

Example: A builder has three basic floor plans: single story, two story, and trilevel. Each of these plans can have either a shake roof or a wood shingle roof. In addition, the plans are available with or without fireplaces. Determine the number of combinations of plans and show these plans in a box diagram.

Let the basic floor plans be represented by the set $F = \{1, 2, 3\}$, the roofing material by the set $R = \{S, W\}$, and the fireplace option by the set $O = \{f, f'\}$. The combination of plans is represented by the set $\{F \times R \times O\}$. The box diagram for the set is shown below.

F

		1	2	3
R	S	S, 1	S, 2	S, 3
	W	W, 1	W, 2	W, 3

$F \times R$

O f	f, S, 1	f, S, 2	f, S, 3	f, W, 1	f, W, 2	f, W, 3
f'	f', S, 1	f', S, 2	f', S, 3	f', W, 1	f', W, 2	f', W, 3

There are twelve combinations of plans.

Example: An advertising agency is placing ads for three products. The media available for advertising are radio, television, and newspaper. The ads will be written by either the agency or the sponsor. Develop the possible combinations with the aid of a box diagram.

Let the products be represented by the set $P = \{1, 2, 3\}$, the media by the set $M = \{R, T, N\}$, and the source of the advertisement by the set $S = \{A, A'\}$. The box diagram for $P \times M \times S$ is constructed as follows:

M

		R	T	N
	1	1, R	1, T	1, N
P	2	2, R	2, T	2, N
	3	3, R	3, T	3, N

$M \times P$

	1, R	1, T	1, N	2, R	2, T	2, N	3, R	3, T	3, N
A	A, 1, R	A, 1, T	A, 1, N	A, 2, R	A, 2, T	A, 2, N	A, 3, R	A, 3, T	A, 3, N
A′	A′, 1, R	A′, 1, T	A′, 1, N	A′, 2, R	A′, 2, T	A′, 2, N	A′, 3, R	A′, 3, T	A′, 3, N

S labels the two rows (A and A′).

The total number of members formed by the Cartesian product of two sets is given by the product of the number of elements in each set. Thus if set A contains five elements and set B contains four elements, then the Cartesian product $A \times B$ contains 5(4) = 20 elements. The rule applies to the Cartesian product of more than two sets. For the three sets A, B, and C, containing N_1, N_2 and N_3 elements respectively, the Cartesian product of the sets $A \times B \times C$ contains $N_1 \cdot N_2 \cdot N_3$ elements.

Example: Set A has 10 elements, set B has 6 elements, set C has 12 elements, and set D has 3 elements. Determine the number of elements in the Cartesian product of $A \times B \times C \times D$.

The number of elements is given by product of the number of elements in each set. Thus the Cartesian product contains $10 \cdot 6 \cdot 12 \cdot 3$, or 2160 elements.

1.6 Applications to Logic†

The algebra of sets can be applied to problems of logic. To illustrate, assume that we receive the following guidelines concerning the allocation of expenditures in a firm:

> *Advertising expenditures are to be directed toward men who are college graduates or are over thirty years old, but not to college graduates under thirty years old.*

To simplify this unnecessarily complicated directive, let A represent men who are college graduates and B represent men who are over thirty years old. The set $A \cup B$ then represents men who are college graduates or over thirty years old. Similarly, the set $A \cap B'$ represents college graduates under thirty years old. Since advertising expenditures are not to be directed toward college graduates who are under thirty years old, the set describing the allocation of advertising expenditures is

$$(A \cup B) \cap (A \cap B')'$$

---

† This section can be omitted without loss of continuity.

This set can be simplified by using either the laws of set algebra or a Venn diagram. To simplify the set using the laws of set algebra, we perform the following steps:

Given: $(A \cup B) \cap (A \cap B')'$
1. Postulate 14: $(A \cup B) \cap (A' \cup B)$
2. Postulate 5: $B \cup (A \cap A')$
3. Postulate 12: $B \cup \phi$
4. Postulate 9: B

Interestingly, the analysis shows that the advertising guide lines can be reduced to the simple statement, "Advertising expenditures should be directed to men who are over thirty years old." The reader should verify that the same result can be shown through use of the Venn diagram.

Example: Simplify the following edict made by the president of Amalgamated Industries:

> *Due to recent cutbacks in the major divisions of our firm, employees who are over sixty years of age or have over thirty years of service with the company are eligible for early retirement. This excludes, however, individuals who have thirty years of service but are under sixty.*

To simplify this policy statement, let A represent the set of individuals over sixty years old and B represent the set of individuals who have over thirty years of service with the company. Those individuals who are over sixty years of age or have over thirty years with the company are described by the set $A \cup B$. Individuals under sixty years of age who have over thirty years of service are given by the set $A' \cap B$. Those eligible for early retirement are thus described by the set $(A \cup B) \cap (A' \cap B)'$. This set is simplified as shown in the preceding example to give $(A \cup B) \cap (A' \cap B)' = A$. The analysis shows that regardless of the number of years of service, only individuals over sixty years of age are eligible for early retirement.

Example: The president of Amalgamated Industries was told that his policy on early retirement specifically included all individuals over sixty years of age, regardless of the length of service to the company. Upon hearing this interpretation, he revised the policy as follows:

> *Due to recent cutbacks in the major divisions of our firm, employees who are over sixty years of age and have over thirty years of service with the company are eligible for early retirement. This specifically excludes individuals who are over sixty but have less than thirty years of service with the company.*

Simplify this policy statement, using a Venn diagram.

The Venn diagram describing the president's policy is shown below. Sets

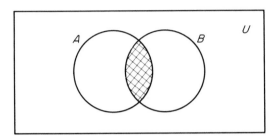

A and B are defined as before. Individuals who are over sixty years of age and have over thirty years of service are described by the set $A \cap B$. Those who are over sixty years of age but have less than thirty years of service are described by the set $A \cap B'$. The individuals described by the president as eligible for early retirement are described by the intersection of these sets, $(A \cap B) \cap (A \cap B')'$. This set can be reduced by using set algebra to the set $A \cap B$, the set shown by the crosshatched area in the Venn diagram.

These examples have used the operations of intersection, union, and complementation together with the laws of set algebra to clarify certain complicated statements. It can be seen that these operators are employed in the logical analysis of statements in much the same fashion as they were in the earlier description of sets and set membership.

1.6.1 STATEMENTS AND CONNECTIVES

One of the important concepts in logic is that of the *statement*. A statement is a simple declaration. For instance, "The sun is shining" and "I received an A on the last examination" are simple statements. The distinguishing characteristic of a statement is that the statement makes an assertion. The assertion made by the statement can be either true or false, but not both. It should be emphasized that interrogations (such as "Is the sun shining?" and imperatives (such as "Please turn in your homework.") do not assert anything and, consequently, are not statements.

The set operators are used to combine simple statements to form *compound* statements. The operations of complementation, intersection, and inclusive and exclusive union that were discussed in Sec. 1.4 have analogous counterparts in logic. These counterparts, termed *connectives* in logic, are negation, conjunction, and inclusive and exclusive disjunction.

The *negation* of a statement A is the statement "not A." To illustrate, let

$$A = \text{The sun is shining}$$

The negation of A is

$$A' = \text{The sun is not shining}$$

The *conjunction* of two statements A and B is the statement "A and B." This is denoted by $A \cap B$. For instance, if

$$A = \text{The sun is shining}$$

and

$$B = \text{I will go swimming}$$

then

$$A \cap B = \text{The sun is shining and I will go swimming}$$

Similarly,

$$A' \cap B' = \text{The sun is not shining and I will not go swimming}$$

The *inclusive disjunction* of two statements A and B is the statement A or B or both and is denoted by $A \cup B$. The *exclusive disjunction* of the two statements is A or B but not both and is denoted by $A \underline{\cup} B$. Defining the statement A as

$$A = \text{I will major in engineering}$$

and the statement B as

$$B = \text{I will major in business}$$

then the inclusive disjunction of the two statements is

$$A \cup B = \text{I will major in engineering or business or both}$$

The exclusive disjunction of the two statements is

$$A \underline{\cup} B = \text{I will major in engineering or business but not both}$$

Example: Define the statements A and B as $A =$ "The Dow-Jones stock market averages advanced" and $B =$ "Corporate profits were reported higher during the past quarter." Use the connectives (i.e., operators) negation, conjunction, and inclusive and exclusive disjunction to describe the following compound statements:

1. The Dow-Jones average advanced, and corporate profits were reported higher during the quarter.

   Answer: $A \cap B$.

2. The Dow-Jones average fell, but corporate profits were reported higher during the quarter.

   Answer: $A' \cap B$.

3. Although not reflected by the Dow-Jones average that continued to advance, corporate profits were reported lower during the past quarter.

Answer: $A \cap B'$.

Example: Define the statements A and B as $A = $ "The labor contract is inflationary" and $B = $ "The price increase is inflationary." Use the logical connectives to describe the following compound statements.

1. The labor contract is inflationary, but the price increase is not.

Answer: $A \cap B'$.

2. Either the labor contract or the price increase is inflationary.

Answer: $A \cup B$.

3. Neither the labor contract nor the price increase is inflationary.

Answer: $(A \cup B)'$.

4. Taken together, the labor contract and the price increase are inflationary.

Answer: $A \cap B$.

5. The labor contract could be considered inflationary or the price increase could be considered inflationary, but certainly both would not be considered inflationary.

Answer: $A \underline{\cup} B$.

1.6.2 LOGICAL ARGUMENTS

The principles of set algebra can easily be extended to the analysis of logical arguments. A *logical argument* consists of a series of statements or *premises* followed by a conclusion. The argument is termed *logically true* if the conclusion is the logical consequence of the premises. Conversely, the argument is *logically false* if the conclusion does not follow from the premises.

The statements or premises upon which the argument is based are termed *factual*. The factual statements may themselves be either true or false. Consequently, an argument that is logically true need not be based on correct facts.

To illustrate the concept of a logical argument, consider the following statement:

> *Automobile insurance premiums in states such as Massachusetts that have passed "no fault" insurance laws are lower than the premiums in states such as California that have not passed the "no fault" insurance laws. Consequently, insurance premiums in Massachusetts are lower than those in California.*

To examine the validity of the argument, let

A = Automobile insurance premiums in states that have passed "no fault" laws are lower than the premiums in states that have not passed the laws.

B = Massachusetts has a "no fault" insurance law.

C = California does not have a "no fault" insurance law.

D = Massachusetts insurance premiums are lower than those in California.

Statements A, B, and C are premises. Statement D is the conclusion based upon the three premises. To determine if D is logically true, consider the Venn diagram in Fig. 1.16. The figure shows that B is a subset of A and that

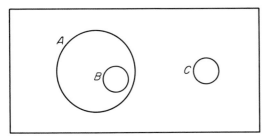

Figure 1.16

C is not a subset of A. Since the insurance premiums in all states that are members of A are lower than those in all states that are not members of A, it logically follows that the insurance premiums in Massachusetts are lower than those in California. The argument, therefore, is logically true.

Example: Determine the validity of the following argument.

Rapid expansion in an economy is often accompanied by inflation. Germany and Japan have experienced rapid economic growth during the past decade. Therefore, Germany and Japan have experienced inflation.

Let:

A = Rapid expansion is often accompanied by inflation

B = Germany and Japan have experienced rapid economic growth

C = Germany and Japan have experienced inflation

To determine the validity of the argument, consider the following Venn diagram. The diagram illustrates that the argument is logically true only if C is a member of the conjunction of A and B. Since there is no factual

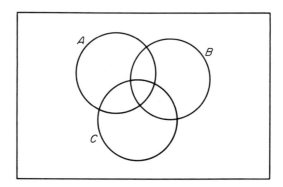

statement to the effect that C is a member of the conjunction of A and B, we conclude that the argument is logically false.

PROBLEMS

1. In a recent editorial, a journalist referred to the Republican voters in an upcoming election. Would it be possible to define this group of voters as a set?

2. In a recent opinion poll, one hundred individuals were questioned concerning a legislative proposal. Both the opinion and the sex of the respondent were noted. Give the subsets that a political analyst might consider important.

3. An economist referred to the set of all family units in the United States whose combined family income is less than $6000. Is the use of the term "set" appropriate in this instance?

4. Let set $A = \{5, 7, 9, 10, 12, 14\}$. List elements in the following sets that are also in A.
   (a) $\{x \mid x = \text{all odd numbers}\}$
   (b) $\{x \mid 2x + 6 = 20\}$
   (c) $\{x \mid x - 3 = \text{an odd integer}\}$

5. List all possible subsets of the set $S = \{a, b, c\}$.

6. A student can take one or more of five courses. How many different course selections are possible?

7. A restaurant offers pickles, ketchup, mustard, onions, lettuce, and tomatoes on its hamburgers. How many different hamburgers are possible?

8. For the sets $U = \{0, 1, 2, 3, 4, 5, 6, 7, 8, 9\}$, $R = \{0, 2, 4, 6, 8\}$, $S = \{2, 3, 4, 5, 6\}$, and $T = \{5, 6, 7, 8, 9\}$, determine the membership of the following:
   (a) $R \cap S$
   (b) $R \cap S \cap T'$
   (c) $R \cap S' \cap T'$
   (d) $R \cup S'$
   (e) $(R \cap S \cap T)'$

9. Simplify the following expressions:
   (a) $(P \cup Q)' \cup (P \cap Q')$
   (b) $(P \cap Q) \cap (P \cup Q)'$
   (c) $(P \cap Q)' \cup P$

10. Kawar Travel Agency handles the winter travel arrangements for all ski clubs in the Los Angeles, San Diego, and Phoenix areas that are affiliated with the United States Ski Association. The billing system for the agency is computerized. From time to time, the agency has need for the information stored in the computer and must call for it by set nomenclature. Following are examples of the type of information which is available.

A = ski clubs in Los Angeles
B = ski clubs in San Diego
C = ski clubs in Phoenix
D = set of ski clubs whose billings
        are more than 45 days old

Develop set nominclature for the following.
   (a) All the ski clubs in California with chapters both in Los Angeles and San Diego that have billings with the agency.
   (b) The billings in Phoenix or San Diego that are more than 45 days old.
   (c) The billings in Phoenix or Los Angeles less than 45 days old.

11. A highway construction crew is composed of men who can operate heavy equipment as follows: bulldozer, 33; crane, 22; cement mixer, 35; mixer and dozer, 14; mixer and crane, 10; dozer and crane, 10; mixer, dozer, and crane, 4. How many men are in the construction crew?

12. An analysis of the membership at a local country club shows that 60 percent of the members are men, 40 percent of the members score below 85, and 50 percent of the individuals belonging to the club are over 30 years old. It was also determined that 50 percent of the men are over

30 years old, 60 percent of the men score below 85, 40 percent of the men over 30 years old score below 85, and that 10 percent of the members who score below 85 are women over 30 years old.

(a) What percentage of the membership are men over 30 years old that score below 85?

(b) What percentage of the membership are men that score below 85?

13. The Johnson Corporation had 500 employees, of which 200 got a raise, 100 got a promotion, and 80 got both.

(a) How many employees got a promotion with no raise?

(b) How many employees got neither a raise nor a promotion?

14. A student surveyed 525 people. He determined that 350 read the newspaper for news, 215 listened to radio, and 140 watched TV. Additionally, 75 read the newspaper and watched television, 40 listened to radio and watched TV, and 100 read the newspaper and listened to radio. If 25 used all three sources of news, how many people utilized none of the three?

15. Define the sets F, C, and S, respectively, as stockholders in Ford, Chrysler, and U.S. Steel. If there are 500 stockholders in Ford, 800 stockholders in Chrysler, 700 stockholders in U.S. Steel, 200 stockholders in both Ford and Chrysler, 100 stockholders in both Ford and U.S. Steel, 100 stockholders in both Chrysler and U.S. Steel, and 50 stockholders in all three, determine numerically the number of stockholders in the following sets:

(a) $F \cup C$

(b) $(F \cap C) \cup (F \cap S)$

(c) $(C \cap S) \cap F'$

(d) $(F \cap C)' \cap (F \cap S)'$

16. A color television set and a black and white television set are being distributed by a company. Both sets offer the option of a remote control unit. Determine the Cartesian product of the two sets.

17. A firm is about to market a new line of hair spray. The product can be called Hair Mist (m), Hair Set (s), or Hair Hold (h). The product can be placed in either the cosmetics section (c) or in the personal care section (p) of a store and can be distributed either through wholesalers (w) or directly to the retailer (r). Determine the possible number of combinations and construct a box diagram to show these combinations.

18. A firm must raise additional capital. Two methods have been suggested, stock or bonds. With either method, the firm has the choice of a public or private offering of the securities. Depending on market conditions, the offering could be undersubscribed or oversubscribed. Determine all possibilities for the offering. How many possibilities must be considered?

19. An automobile manufacturer offers five different models of cars. Each model has nine separate option packages and is available in 14 colors.

If a car dealer were to maintain a complete inventory of cars, what is the minimum number of cars the dealer must stock?

20. Consider the following two statements.

A = "The illegal sale of drugs is rising."
B = "The Department of H. E. W. is concerned with rising drug traffic."

Describe in words the statements represented by the following logical operations.

(a) $A' \cap B$
(b) $A \cap B$
(c) $(A \cup B)'$
(d) $A \cup B$
(e) B'

21. A railroad company executive released the following directive.

All passenger service to cities where revenue is less than $5000 per month or the number of passengers is less than 500 per month will be discontinued. This does not include cities that have less than 500 passengers per month which produce revenue of $5000 per month, or more.

Where will passenger service be discontinued?

SUGGESTED REFERENCES

CANGELOSI, VINCENT E., *Compound Statements and Mathematical Logic* (Columbus, Ohio: Charles E. Merrill Books, Inc., 1967).

FREUND, JOHN E., *College Mathematics with Business Applications* (Englewood Cliffs, N.J.: Prentice-Hall, Inc., 1969) 1.

HANNA, SAMUEL C. and JOHN C. SABER, *Sets and Logic* (Homewood, Ill.: Richard D. Irwin, Inc., 1971).

KEMENY, JOHN G., et al., *Finite Mathematical Structures* (Englewood Cliffs, N.J.: Prentice-Hall, Inc., 1958), 1, 2.

KEMENY, JOHN G., et al., *Finite Mathematics with Business Applications*, 2nd ed. (Englewood Cliffs, N.J.: Prentice-Hall, Inc., 1972), 1, 2.

LIPSCHULTZ, SEYMOUR, *Set Theory and Related Topics*, Schaum's Outline Series (New York, N.Y.: McGraw-Hill Book Company, Inc., 1964).

THEODORE, CHRIS A., *Applied Mathematics: an Introduction* (Homewood, Ill.: Richard D. Irwin, Inc., 1965), 1–4.

Chapter 2

Functions
and Systems
of Equations

For mathematics to be of use in business, it is important that relationships that exist between business variables be formally defined. For instance, the retailer knows that profits are related to the number of units of product sold and the cost of the product. If it is possible for the retailer to state these relationships mathematically, the break-even point can be calculated and profit forecast. In this chapter we discuss the properties of mathematical relationships and offer examples of these properties. One of the most important relationships is the function.

2.1 Functions, Sets, and Relations

A *function* is a mathematical relationship in which the values of a single *dependent variable* are determined from the values of one or more *independent variables*. Before we discuss the general definition and properties of functions, it will be helpful to consider one of the most elementary functional forms, the *linear function*. The functional form of the linear function is

$$f(x) = a + bx \tag{2.1}$$

32

where $f(x)$ is the dependent variable.

x is the independent variable.

a is the value of the dependent variable when x is zero.

b is the coefficient of the independent variable.

The linear function has one dependent and one independent variable. The symbol $f(x)$, read "f of x," represents values of the dependent variable, and x represents values of the independent variable; $f(x)$ varies according to the rule of the function as x varies. For the linear function, the *rule of the function* states that b is to be multiplied by x and this product added to a. This sum determines the value of the dependent variable $f(x)$. Since the value of the dependent variable depends upon the value of the independent variable, $f(x)$ is termed the dependent variable and x is termed the independent variable. The term *variable* refers to a quantity that is allowed to assume different numerical values.

The properties of a linear function can be illustrated by an example. The linear function $f(x) = -4 + 2x$ is graphed in Fig. 2.1.

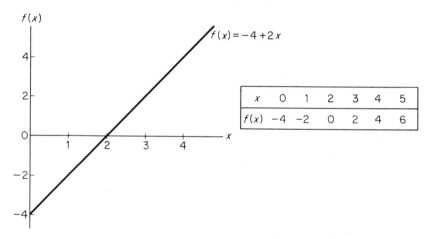

Figure 2.1

For each value of the independent variable x, there is one and only one value of the dependent variable $f(x)$. The value of the dependent variable is calculated by the rule of the function. The rule of the function in the example in Fig. 2.1 states that $f(x)$ is equal to -4 plus $2x$.

Both $f(x)$ and x are termed variables, since they are both permitted to take on different numerical values. The table in Fig. 2.1 lists six possible values of x and $f(x)$. For this linear function, however, there are an infinite number of possible values of the variables. For example, between $x = 0$ and $x = 1$ there are an infinite number of possible values of x. Similarly, there are an infinite number of possible values of x between $x = 1$ and $x = 2$. From the infinite number of possible values, we have selected the six values in the table to illustrate the function.

2.1.1 DOMAIN AND RANGE

The permissible values of the independent variable are termed the *domain* of the function. If, for instance, the analyst is interested in all positive values of the independent variable (including zero), he would state that the domain of the function consists of all positive numbers. The function graphed in Fig. 2.1 could thus be expressed as

$$f(x) = -4 + 2x \qquad \text{for } x \geq 0$$

where $x \geq 0$ is read "x greater than or equal to zero." The domain of this function is shown by $x \geq 0$ to consist of all positive numbers.

In many business situations, the analyst will be concerned with only selected values of the independent variable. If he were concerned with only positive integer values of x, then the domain of the function would be values of the independent variable that are positive integers. The function graphed in Fig. 2.1 could be expressed as

$$f(x) = -4 + 2x \qquad \text{for } x = 0, 1, 2, 3, \ldots$$

This function consists of the set of ordered pairs $(x, f(x))$,

$$\{(0, -4), (1, -2), (2, 0), (3, 2), (4, 4), \ldots\}$$

The function is graphed in Fig. 2.2.

The permissible values of the dependent variable are termed the *range* of the function. The range of the function is those numbers that the dependent variable assumes as the independent variable takes on all values in the domain. In the example illustrated in Fig. 2.1, the domain of the function is all values of x along the continuous line between $x = 0$ and $x = \infty$. The range for this function contains all the numbers along the continuum from $f(x) = -4$ to $f(x) = \infty$. Similarly, if the domain of the function consists of positive integer values, the corresponding range consists of the values calculated by $f(x) = -4 + 2x$ for $x = 0, 1, 2, 3, \ldots$.

To summarize, a linear function is a mathematical relationship in which the values of the dependent variable $f(x)$ are determined from the values of the independent variable according to the rule that $f(x) = a + bx$. For the linear function, as for all functions, one and only one value of the dependent variable is possible for each value of the independent variable. The permissible values of the independent variable are termed the *domain* of the function, and the permissible values of the dependent variable are termed the *range* of the

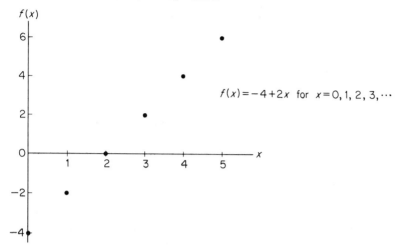

Figure 2.2

function. The concepts of function, domain, and range are further illustrated by the following examples.

Example: Harvey West, an analyst for Pacific Soft-Drink Company, is attempting to develop a cost function for a diet cola. From the accounting department he has learned that fixed costs are $10,000 and variable costs are $0.03 per bottle. Present capacity limitations are 50,000 bottles per month, or 600,000 bottles per year. Mr. West wishes to develop the cost function and to specify the domain and range of the function.

We shall represent the number of bottles of the diet cola by x and the total cost by $f(x)$. Since total cost is composed of fixed and variable costs, the cost function on a yearly basis is

$$f(x) = 10,000 + 0.03x$$

The domain of the function is $x = 0$ to $x = 600,000$. The corresponding range is $f(x) = \$10,000$ to $f(x) = \$28,000$.

Example: Bill Short owns a small auto repair shop. Mr. Short acts as supervisor and employs three auto repairmen. His shop is operated on a 40-hour per week basis, and his employees are guaranteed 40 hours of pay each week. Weekly labor costs consist of a fixed component and a sum that varies with the amount of overtime during the week. The weekly salaries of Mr. Short and the three repairmen total $750. The overtime rate is $5.00 per hour for the employees. Mr. Short does not draw overtime pay. The maximum amount of overtime per employee is 20 hours per week. Determine the cost function and the domain and range of the function.

The cost function consists of the sum of the weekly salaries and the overtime pay. If we represent the overtime hours per week by x and the total weekly salary by $f(x)$, then the cost function is

$$f(x) = 750 + 5x$$

The domain of the function is $x = 0$ to $x = 60$, and the corresponding range of the function is $f(x) = \$750$ to $f(x) = \$1,050$.

2.1.2 FUNCTIONS AND SET TERMINOLOGY

The language of sets can be usefully employed to define and identify the essential properties of a function. Expressed in set notation, a linear function is the set of ordered pairs of numbers $(x, f(x))$ given by

$$\{(x, f(x)) \mid f(x) = a + bx\} \tag{2.2}$$

The expression is read "that set of ordered pairs $(x, f(x))$ such that $f(x) = a + bx$."

The functional relationship between x and $f(x)$ has been described by the defining property method of specifying sets. This method consists of stating a rule within the braces that allows one to determine set membership. The rule must conform to the requirement that for each value of x there can be only one value of $f(x)$. The members of the set consist of the ordered pairs $(x, f(x))$. The permissible values of x are termed the *domain* of the function, and the corresponding values of $f(x)$, determined from the rule of the function, are termed the *range* of the function.

An alternative method of specifying set membership is the roster method. This method consists of listing within the braces all members of the set. A set of ordered pairs

$$\{(x_1, f(x_1)), (x_2, f(x_2)), (x_3, f(x_3))\} \tag{2.3}$$

is termed a function provided that for each value of the independent variable there exists only one value of the dependent variable. The values of the first component of each ordered pair make up the domain of the function, while the values of the second component of each ordered pair make up the range of the function.

Example: Use set terminology to describe the linear function

$$f(x) = 6 - 3x, \qquad \text{for } -10 \le x \le 10$$

The function consists of the set of ordered pairs

$$\{(x, f(x)) \mid f(x) = 6 - 3x \text{ for } -10 \le x \le 10\}$$

For each value of x in the domain $-10 \le x \le 10$, the value of $f(x)$ is determined by the rule of the function, $f(x) = 6 - 3x$. The domain of the function is the continuum $-10 \le x \le 10$, and the range of the function is the continuum $-24 \le f(x) \le 36$.

Example: A probability function is a functional relationship between the value of a variable and the probability of occurrence of that value. If we define the variable as the number of heads occurring in two tosses of a fair coin, then the probability function for the experiment of tossing the coin is given by the set of ordered pairs $(x, p(x))$,

$$\{(0, \tfrac{1}{4}), (1, \tfrac{1}{2}), (2, \tfrac{1}{4})\}$$

where x represents the number of heads and $p(x)$ represents the probability of this number occurring. The domain of the function (the possible number of heads) is $x = 0, 1, 2$, and the range of the function (the associated probabilities) is $p(x) = \tfrac{1}{4}$ and $\tfrac{1}{2}$. Notice that we have used the roster method of specifying set membership to list the elements of the set. This method of specifying a functional relationship can be employed only when a limited number of ordered pairs are included in the function.

Example: Verify that the following set is a function. Give the domain and range of the function.

$$\{(0, 3), (1, 2), (3, 3), (4, 2), (5, -1)\}$$

The roster method is used to specify the ordered pairs $(x, f(x))$ that comprise the set. For each value of x there is only one value of $f(x)$; consequently, the set is a function. The domain of the function is $x = 0, 1, 3, 4, 5$ and the range of the function is $f(x) = -1, 2, 3$. This example illustrates that for each value of x there is only one value of $f(x)$. This value of $f(x)$ need not be unique, however. In this function $f(x) = 3$ when $x = 0$ and $f(x) = 3$ when $x = 3$.

Example: State why the following set is not a function.

$$\{(0, 0), (0, 1), (2, 1), (3, 2)\}$$

This set does not qualify as a function, since the first two ordered pairs $(0, 0)$ and $(0, 1)$ indicate that $f(x)$ is 0 or 1 when $x = 0$.

The Cartesian product of two or more sets can be used in defining and illustrating functions. The Cartesian product of two sets A and B is the set of ordered pairs (a, b), where $a \in A$ and $b \in B$. Utilizing set notation, we find that the Cartesian product $A \times B$ is the set

$$A \times B = \{(a, b) \mid a \in A, b \in B\} \tag{2.4}$$

The elements of the set A are termed the *domain of the Cartesian product* and the elements of the set B are termed the *range of the Cartesian product*.

For the set $A \times B$, a subset in which elements of A occur only once is a function. The domain of the function consists of elements in A and the range of the function of elements in B. Although the elements in A may occur only once in the function, there is no requirement that all elements in A be included in the function. Elements in B can occur zero, once, or repeatedly in the function.

Example: Given the set $A = \{a, b, c\}$ and the set $B = \{1, 2\}$, the Cartesian product of A and B is the set

$$A \times B = \{(a, 1), (a, 2), (b, 1), (b, 2), (c, 1), (c, 2)\}$$

Examples of subsets that are functions include $\{(a, 1), (b, 1), (c, 1)\}$, $\{(a, 1), (b, 2), (c, 1)\}$, and $\{(a, 2), (b, 1), (c, 1)\}$. The subset $\{(a, 1), (a, 2), (b, 1)\}$ is, however, not a function.

Example: Set $A = \{\text{boy, girl}\}$ and $B = \{\text{Yale, Princeton, Vassar, Smith}\}$. The Cartesian product of A and B is

$$A \times B = \{(\text{boy, Yale}), (\text{boy, Princeton}), (\text{boy, Vassar}), (\text{boy, Smith}),$$
$$(\text{girl, Yale}), (\text{girl, Princeton}), (\text{girl, Vassar}), (\text{girl, Smith})\}$$

Two subsets of the Cartesian product that satisfy the requirements of a function are $\{(\text{boy, Yale}), (\text{girl, Vassar})\}$ and $\{(\text{boy, Princeton}), (\text{girl, Smith})\}$.

2.1.3 RELATIONS

A subset of the Cartesian product in the preceding example which is not a function but nevertheless is of interest is the subset

$$\{(\text{boy, Yale}), (\text{boy, Princeton}), (\text{girl, Vassar}), (\text{girl, Smith})\}$$

This subset fails to satisfy the requirements for a function since elements in the domain are associated with more than one element in the range. The subset of a Cartesian product in which elements in the domain and range may both appear repeatedly is called a *relation*. A relation in $A \times B$ is thus a subset of $A \times B$. The difference between a relation and a function is that a relation may have elements in the domain associated with more than one element in the range. A function consists of ordered pairs in which each element in the domain is associated with one and only one element in the range. These concepts are illustrated by the following examples.

Example: Let $S = \{\text{all stocks traded on the New York Stock Exchange on a Friday}\}$ and let $A = \{+, 0, -\}$ represent the stock closing up, closing unchanged, or closing down. The Cartesian product $S \times A$ is the set of all possible movements in the stock prices. From the Cartesian product, the relation that describes the possible stock movements in a portfolio that contains Ford and IBM is

$$\{(\text{Ford, }+), (\text{Ford, }0), (\text{Ford, }-), (\text{IBM, }+), (\text{IBM, }0), (\text{IBM, }-)\}.$$

A subset of the relation is the function which describes the actual price movement for a given Friday. This function is

$$\{(\text{Ford, }+), (\text{IBM, }-)\}.$$

Notice that in the relation elements in the domain may appear more than once and may be associated with more than one element in the range. In the function, elements in the domain are associated with only one element in the range.

Example: Develop the relation between x and y such that x is greater than y and both x and y are positive integers.

The set I consists of all positive integers, i.e.,

$$I = \{1, 2, 3, 4, \ldots\}$$

The Cartesian product, $I \times I$, represents all integer pairs (x, y) in the first quadrant. The relation

$$R = \{(x, y) \mid x > y \text{ and } x, y \in I\}$$

is a subset of the Cartesian product $I \times I$. This relation is shown in Fig. 2.3.

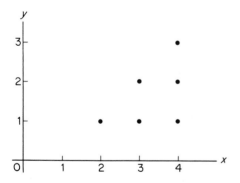

Figure 2.3

Only a few of the infinite number of ordered pairs that comprise the relation are shown in this figure.

A relation that will be of major importance in the chapters on linear programming is the linear inequality. A linear inequality is a relation of the form

$$y \geq kx + c \qquad\qquad (2.5)$$

or alternatively of the form

$$y \leq kx + c \qquad\qquad (2.6)$$

An inequality is a relation rather than a function, since for each value of x in the domain there is more than one value of y in the range. In fact, for any value of x in the linear inequality $y \geq kx + c$ there are an infinite number of values of y that satisfy the requirement that $y \geq kx + c$.

As an example of a linear inequality, consider $y \geq 2x - 3$. This relation is shown by the shaded area in Fig. 2.4. Using set notation, we can express

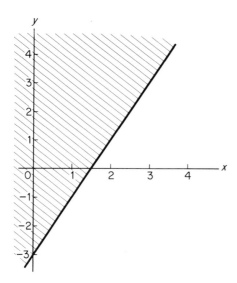

Figure 2.4

the linear inequality as $\{(x, y) \mid y \geq 2x - 3\}$.

Example: A university limits class enrollment to sixty-four students per teacher and cancels any class in which enrollment is six or less. Develop relations that describe the limitations.

If we represent the faculty size as x and the student enrollment as y, then the relation which describes the limitation is

$$\{y \leq 64x \text{ and } y \geq 7x \text{ for } x, y \in I\}$$

The relation is a subset of the Cartesian product of $I \times I$, where I is the set of positive integers.

Example: Show the inequality $y \leq x - 1$ on a graph.

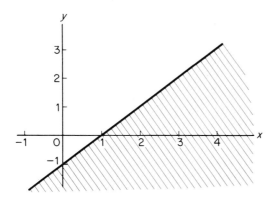

The inequality is shown by the shaded area.

Both functions and relations have been shown to be subsets of Cartesian products. In examples such as on p. 38 concerning the set $A = \{$boy, girl$\}$ and $B = \{$Yale, Princeton, Vassar, Smith$\}$, the Cartesian product was obvious. We have, however, not indicated the Cartesian product of which the function $f(x) = 2x + 3$ or the relation $y \leq 6x - 4$ are subsets. If x is unrestricted, the domain for both the function and the relation is $-\infty \leq x \leq \infty$.

The continuum $-\infty \leq x \leq \infty$ is described by the set of real numbers $R = \{-\infty \leq x \leq \infty\}$. The range of both the function and the relation is similarly $R = \{-\infty \leq f(x) \leq \infty\}$. The Cartesian product $R \times R$ is the set of all ordered pairs of real numbers, i.e.,

$$R \times R = \{(x, f(x)) \mid x \in R \text{ and } f(x) \in R\}$$

or alternatively

$$R \times R = \{(x, y) \mid x \in R \text{ and } y \in R\}$$

Both the relation and the function are subsets of the Cartesian product of real numbers.

Example: Verify that the mathematical relationship

$$y^2 = x$$

is a relation rather than a function.

If $y^2 = x$, then $y = \pm\sqrt{x}$ and for each value of x in the domain $0 \leq x \leq \infty$, there are two values of y. This violates the requirement that there exist only one value of the dependent variable for each value of the independent variable. Consequently, $y = \pm\sqrt{x}$ is a relation rather than a function. This relation is graphed in **Fig. 2.5.**

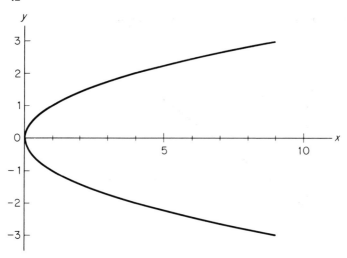

Figure 2.5

2.2   Establishing and Recognizing Functions

Numerous types of functions are used to describe functional relationships between business variables. The analyst should be able to recognize various types of functions and, when appropriate, establish functions that model the relationships that are observed to exist between business variables. Our purpose in this section is to describe the linear and quadratic functions and to illustrate selected applications of these functions. Methods of establishing these functions are presented and illustrated.

2.2.1   LINEAR FUNCTIONS

The linear function was defined earlier as

$$f(x) = a + bx \qquad (2.1)$$

where $f(x)$ is the dependent variable.

  x is the independent variable.
  a is the value of the dependent variable where x is zero.
  b is the coefficient of the independent variable.

As an example of the linear function, the function

$$f(x) = 8 - 2x \qquad \text{for } -2 \le x \le 8$$

is graphed in Fig. 2.6. Values of $f(x)$ are given on the vertical axis and values

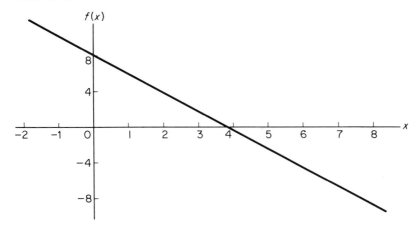

Figure 2.6

of x on the horizontal axis. The linear function plots as a straight line that intersects the vertical axis at $(0, 8)$, i.e., at $x = 0$ and $f(x) = 8$.

The slope of a linear function is defined as the change in the value of the dependent variable divided by the change in the value of the independent variable. For a linear function, the slope is given by b, the coefficient of x. It can also be calculated by determining the change in the dependent variable for a one-unit change in the independent variable. The slope of the function $f(x) = 8 - 2x$ is $b = -2$.

Example: Graph the linear function $f(x) = -3 + 1.5x$ for $-5 \le x \le 6$. Give the intercept and slope of the function.

A linear function can be graphed from two data points. One obvious data point is the ordered pair $(0, -3)$. This data point is the intercept of the function, i.e., the value of the dependent variable when the independent variable is 0. Since there are an infinite number of values of x in the domain $-5 \le x \le 6$, there are an infinite number of ordered pairs $(x, f(x))$ that could be used for a second data point. One of these ordered pairs is the data point $(4, 3)$. The graph of the function is shown on p. 44.

The intercept of the function is $f(x) = -3$ and the slope is $b = 1.5$.

One of the many uses of the linear function is in describing linear revenue and cost models. For certain products, the revenue generated from the sales of the product is a linear function of the number of units of the product sold. If, for instance, a firm is capable of supplying up to 1000 units of a product at a price of \$10 per unit, the revenue function for this product would be

$$R = 10x, \quad \text{for } 0 \le x \le 1000$$

If the cost of producing the product is composed of a fixed cost and a variable

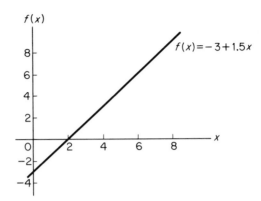

cost, the total cost of production is $TC = FC + VC$. Provided that variable cost is a linear function of the number of units manufactured and the fixed cost is constant for the specified domain of the cost function, the total cost function becomes $TC = FC + cx$, where c is the variable cost per unit of production. If fixed costs are \$1000 and variable costs are \$8, the total cost function is

$$TC = 1000 + 8x, \quad \text{for } 0 \leq x \leq 1000$$

The revenue and cost functions for the product are graphed in Fig. 2.7.

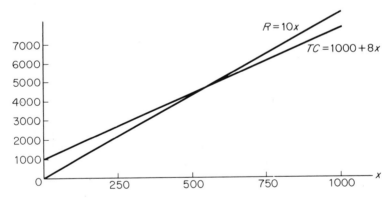

Figure 2.7

The breakeven level of production is the number of units for which revenue equals total cost. This quantity is equal to the value of x at which the revenue and total cost curves intersect. The quantity is calculated by equating R with TC.

$$R = TC$$
$$10x = 1000 + 8x$$
$$x = 500$$

The breakeven level of production is $x = 500$.

Example: The anticipated price of a product is $15. The fixed cost of manufacturing the product is $10,000, and the variable cost is $10 per unit. Plant limitations restrict production to 5000 units. Develop revenue and total cost functions and calculate breakeven production.

The revenue function is $R = 15x$ for $0 \leq x \leq 5000$, and the total cost function is $TC = 10,000 + 10x$ for $0 \leq x \leq 5000$. Breakeven is found by equating the two functions.

$$R = TC$$
$$15x = 10,000 + 10x$$
$$5x = 10,000$$
$$x = 2000$$

Example: Profit is defined as revenue minus total cost. Develop the profit function for the product in the preceding example and verify that profit is 0 at breakeven sales of $x = 2000$.

Revenue is $R = 15x$, and total cost is $TC = 10,000 + 10x$. Profit is thus

$$P = R - TC$$
$$P = 15x - (10,000 + 10x)$$
$$P = 5x - 10,000$$

Substituting 2000 for x shows that profit is 0 at breakeven.

It is often necessary to determine the function that passes through data points. In these instances, the analyst must determine the appropriate type of function and the parameters of the function (e.g., the coefficients a and b in the linear function $f(x) = a + bx$). To illustrate the method for determining the parameters of the linear function, consider the example of Bud Brewer.

Example: Bud Brewer is attempting to predict sales for his company for the coming year. Sales for the past five years are plotted on the graph shown on p. 46. On the basis of his knowledge of the market and the historical data, Bud believes that a linear function that passes through the data points for year 1 and year 5 will provide the best forecast for sales in the coming year (year 6).

The procedure for establishing functions through data points requires substituting the data points into the general form of the function and solving

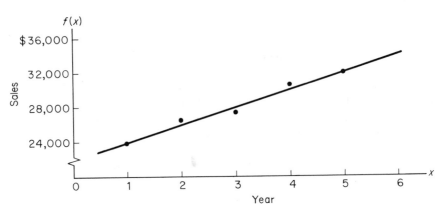

the resulting equations simultaneously for the parameters of the function. Two equations are required for a linear function. From Bud's statements, the two data points are the ordered pairs ($x = 1, f(x) = \$24{,}000$) and ($x = 5$, $f(x) = \$32{,}000$). These data points are each substituted into the general form of the linear function

$$f(x) = a + bx \qquad (2.1)$$

where x is the independent variable, time, and $f(x)$ is the dependent variable, sales. The substitution gives the two equations

$$32{,}000 = a + b(5)$$
$$24{,}000 = a + b(1)$$

These two equations are solved simultaneously by the method of substitution.† Substituting $a = 32{,}000 - b(5)$ from the first equation into the second equation gives

$$24{,}000 = 32{,}000 - b(5) + b(1)$$

or

$$4b = 8000$$

and

$$b = 2000$$

From the first equation we obtain

$$a = 32{,}000 - 5b = 22{,}000$$
$$f(x) = 22{,}000 + 2000x$$

Sales in year 6 are determined by evaluating the function for $x = 6$.

$$f(6) = 22{,}000 + 2000(6) = \$34{,}000$$

---

† The method of subtraction is discussed in Appendix A.

Since Bud Brewer is interested in predicting sales based upon this function for year 6 only, he has implied that the domain of the function is $x = 1$ to 6. The corresponding range of the function is $f(x) = \$24,000$ to $f(x) = \$34,000$. The forecast for year 6 is $f(6) = \$34,000$.

Example: Joe Findley is a stockholder in Electronic Products, Inc. In studying the annual report to stockholders, Joe notes that Electronic Products' computer is being depreciated on a straight-line basis. This means that the depreciation charge is constant each year. If the book value of the computer was \$1 million in year 2 and \$700,000 in year 5, determine the year in which the computer is fully depreciated, i.e., in which the book value will equal the scrap value. The scrap value of the computer is \$200,000.

The function that describes the depreciation and book value can be determined by substituting the data points (2, \$1,000,000) and (5, \$700,000) into the linear function

$$B(t) = a + bt$$

where $B(t)$ is the book value in year t, a is the original book value, and b is the yearly straight-line depreciation. The two equations are

$$1,000,000 = a + b(2)$$
$$700,000 = a + b(5)$$

These equations are solved simultaneously to give $a = \$1,200,000$, $b = -\$100,000$ and

$$B(t) = 1,200,000 - 100,000t$$

The year in which the computer is fully depreciated is determined by equating the function with \$200,000 and solving for t.

$$200,000 = 1,200,000 - 100,000t$$

The solution is $t = 10$ years. The depreciable life of the computer is ten years.

2.2.2 QUADRATIC FUNCTIONS

The *quadratic* function is a second function which has important applications in business and economics. The general form of the quadratic function is

$$f(x) = a + bx + cx^2 \tag{2.7}$$

where $f(x)$ is the dependent variable, x is the independent variable, and a, b, and c are the parameters of the function. The shape of the quadratic function is determined by the magnitude and signs of the parameters a, b, and c. Common forms of the quadratic function are shown in Fig. 2.8.

From the diagrams in Fig. 2.8, it can be seen that the value of a positions the function vertically. In Fig. 2.8(a), for instance, if a were negative rather than positive, the function would intercept the vertical axis below the hori-

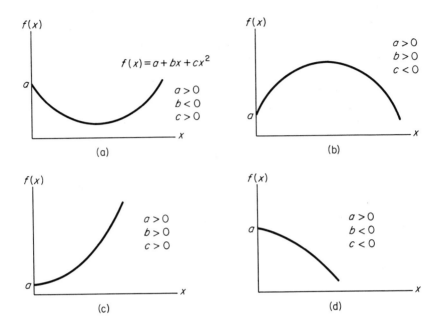

$$f(x) = a + bx + cx^2$$

(a) $a > 0$, $b < 0$, $c > 0$

(b) $a > 0$, $b > 0$, $c < 0$

(c) $a > 0$, $b > 0$, $c > 0$

(d) $a > 0$, $b < 0$, $c < 0$

Figure 2.8

zontal axis. This would have the effect of moving the entire function downward. The values of b and c determine the general shape of the function. For b negative and c positive, the function will have the concave shape illustrated by Fig. 2.8(a). For b positive and c negative, the function will be convex as shown in Fig. 2.8(b). If b and c are both positive, the function has the shape illustrated by Fig. 2.8(c). Conversely, if b and c are both negative, the function appears as illustrated by Fig. 2.8(d).

Three data points are necessary to establish a quadratic function. The procedure in establishing a quadratic function is to substitute the data points into the general form of the quadratic function and solve the resulting three equations simultaneously for the parameters a, b, and c. This procedure is illustrated in Appendix A and by the following two examples.

Example: Bill Cook owns a small machine shop. Because of the shop's limited capacity, average costs are related to the volume of output. Cost studies undertaken by Bill show that the average cost of producing 90 units of output is $850 per unit, the average cost of producing 100 units is $800 per unit, and the average cost per unit of producing 120 units is $825 per unit. Determine the functional relationship between average cost and output.

The three data points, when plotted, show that a function that decreases and then increases describes the cost-output data. This type of functional relationship is described by the quadratic function

$$f(x) = a + bx + cx^2 \qquad (2.7)$$

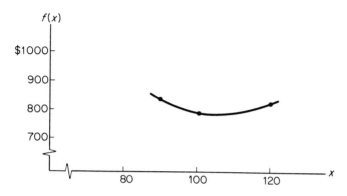

The procedure for determining the function is to substitute the data points into the general form of the quadratic function, Eq. (2.7), and solve the resulting three equations simultaneously for the parameters a, b, and c.† The three equations are

$$850 = a + b(90) + c(90)^2$$
$$800 = a + b(100) + c(100)^2$$
$$825 = a + b(120) + c(120)^2$$

The solution of the three simultaneous equations is $a = 3175$, $b = -44.58$, and $c = 0.2083$. The average cost function is

$$f(x) = 3175 - 44.58x + 0.2083x^2$$

Example: James Mason is concerned with the effect of fatigue on the productivity of skilled machinists. He has observed that productivity, which he defines as standard units of output per man-hour, increases as the number of hours that an individual works increases up to approximately 40 hours per week and then begins to decline as fatigue begins to become a factor. Mason believes that three representative observations of the effect of man-hours on productivity are 9.3 units of output during the 35th hour of work, 9.5 units during the 40th hour, and 8.5 units during the 45th hour.

---

† These equations can be solved simultaneously by the method of substitution, the method of subtraction, or by using matrix algebra. The method of subtraction is discussed in Appendix A, and matrix algebra is discussed in Chapter 3.

The observations are plotted below. Productivity as a function of the hour of work increases and then decreases. The type of a relationship shown in the illustration is described by a quadratic function.

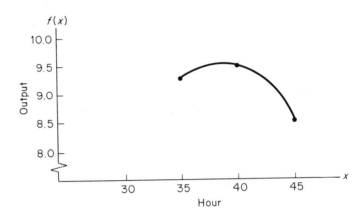

We substitute the data points in the general form of the quadratic function, Eq. (2.7), and obtain

$$9.3 = a + b(35) + c(35)^2$$
$$9.5 = a + b(40) + c(40)^2$$
$$8.5 = a + b(45) + c(45)^2$$

Solving the three equations simultaneously for a, b, and c gives $a = -25.7$, $b = 1.84$, and $c = -0.024$. The productivity function is

$$f(x) = -25.7 + 1.84x - 0.024x^2$$

2.2.3 POLYNOMIAL FUNCTIONS

Linear and quadratic functions belong to the class of functions termed *polynomial* functions. The general form of the polynomial function is

$$f(x) = a_0 + a_1x + a_2x^2 + a_3x^3 + \ldots + a_nx^n \qquad (2.8)$$

where $a_0, a_1, a_2, a_3, \ldots, a_n$ are parameters and n is a positive integer. The parameters may be positive, negative, or zero. The polynomial function is linear if $n = 1$ and quadratic if $n = 2$. This can be verified by comparing Eq. (2.8) for $n = 1$ with the general form of the linear function (2.1), and by comparing Eq. (2.8) for $n = 2$ with the general form of the quadratic function (2.7).

A polynomial function in which the largest exponent is $n = 3$ is termed a *cubic* function. The cubic function has the general form

$$f(x) = a_0 + a_1x + a_2x^2 + a_3x^3 \qquad (2.9)$$

The general shape of the cubic function is shown in Fig. 2.9.

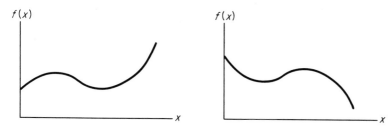

Figure 2.9

Although the two illustrations in Fig. 2.9 do not represent all possible shapes for the cubic function, they do illustrate a distinguishing characteristic of the cubic function. The reader will remember that the quadratic function was described as convex or concave. Referring to Fig. 2.9, we see that the cubic function changes from convex to concave or from concave to convex. An example of a cubic function that changes from convex to concave as the value of the independent variable x increases is

$$f(x) = 5 + 2.5x - x^2 + 0.1x^3$$

This function is plotted in Fig. 2.10.

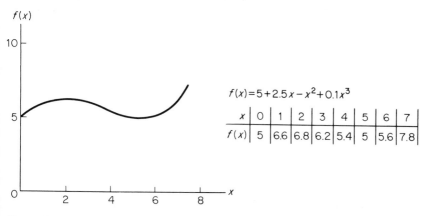

$f(x) = 5 + 2.5x - x^2 + 0.1x^3$

x	0	1	2	3	4	5	6	7
$f(x)$	5	6.6	6.8	6.2	5.4	5	5.6	7.8

Figure 2.10

The number of data points required to establish a polynomial function is equal to $n + 1$, where n is the largest exponent in the function. Thus, two data points are required for the linear function, three data points are required for the quadratic function, and four data points are required for the cubic function. The procedure in establishing the function is to substitute each data point into the general form of the polynomial function, and solve the resulting $n + 1$ equations simultaneously for the $n + 1$ parameters. This procedure has been illustrated for the linear and quadratic functions.†

2.3 Multivariate Functions

We have stated that a function has only one dependent variable but may have one or more independent variables. Functions in which the single dependent variable is related to more than one independent variable are termed *multivariate* functions. An example of a multivariate function is

$$f(x_1, x_2) = 2x_1 + 3x_1x_2 + 6x_2$$

where $f(x_1, x_2)$ is the dependent variable, x_1 is an independent variable, and x_2 is a second independent variable. Multivariate functions are difficult to plot, since they require one axis (or dimension) for each variable. A function with two independent variables and one dependent variable, for instance, requires three dimensions for plotting. Multivariate functions with three independent variables cannot be plotted, since four dimensions are required. In spite of the fact that multivariate functions with more than two independent variables cannot be plotted, they are quite necessary to describe many business and economic models. Multivariate functions with two independent variables are illustrated by the following examples.

Example: Sales of the Southern Distributing Company consist of sales made through retail and wholesale outlets. Profit per dollar of sales at retail is \$0.15 and profit per dollar of sales at wholesale is \$0.05. Determine the profit function.

If retail sales in dollars are represented by x_1, wholesale sales in dollars by x_2, and profit by $P(x_1, x_2)$, the profit function is

$$P(x_1, x_2) = 0.15x_1 + 0.05x_2$$

Example: Assume that the Neshay Candy Company makes net profit of \$0.05 per almond bar and \$0.07 per chocolate bar. If net profit P is represented as a function of monthly sales in units of the almond bar x_1 and the chocolate bar x_2, determine the appropriate functional relationship and calculate net profit for a month in which $x_1 = 10,000$ and $x_2 = 20,000$.

---

† An example of establishing a cubic function is given in Appendix A.

$$P = 0.05x_1 + 0.07x_2$$
$$P = 0.05(10,000) + 0.07(20,000)$$
$$P = \$1,900$$

Multivariate functions can be established through data points by using the same technique described for polynomial functions. This technique involves substituting the appropriate data points into the specified general form of the multivariate function and solving the resulting equations simultaneously for the parameters of the function. The technique is illustrated by the following example.

Example: Research has demonstrated that the output Y of a firm is related to labor L and capital K. The general form of the function is

$$Y = aL + bK + c(LK)$$

It has been determined that: $Y = 100,000$ when $L = 9.0$ and $K = 4.0$; $Y = 120,000$ when $L = 10.0$ and $K = 5.0$; and $Y = 150,000$ when $L = 11.5$ and $K = 7.0$. The domains of L and K are $9.0 \leq L \leq 11.5$ and $4.0 \leq K \leq 7.0$. Determine the values of a, b, and c.

The three data points are substituted into the general form of the function to give

$$100,000 = a(9.0) + b(4.0) + c(36.0)$$
$$120,000 = a(10.0) + b(5.0) + c(50.0)$$
$$150,000 = a(11.5) + b(7.0) + c(80.5)$$

These equations are solved simultaneously to give $a = -60,000$, $b = 304,000$, and $c = -16,000$. The function is

$$Y = -60,000L + 304,000K - 16,000(LK) \qquad \text{for } 9.0 \leq L \leq 11.5$$
$$4.0 \leq K \leq 7.0.$$

2.4 Systems of Equations

A function was defined earlier in this chapter as a mathematical relationship in which the value of a single dependent variable is determined from the values of one or more independent variables. In certain instances, it is not appropriate to consider one variable as "dependent" upon other variables. Although the variables are explicitly related and the requirements for a function are satisfied, the relationship is referred to as an *equation* rather than a function. Expressed in set notation, a linear equation with two variables is the set of ordered pairs (x, y) given by

$$\{(x, y) \mid ax + by = c\} \qquad (2.10)$$

In comparing the definition of a linear equation given by Eq. (2.10) with that of a linear function given by Eq. (2.2), the primary difference is that neither variable is distinguished in Eq. (2.10) as the dependent or the independent variable.

A *system of equations* consists of one or more equations with one or more variables. As an example, a system of two equations with two variables is

$$2x + 4y = 28$$
$$3x - 2y = 10$$

Similarly, a system of two equations with three variables is

$$20x + 15y - 10z = 20$$
$$30x - 25y + 35z = 85$$

The concept of a system of equations is illustrated by the following examples.

Example: A firm produces two products, A and B. Each unit of product A requires two man-hours of labor and each unit of product B requires three man-hours of labor. Develop the equation describing the relationship between the two products, assuming that 110 man-hours of labor will be used.

The number of units of product A is represented by x and the number of units of product B is represented by y. The equation describing the relationship between the two variables is

$$2x + 3y = 110$$

Example: Products A and B discussed in the preceding example require machine time as well as labor time. Each unit of product A requires one hour of machine time, and each unit of product B requires 0.5 hour of machine time. Develop the equation describing the relationship between the two products, assuming that thirty-five hours of machine time will be used.

If we again represent the number of units of product A by x and the number of units of product B by y, the equation is

$$x + 0.5y = 35$$

2.4.1 SOLUTION SETS FOR SYSTEMS OF EQUATIONS

The preceding examples illustrate a system of two equations with two variables. The equation that relates units of products A and B with man-hours can be described by the set

$$L = \{(x, y) \mid 2x + 3y = 110\}$$

Similarly, the equation that relates units of products A and B with machine time can be described by the set

$$M = \{(x, y) \mid x + 0.5y = 35\}$$

There are an infinite number of ordered pairs (x, y) that are members of set L. For instance, $(0, 36.67)$, $(1, 36)$, $(2, 35.33)$, $(3, 34.67)$ are four ordered pairs which are solutions to the equation $2x + 3y = 110$. There are also an infinite number of ordered pairs (x, y) that are members of set M. These include $(0, 70)$, $(1, 68)$, $(2, 66)$, $(3, 64)$, ... The ordered pairs (x, y) that are members of the set L are termed *solutions* to the equation $2x + 3y = 110$. Similarly, the ordered pairs that are members of the set M are solutions to the equation $x + 0.5y = 35$. The sets L and M are referred to as *solution sets*.

The sets L and M describe the labor and machine constraints on the production of products A and B. The intersection of sets L and M contains those elements (x, y) that are members of both L and M. That is,

$$L \cap M = \{(x, y) \mid 2x + 3y = 110 \text{ and } x + 0.5y = 35\}$$

The elements (x, y) represent the quantities of products A and B that are solutions for both the labor and the machine constraints.

The method of determining the solution set for a system of two equations is to determine the values of the variables that are solutions to both equations. These values are determined by solving the equations simultaneously. One of the methods commonly used in algebra for solving equations simultaneously is the method of substitution. This method involves solving for one variable in terms of the remaining variables and substituting this expression for the variable in the remaining equations. The two equations are

$$2x + 3y = 110$$
$$x + 0.5y = 35$$

Using the method of substitution, we solve for x in the second equation and obtain $x = 35 - 0.5y$. The expression $35 - 0.5y$ is substituted for x in the first equation to give

$$2(35 - 0.5y) + 3y = 110$$
$$2y = 40$$
$$y = 20$$

Substituting $y = 20$ in the second equation gives $x = 25$. The values of x and y that are solutions to both equations are $L \cap M = \{(25, 20)\}$, or 25 units of product A and 20 units of product B.

Example: A firm uses three processes to produce three products. Product 1 requires 2 hours of process A time, 4 hours of process B time, and 6 hours of process C time. Product 2 requires 1 hour of process A time and 2 hours of process B and C time. Product 3 requires 3 hours of process A time and 4 hours of process B and C time. One hundred hours of process A time, 160 hours of process B time, and 190 hours of process C time have been allocated to the production of the products. Determine the quantities of products 1, 2, and 3 that can be manufactured.

We let x_1, x_2, and x_3 represent the number of units of products 1, 2, and 3. The process A equation is given by

$$A = \{(x_1, x_2, x_3) \mid 2x_1 + x_2 + 3x_3 = 100\}$$

The process B equation is

$$B = \{(x_1, x_2, x_3) \mid 4x_1 + 2x_2 + 4x_3 = 160\}$$

The process C equation is

$$C = \{(x_1, x_2, x_3) \mid 6x_1 + 2x_2 + 4x_3 = 190\}$$

There are an infinite number of ordered triplets (x_1, x_2, x_3) that are solutions to each equation. We require, however, the set of ordered triplets (x_1, x_2, x_3) that are solutions to all three equations. This is the solution set for the system of simultaneous equations. The solution set is determined by solving the three equations simultaneously for x_1, x_2, and x_3. The solution set is $A \cap B \cap C = \{(15, 10, 20)\}$, or $x_1 = 15$ units, $x_2 = 10$ units, and $x_3 = 20$ units.

The solution set for $A \cap B \cap C$ can be determined by using any one of the methods introduced in algebra. Using the method of substitution, we solve the first equation for x_1 to obtain $x_1 = -0.5x_2 - 1.5x_3 + 50$. Substituting this quantity for x_1 in the second and third equations gives

$$4(-0.5x_2 - 1.5x_3 + 50) + 2x_2 + 4x_3 = 160$$

and

$$6(-0.5x_2 - 1.5x_3 + 50) + 2x_2 + 4x_3 = 190$$

These two equations reduce to

$$2x_3 = 40$$

and

$$x_2 + 5x_3 = 110$$

These two equations are solved for x_2 and x_3 to give $x_2 = 10$ and $x_3 = 20$. Substituting these values in equation A gives $x_1 = 15$.

2.4.2 CONSISTENT AND INCONSISTENT SYSTEMS OF LINEAR EQUATIONS

The solution set for the system of linear equations in the preceding two examples had only a single element. In the first example the element was $L \cap M = \{(25, 20)\}$ and in the second example the element was $A \cap B \cap C = \{(15, 10, 20)\}$. These examples illustrate the case in which a *unique* solution exists for the system of equations. A unique solution exists when there is only one element in the intersection of the sets describing the individual equations.

A unique solution is one of three possibilities for the solution to a system of linear equations. A system of linear equations may have: (1) no solution; (2) exactly one solution (unique); or (3) an infinite number of solutions. A system that has no solution is termed *inconsistent*. A system that has one or more solutions is said to be *consistent*. These three possibilities are illustrated by Fig. 2.11.

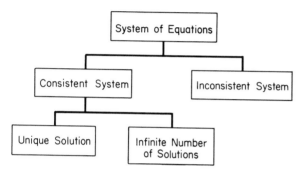

Figure 2.11

The concepts presented in Fig. 2.11 can be explained by using set notation. Assume that $f(x_1, x_2, x_3)$, $g(x_1, x_2, x_3)$, and $h(x_1, x_2, x_3)$ represent three different equations. The set of ordered triplets (x_1, x_2, x_3) that are solutions to each equation is described by the sets

$$F = \{(x_1, x_2, x_3) \mid f(x_1, x_2, x_3)\}$$

$$G = \{(x_1, x_2, x_3) \mid g(x_1, x_2, x_3)\}$$

$$H = \{(x_1, x_2, x_3) \mid h(x_1, x_2, x_3)\}$$

The solution to the system of equations is defined as all ordered triplets (x_1, x_2, x_3) that are common to the three equations. This solution set is the set of ordered triplets contained in the intersection of F, G, and H, i.e., $F \cap G \cap H$. If this intersection is null (empty), the system of equations is inconsistent. If, however, the intersection contains one or more elements, the system is consistent. For those cases in which the intersection contains exactly one element, this element is the unique solution to the system of equations. If the intersection contains more than one element, the system of equations is consistent, but the solution is not unique.

Graphs of equations are quite useful as an aid to understanding the solution sets for systems of equations. We shall show a system of equations that is consistent and has a unique solution, a system of equations that is consistent with an infinite number of solutions, and a system of equations that is inconsistent and consequently has no solution.

2.4.3 CONSISTENT SYSTEM WITH A UNIQUE SOLUTION

As an illustration of a consistent system of equations with a unique solution, consider the equations $2x + y = 8$ and $3x - 2y = -2$. These equations are shown in Fig. 2.12.

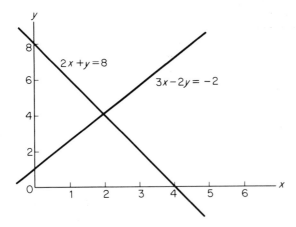

Figure 2.12

The equations can be described by the sets

$$A = \{(x, y) \mid 2x + y = 8\}$$
$$B = \{(x, y) \mid 3x - 2y = -2\}$$

The set A contains all ordered pairs (x, y) that are solutions to the equation $2x + y = 8$. These ordered pairs are shown in Fig. 2.12 by the linear equation $2x + y = 8$. The set B contains all ordered pairs (x, y) that are solutions to the equation $3x - 2y = -2$. These ordered pairs are shown by the linear equation $3x - 2y = -2$. The solution set for the system of equations contains all ordered pairs that are solutions to both A and B. This set is shown in Fig. 2.12 by the intersection of the two equations. The ordered pairs that are common to both equations are

$$A \cap B = \{(x, y) \mid 2x + y = 8 \text{ and } 3x - 2y = -2\}$$

The values of the variables x and y are determined by solving the two equations simultaneously for x and y. Substituting $y = 8 - 2x$ from the first equation for y in the second equation gives

$$3x - 2(8 - 2x) = -2$$
$$7x = 14$$
$$x = 2$$

Substituting $x = 2$ in the first equation gives $y = 4$. The ordered pair $(2, 4)$ is the unique solution to the system of equations.

Example: Determine the unique solution to the following system of three linear equations and three variables.

$$A = \{(x, y, z) \mid 2x + y - 2z = -1\}$$
$$B = \{(x, y, z) \mid 4x - 2y + 3z = 14\}$$
$$C = \{(x, y, z) \mid x - y + 2z = 7\}$$

A unique solution exists for the system of equations if there is an ordered triplet (x, y, z) that is a common solution to all three equations. Solving the three equations simultaneously gives $x = 2$, $y = 3$, and $z = 4$. The solution set for the system of equations is $A \cap B \cap C = \{(2, 3, 4)\}$.

Example: Verify that the following system of three equations with two variables is consistent and has a unique solution.

$$A = \{(x, y) \mid 2x + y = 9\}$$
$$B = \{(x, y) \mid x - y = 3\}$$
$$C = \{(x, y) \mid x + 2y = 6\}$$

The system of three equations with two variables is consistent and has a unique solution if there is a single ordered pair (x, y) that is a common solution to all three equations. Solving the first two equations simultaneously gives $A \cap B = \{(4, 1)\}$ as the unique solution to the first two equations. To determine if $(4, 1)$ is also a solution to the third equation, the ordered pair is substituted into the third equation. Since $4 + 2(1) = 6$, the ordered pair $(4, 1)$ is a solution for the third equation. The solution for the system of equations is $A \cap B \cap C = \{(4, 1)\}$. The system of equations is consistent with a unique solution.

The equations described by the sets A, B, and C are shown in Fig. 2.13. The unique solution $x = 4$, $y = 1$ occurs at the intersection of the three equations.

A system of three equations with two variables has a unique solution when one of the equations can be expressed as a linear combination of the remaining two equations. If we use the sets A, B, and C to represent the equa-

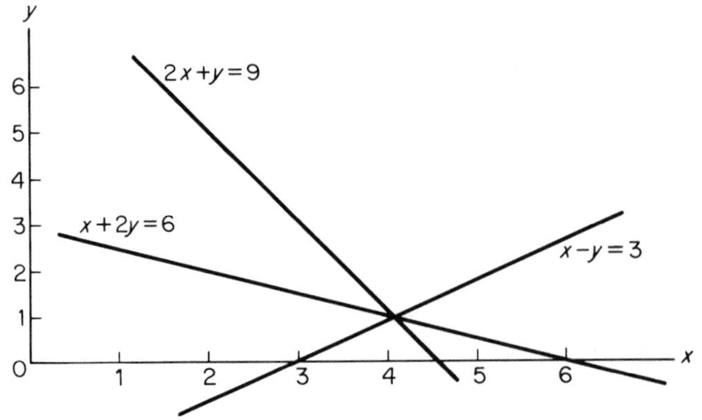

Figure 2.13

tions, a linear combination exists if there are numbers a and b such that $aA + bB = C$.

To determine if a linear combination exists in the preceding example, we first write the equations as $2x + y - 9 = 0$, $x - y - 3 = 0$, and $x + 2y - 6 = 0$. A linear combination exists if there are numbers a and b such that

$$a(2x + y - 9) + b(x - y - 3) = x + 2y - 6$$

In this example equation C can be formed by $1A - 1B$. This is shown by

$$1(2x + y - 9) - 1(x - y - 3) = x + 2y - 6$$

The appropriate numbers are thus $a = 1$ and $b = -1$, and equation C is a linear combination of equations A and B.†

2.4.4 CONSISTENT SYSTEM WITH AN INFINITE NUMBER OF SOLUTIONS

A system of equations has been defined as consistent if one or more solutions exist for the system of equations. As an illustration of a system of equations for which there is more than one solution, consider the equations

$$A = \{(x, y, z) \mid 2x - 4y + 4z = 20\}$$
$$B = \{(x, y, z) \mid 3x + 4y - 2z = 30\}$$

The solution to the system of two equations and three variables consists of

---

† In general, a unique solution to a system of m equations with n variables exists if $m - n$ of the equations can be expressed as a linear combination of the remaining n equations. The technique for determining the existence of a unique solution to a system of equations can be easily demonstrated by the use of matrix algebra. Consequently, we shall postpone further discussion of the requirements for a unique solution until Chapter 4.

all ordered triplets (x, y, z) that are solutions to both equations. From introductory algebra, we remember that it is impossible to determine a unique solution for three variables with two equations. We can, however, determine ordered triplets that are solutions to both equations. The procedure is to arbitrarily specify the value for one of the variables and to solve the two equations simultaneously for the remaining two variables. For instance, if the value of x is specified as 0, the solution set is $x = 0$, $y = 20$, and $z = 25$. These values are determined by solving the two equations

$$2(0) - 4y + 4z = 20$$

$$3(0) + 4y - 2z = 30$$

for y and z. Similarly, if $x = 5$ the equations can be solved simultaneously to obtain $y = 10$ and $z = 12.5$. Since an infinite number of values of x, y, or z could arbitrarily be specified, there are an infinite number of solutions to the system of two equations.

Example: Verify that the following system of two equations with three variables is consistent.

$$A = \{(x, y, z) \mid 2x + 3y + 4z = 20\}$$

$$B = \{(x, y, z) \mid 5x - 4y + 3z = 15\}$$

The system of equations is consistent if one or more solutions exist for the system. Equating $x = 0$ gives $y = 0$ and $z = 5$. Equating $x = 5$ gives $y = 2.8$ and $z = 0.4$. Since an infinite number of values of x, y, or z could be specified, the solution set $A \cap B$ contains an infinite number of triplets (x, y, z), and the system of equations is consistent.

Example: Verify that the following system of equations is consistent and has an infinite number of solutions.

$$A = \{(x, y, z) \mid 2x + 3y - 2z = 40\}$$

$$B = \{(x, y, z) \mid 3x - 2y + z = 50\}$$

$$C = \{(x, y, z) \mid x - 5y + 3z = 10\}$$

The system of three equations with three variables would appear at initial inspection to be consistent with a unique solution. When attempting to determine the solution, however, we find that this is not the case. Substituting the expression for x from the third equation in the first equation gives

$$2(10 - 3z + 5y) + 3y - 2z = 40$$

$$13y - 8z = 20$$

Substituting the same expression for x in the second equation gives

$$3(10 - 3z + 5y) - 2y + z = 50$$

$$13y - 8z = 20$$

Using the method of substitution, we would normally solve the two equations simultaneously for y and z. The equations that resulted from the original substitution are, however, the same. Consequently, there are an infinite number of ordered triplets (x, y, z) that are members of the solution set $A \cap B \cap C$. To obtain a solution to this system of equations, the value of one of the variables must be arbitrarily specified. If z is specified, the complete solution to the system of equations is

$$z = \text{specified}$$

$$y = \frac{20 + 8z}{13}$$

$$x = 10 - 3z + 5y$$

For instance, if $z = 0$, then $y = \frac{20}{13}$ and $x = \frac{230}{13}$. One solution is thus the ordered triplet $(\frac{230}{13}, \frac{20}{13}, 0)$. If $z = 1$, then $y = \frac{28}{13}$ and $x = \frac{231}{13}$. A second solution is $(\frac{230}{13}, \frac{28}{13}, 1)$. It is obvious that an infinite number of values of z could be specified and therefore an infinite number of solutions are possible.

Example: Verify that the equation $x - y = 6$ represents a consistent system of equations with an infinite number of solutions.

A system of equations is termed consistent if there are one or more solutions for the system of equations. The equation $x - y = 6$ is a system of one equation with two variables. The solution set consists of an infinite number of ordered pairs (x, y) that are solutions to the equation. One of the variables must be specified to determine a solution. For instance, if $x = 0$, then $y = -6$ and $(0, -6)$ is a solution. Similarly, if $x = 1$, then $y = -5$ and $(1, -5)$ is another solution. We thus conclude that the equation $x - y = 6$ is a consistent system of equations with an infinite number of solutions.

2.4.5 INCONSISTENT SYSTEM OF EQUATIONS

A system of equations is termed inconsistent when the solution set for the system of equations is null. As an example, consider the two equations

$$A = \{(x, y) \mid x + y = 6\}$$

$$B = \{(x, y) \mid 2x + 2y = 8\}$$

These equations are shown in Fig. 2.14. The solution to the system of two equations consists of the ordered pairs (x, y) that are common to both set A and set B, i.e., $A \cap B$. The equations described by the sets A and B are, however, parallel. Consequently, there is no ordered pair (x, y) that is a

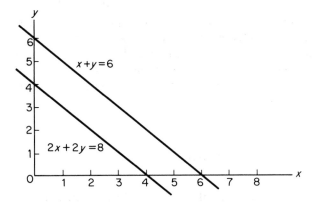

Figure 2.14

common solution to both equations. The solution set for the system of two equations with two variables is thus null, i.e., $A \cap B = \phi$.

Another example of an inconsistent system of equations is

$$A = \{(x, y) \mid x + y = 8\}$$
$$B = \{(x, y) \mid 2x - y = 2\}$$
$$C = \{(x, y) \mid x - 2y = -2\}$$

These equations are shown in Fig. 2.15. The solution to the system of equa-

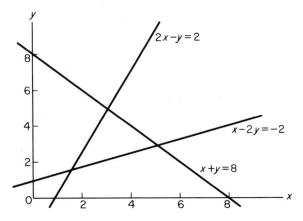

Figure 2.15

tions consists of the ordered pairs (x, y) that are common solutions to all three equations. Figure 2.15 shows that there are no ordered pairs (x, y) that are members of the solution set $A \cap B \cap C$. This can be verified algebraically by determining the solution to any two of the three equations and determining if this ordered pair is a solution to the third equation. For instance, the solution set for equation A and B is $A \cap B = \{(3\frac{1}{3}, 4\frac{2}{3})\}$. The ordered pair $(3\frac{1}{3}, 4\frac{2}{3})$ is, however, not a solution for equation C. Since there is no ordered pair that is a common solution for all three equations, we conclude that the equations are inconsistent.

PROBLEMS

1. Plot the following functions and specify the slope of each function.
   (a) $f(x) = 4$ (b) $f(x) = x + 3$
   (c) $f(x) = 2x + 3$ (d) $f(x) = 4 - 3x$
   (e) $f(x) = -0.5x + 2$ (f) $f(x) = -x$
   (g) $f(x) = -3x - 4$ (h) $f(x) = \dfrac{x}{3} - 2$

2. Establish linear functions through the data points $(x, f(x))$.
   (a) $(3, 4)$ and $(5, 2)$ (b) $(9, 4)$ and $(4, -1)$
   (c) $(8, 2)$ and $(5, 5)$ (d) $(4, -3)$ and $(-5, 6)$
   (e) $(8, 4)$ and $(4, -3)$ (f) $(-4, -5)$ and $(4, 5)$
   (g) $(3, 3)$ with $b = 0.5$ (h) $(-5, 7)$ with $b = 0.75$
   (i) $(\frac{3}{4}, \frac{6}{5})$ with $b = -\frac{3}{2}$ (j) $(\frac{4}{3}, 70)$ with $b = \frac{3}{4}$
   (k) $(0.8, 0.5)$ with $b = 0.5$ (l) $(3, -4)$ with $b = 5$

3. Establish linear functions for the following problems.
   (a) Costs are \$3000 when output is 0 and \$4000 when output is 1500 units.
   (b) Fixed costs are \$75,000 and variable costs are \$3.50 per unit.
   (c) Costs are \$5000 for 150 units and \$6000 for 250 units.
   (d) If 1000 units can be sold at a price of \$2.50 per unit and 2000 units can be sold at a price of \$1.75 per unit, determine price as a function of the number of units sold.
   (e) If 1500 units can be sold at a price of \$10 per unit and 2500 units can be sold at a price of \$9 per unit, determine price as a function of the number of units sold.
   (f) A firm has 250 employees and expects to add to their work force by 10 employees per year.
   (g) An individual now earning \$10,000 per year expects raises of \$500 per year.
   (h) A dealer advertises that a car worth \$3000 will be reduced in price by \$100 per month until sold.

4. For the sets A and B defined below, determine $A \cap B$.
$$A = \{x \mid x^2 - x - 2 \leq 0\}$$
$$B = \{x \mid x + 1 > 0\}$$

5. For the sets A and B, find $A \cap B$ and $A \cup B$.
$$A = \{-5, -4, -3, -2, -1, 0, 1, 2, 3, 4\}$$
$$B = \{x \mid x^2 - 9x + 20 = 0\}$$

6. For the set A in problem 5, list the elements in the following sets that are members of A.
   (a) $\{x \mid x^2 = 9\}$
   (b) $\{x \mid -3 < 2x < 4\}$

7. Verify that the following set is a function. Give the domain and range of the function.
$$\{(3, 7), (4, 10), (5, 8), (0, 6)\}$$

8. Is the following set a function or relation?
$$\{(A, 1), (A, 2), (B, 1), (B, 2)\}$$

9. Develop inequalities for the following problems and give the domain of the variables.
   (a) Two products can be manufactured by using the turret lathe. One product requires 2 hours per unit on the lathe and the other product requires 3 hours per unit. The maximum capacity of the lathe is 80 hours per week.
   (b) An airplane has a useful load of 1800 pounds. Gasoline weighs 6 pounds per gallon, and the fuel cells have a maximum capacity of 140 gallons. The standard weight for passengers and crew is 170 pounds, and a maximum of six people can be carried on the airplane.
   (c) A construction equipment leasing company is considering the purchase of dump trucks, road graders, and bulldozers at a cost of $20,000 per truck, $60,000 per grader, and $80,000 per dozer. A maximum of $1.2 million has been authorized for the equipment.
   (d) An individual must have a minimum of 300 units of a special dietary supplement each day. This supplement is available in two commercial products, A and B. Each ounce of A contains 25 units of the supplement and each ounce of B contains 40 units.

10. A firm is interested in determining the breakeven production for a new product. The cost of introducing the product is estimated to be $40,000. This cost includes initial advertising and promotion as well as the fixed cost necessary for one year of production. The variable cost per unit is $35, and the proposed selling price is $60.
    (a) Determine breakeven production.
    (b) Determine the profit on sales of 2000 units.

11. The sole proprietor of a rug-cleaning business has determined that he must work at least as many hours as it takes to cover his fixed cost of $45 per day. He averages $9 per hour and can work no more than 12 hours on any one day. Determine the profit function and the domain and range of the function. Also, determine his breakeven point in hours.

12. The Handley Company plans to produce and sell an item. The cost of production includes fixed cost of $50,000 and variable cost of $10 per item. They plan to sell the item for $25. How many items must be sold to obtain a profit of $100,000?

13. The profit function of Graham Manufacturing Company is described by a quadratic function. The following data points were estimated by the chief accountant: output of zero units results in a loss of $10 million, output of 6000 units results in a profit of $8 million, and output of 8000 units results in a profit of $6 million. Determine the quadratic function that describes profit as a function of the number of units produced.

14. The average per unit cost of a certain assembly manufactured by Milwaukee Electronics Corporation is $400 for the 300th unit, $250 for the 500th unit, and $550 for the 600th unit. Determine the quadratic function that describes the relationship between average cost per unit and output.

15. In a controlled experiment the productivity of a work group was found to be related to the number of workers assigned to the group. Specifically, the analyst found that with a group of six workers the productivity was eight units per man. When the group size was increased to eight workers, the productivity rose to 12 units per man. An additional increase in group size to 10 workers resulted in an increase in productivity to 14 units per man. Determine the quadratic function that relates productivity to the group size.

16. Profit derived from gas sales at a service station depends upon the quantities sold of three types of gas; Regular, Unleaded, and Premium.
    (a) State the profit function for total profit on gallons sold of the three types of gas.

    It was determined from past experience that the profit derived from certain sales combinations is as follows:

    Profit = {($440, 5000 gal. R, 1000 gal. U, 4000 gal. P),
    ($245, 3000 gal. R, 500 gal. U, 2000 gal. P),
    ($ 93, 1000 gal. R, 100 gal. U, 1000 gal. P)}

    (b) Determine the profit per gallon for each type of gas by substituting these data points into the profit function.

17. In a certain company, cost per unit is constant at $100 from 0 to 1000 units produced per month. From 1001 to 3000 units, costs decrease

linearly to $50 per unit at 3000 units. Costs then increase by $0.01 per unit produced up to 4000 units. Determine the functions describing the cost per unit for the product.

18. The total cost of four milk shakes and two coffees is $1.40. The cost of one milk shake and three coffees is $0.60. Find the price of a milk shake and a coffee.

19. A firm uses two machines in the manufacture of two products. Product A requires three hours on machine 1 and five hours on machine 2. Product B requires six hours on machine 1 and four hours on machine 2. If 36 hours of machine 1 time and 42 hours of machine 2 time are to be scheduled, determine the product mix that fully utilizes the machine time.

20. A company packages fruit in fancy boxes for sale as Christmas presents. One box contains six apples, four pears, and four oranges and sells for $1.68. A second box contains five apples, four pears, and six oranges and sells for $1.88. A third box contains eight apples, six pears, and ten oranges and sells for $3.02. Assuming that the cost of fruit is constant, determine the per unit cost of each type of fruit.

21. Determine whether the following consistent system of equations has a unique or an infinite number of solutions.

$$3x + 2y + 7z - 5 = 0$$
$$4x - y + 5z + 4 = 0$$
$$7x + y + 12z - 1 = 0$$

22. Verify by graphing that the following system of linear inequalities has an infinite number of solutions.

$$4x_1 + 5x_2 \leq 20$$
$$6x_1 + 3x_2 \geq 18$$

23. Specify whether the following systems of equations are consistent or inconsistent. For those that are consistent, state whether there is a unique or an infinite number of solutions.

(a) $5x + 7y = 70$
    $x + y = 12$

(b) $6x + 15y = 90$
    $4x + 5y = 40$
    $11x + 8y = 87$

(c) $x + y = 10$
    $8x + 11y = 88$
    $7x + 14y = 98$

(d) $2x_1 + 3x_2 + 8x_3 = 56$
    $4x_1 - 3x_2 + 5x_3 = 48$
    $-2x_1 + 6x_2 + 3x_3 = 8$

24. Determine if the following system of equations has a consistent, unique solution.

$$x + y = 4$$

$$2x + y \geq 6$$
$$2x^2 + 3xy + y^2 = 24$$

SUGGESTED REFERENCES

BURNS, CARL M., et al., *Algebra: An Introduction for College Students* (Menlo Park, Ca.: Cummings Publishing Company, Inc., 1972).

FREUND, JOHN E., *College Mathematics with Business Applications* (Englewood Cliffs, N.J.: Prentice-Hall, Inc., 1969), 3, 4.

MOORE, GERALD E., *Algebra*, The Barnes and Noble College Outline Series (New York, N.Y.: Barnes and Noble, Inc., 1970).

NIELSEN, KAJ L., *Algebra: A Modern Approach*, The Barnes and Noble College Outline Series (New York, N.Y.: Barnes and Noble, Inc., 1969).

NIELSEN, KAJ L., *College Mathematics*, The Barnes and Noble College Outline Series (New York, N.Y.: Barnes and Noble, Inc., 1958), 1–6.

RICH, BARNETT, *Elementary Algebra*, Schaum's Outline Series (New York, N.Y.: McGraw-Hill Book Company, Inc., 1960).

THEODORE, CHRIS A., *Applied Mathematics: an Introduction* (Homewood, Ill.: Richard D. Irwin, Inc., 1965), 5–7.

Chapter 3

Matrices
and
Matrix Algebra

Matrices and matrix algebra have begun to play an increasingly important role in modern techniques for quantitative and statistical analysis of business decisions. These tools have also become quite important in the functional business areas of accounting, production, finance, and marketing. Since the matrix provides a method of representing large quantities of data, it has become an important working tool of the computer programmer and program analyst.

Before beginning a formal discussion of matrices and matrix algebra, let us define a *matrix* as an array or group of numbers. To illustrate, the quantity of inventory by product line at three local car dealers can be described by a matrix. This matrix might have the form shown below.

$$
\begin{array}{ccc}
\text{Ford} & \text{Chevrolet} & \text{Plymouth}
\end{array}
$$

$$
\begin{pmatrix}
6 & 8 & 5 \\
10 & 10 & 8 \\
3 & 4 & 2 \\
4 & 5 & 3
\end{pmatrix}
\begin{array}{l}
\text{Compact} \\
\text{Sedan} \\
\text{Convertible} \\
\text{Station wagon}
\end{array}
$$

The matrix shows that the Ford dealer has an inventory of 6 compacts, 10 sedans, 3 convertibles, and 4 station wagons. Similarly, the inventories of the Chevrolet and Plymouth dealers can be read directly from the matrix.

An important characteristic of a matrix is that the position within the matrix of each number as well as the magnitude of the number is important. For instance, the fact that the local Chevrolet dealer has four convertibles in inventory is described by the entry in the matrix at the intersection of the convertible row and the Chevrolet column. The position of the number is important in that a specific location within the matrix is reserved for the inventory of Chevrolet convertibles. The magnitude of the number is important, since it gives the number of Chevrolet convertibles in inventory.

Certain algebraic operations can be performed on matrices. One such operation might involve determining the value of the inventories of the car dealers. The value of the inventories could be found by multiplying the inventory matrix by a price matrix. This would be an example of matrix multiplication. Other examples of matrix algebra might involve the addition or subtraction of two or more matrices or the multiplication of each of the numbers in the matrix by a common number.

This chapter provides an introduction to matrices, matrix algebra, and selected business applications of matrices. The discussion is extended by the introduction of additional topics in Chapter 4. It should be emphasized that the procedure introduced in this chapter for determining solutions to simultaneous equations will be quite important in the chapters on linear programming.

3.1 Vectors

A *vector* is defined as a row or column of numbers in which the position of each number within the row or column is of importance. A vector is a special case of a matrix in that it has only a single row or column rather than multiple rows and columns. Column vectors are written vertically and row vectors horizontally. Examples of column vectors are $\begin{pmatrix} 2 \\ 1 \end{pmatrix}$, $\begin{pmatrix} -3 \\ 6 \end{pmatrix}$, $\begin{pmatrix} 1 \\ 0 \\ 2 \\ 3 \end{pmatrix}$, and $\begin{pmatrix} 3 \\ -4 \\ 7 \end{pmatrix}$.

Examples of row vectors are $(3, 2)$, $(6, 0, 4)$, and $(-6, 4, \frac{1}{2}, 7)$.

To illustrate that the position of the number within the vector is important, the row vector $(3, 2)$ and the row vector $(2, 3)$ have the same components but are not equal. Vectors are equal only if they are exactly the same. Thus, for two vectors to be equal, each component of the first vector must be exactly the same and occupy the same position as each component of the second vector. Row vectors can, therefore, equal other row vectors but not column vectors. Similarly, column vectors can equal other column vectors but not row vectors.

Example: Are the vectors $(6, 4)$ and $\begin{pmatrix} 6 \\ 4 \end{pmatrix}$ equal?

Since $(6, 4)$ is a row vector and $\begin{pmatrix} 6 \\ 4 \end{pmatrix}$ a column vector, the two vectors are not equal.

Example: Are the vectors $(6, 1, 0)$ and $(6, 1)$ equal?

Since the first vector has the element 0 while the second does not, the two row vectors are not equal.

Both row and column vectors are designated by capital letters. The numbers that comprise the vector are called the *elements* or *components* of the vector. We could thus write $A = (6, 10, 12)$, where A designates the row vector with components or elements 6, 10, and 12.

3.2 Vector Algebra

The basic rules for the addition, subtraction, and multiplication of vectors are analogous to the rules of ordinary algebra. The concept of division is not directly extendable to either vectors or matrices. The matrix counterpart of division will be treated in the discussion of matrix algebra.

3.2.1 VECTOR ADDITION AND SUBTRACTION

Row or column vectors may be added or subtracted to other like row or column vectors by adding or subtracting components. If, for instance, $A = (6, 10, 12)$ and $B = (4, 6, -3)$, then

$$A + B = (6 + 4, 10 + 6, 12 - 3) = (10, 16, 9)$$

and

$$A - B = (6 - 4, 10 - 6, 12 + 3) = (2, 4, 15)$$

Similarly, for the column vectors $G = \begin{pmatrix} 3 \\ 2 \end{pmatrix}$ and $H = \begin{pmatrix} -1 \\ 3 \end{pmatrix}$,

$$G + H = \begin{pmatrix} 3 \\ 2 \end{pmatrix} + \begin{pmatrix} -1 \\ 3 \end{pmatrix} = \begin{pmatrix} 2 \\ 5 \end{pmatrix}$$

It is important to recognize that the addition or subtraction of vectors can take place only if the vectors are of the same type; i.e., row vectors may be added to or subtracted from row vectors and column vectors may be added to or subtracted from column vectors. Furthermore, the vectors to be added or subtracted must have the same number of components. The vector $A = (6, 10, 12)$ cannot be added to the vector $E = (3, 2)$ nor the vector $F = (5, 3, 7, 10)$. Vectors that are the same type and have the same number of components are said to be *conformable* for addition or subtraction.

The commutative and associative laws of addition apply to vector addition. The reader may remember that the *commutative law* of addition states that $a + b = b + a$. If this is applied to vector addition, the commutative law of addition states that $A + B = B + A$. This is easily verified, since for $A = (6, 10, 12)$ and $B = (4, 6, -3)$, $A + B = (10, 16, 9)$ and $B + A = (10, 16, 9)$.

The *associative law* of addition states that $a + (b + c) = (a + b) + c$. Applying this law to the vectors $A = (6, 10, 12)$, $B = (4, 6, -3)$, and $C = (-4, 8, 6)$, we have

$$A + (B + C) = (A + B) + C$$
$$(6, 10, 12) + (4 - 4, 6 + 8, -3 + 6) = (6 + 4, 10 + 6, 12 - 3) + (-4, 8, 6)$$
$$(6, 10, 12) + (0, 14, 3) = (10, 16, 9) + (-4, 8, 6)$$
$$(6 + 0, 10 + 14, 12 + 3) = (10 - 4, 16 + 8, 9 + 6)$$
$$(6, 24, 15) = (6, 24, 15)$$

This verifies that the associative law applies to vectors.

Example: A student has 96 semester credits of which 24 units are A, 36 units are B, 30 units are C, 6 units are D, and 0 units are F. If he receives 6 units of A, 6 units of B, 3 units of C, 0 units of D, and 3 units of F, determine the vector that describes his grades.

Let $G = (24, 36, 30, 6, 0)$ represent grades excluding the current semester and $S = (6, 6, 3, 0, 3)$ represent the current semester's grades. Then $G + S = (30, 42, 33, 6, 3)$ is the vector that describes his grades including the current semester.

3.2.2 VECTOR MULTIPLICATION

Vector multiplication includes the multiplication of a vector by a vector or a vector by a scalar. A *scalar* is any number, such as 3, or -10, or 1.10, etc. A vector may be multiplied by a scalar simply by multiplying each component of the vector by the scalar and listing the resulting products as a new vector. To demonstrate, let $X = \begin{pmatrix} 3 \\ -6 \\ 2 \end{pmatrix}$ and $a = -3$. The product of the vector X and the scalar a is

$$aX = -3 \begin{pmatrix} 3 \\ -6 \\ 2 \end{pmatrix} = \begin{pmatrix} -9 \\ 18 \\ -6 \end{pmatrix}$$

Similarly, if we multiply the row vector $A = (6, 10, 12)$ by the scalar $b = 1.5$, we obtain

$$bA = 1.5(6, 10, 12) = (9, 15, 18)$$

Example: The vector $P = (\$10, \$8, \$6)$ represents the prices of three items. If prices increase by 10 percent, determine the new price vector.

The new price vector is formed by the product of the scalar 1.10 and the vector P. Thus, $1.10P = 1.10(\$10, \$8, \$6) = (\$11, \$8.80, \$6.60)$.

Two vectors may be multiplied together, provided that: (1) one of the vectors is a row vector and the other a column vector; and (2) the row vector and column vector each contain the same number of components. To illustrate, consider the row vector $R = (r_1, r_2, r_3)$ and the column vector $C = \begin{pmatrix} c_1 \\ c_2 \\ c_3 \end{pmatrix}$.

Since R is a row vector and C a column vector, and since each vector has three components, it is possible to determine the product of R and C. The vector multiplication is made by placing the row vector to the left of the column vector. The first component of the row vector is then multiplied by the first component of the column vector, the second component of the row vector is multiplied by the second component of the column vector, etc. These products are summed and the resulting sum represents the product of the two vectors. The product of R and C is

$$RC = (r_1, r_2, r_3) \begin{pmatrix} c_1 \\ c_2 \\ c_3 \end{pmatrix} = r_1c_1 + r_2c_2 + r_3c_3 \qquad (3.1)$$

The result is termed the *inner product* of R and C and is a scalar.

Example: Determine the inner product of $A = (6, 10, 12)$ and $X = \begin{pmatrix} 3 \\ -6 \\ 2 \end{pmatrix}$.

$$AX = 6 \cdot 3 + 10(-6) + 12 \cdot 2 = -18$$

Example: At the end of a 64-unit MBA program a student has 30 units of A, 20 units of B, 10 units of C, 4 units of D, and 0 units of F. If an A is worth 4 grade points, a B is worth 3 grade points, a C is worth 2 grade points, a D is worth 1 grade point, and an F is worth 0 grade points, determine the student's total number of grade points.

Let

$$A = (30, 20, 10, 4, 0) \text{ and } G = \begin{pmatrix} 4 \\ 3 \\ 2 \\ 1 \\ 0 \end{pmatrix}$$

The total number of grade points is given by the inner product of A and G. Thus,

$$AG = (30, 20, 10, 4, 0) \begin{pmatrix} 4 \\ 3 \\ 2 \\ 1 \\ 0 \end{pmatrix} = 120 + 60 + 20 + 4 + 0 = 204$$

The *commutative law* of multiplication states that $a \cdot b = b \cdot a$. Although this result is true in ordinary algebra, it is not true that $AB = BA$ in vector algebra. If A is a row vector and B a column vector, then AB is a scalar. As we demonstrate later in this chapter, the product of the column vector B and the row vector A, i.e., BA, is a matrix. It will become obvious that the scalar AB is not equal to the matrix BA; thus, the commutative law of multiplication does not apply to vector multiplication.

The *distributive law* of algebra applies to vector algebra. The distributive law states that $a(b + c) = ab + ac$. Applying this to vector algebra gives the result that $P(B + E) = PB + PE$, where P is a row vector and B and E are column vectors. The distributive law is demonstrated in the following example.

Example: A retailer stocks four brands of a product. The costs of these four brands are given by the row vector $P = (60, 70, 90, 100)$. The beginning inventory of the product is given by the column vector $B = \begin{pmatrix} 30 \\ 20 \\ 15 \\ 10 \end{pmatrix}$, and the ending inventory by $E = \begin{pmatrix} 20 \\ 15 \\ 10 \\ 5 \end{pmatrix}$. Assuming no purchases of inventory, determine the cost of goods sold during the period.

To illustrate the distributive law, determine $PB - PE$ and $P(B - E)$.

$$PB - PE = (60, 70, 90, 100) \begin{pmatrix} 30 \\ 20 \\ 15 \\ 10 \end{pmatrix} - (60, 70, 90, 100) \begin{pmatrix} 20 \\ 15 \\ 10 \\ 5 \end{pmatrix}$$

$$PB - PE = (60 \cdot 30 + 70 \cdot 20 + 90 \cdot 15 + 100 \cdot 10)$$
$$- (60 \cdot 20 + 70 \cdot 15 + 90 \cdot 10 + 100 \cdot 5)$$

$$PB - PE = \$1900$$

and

$$P(B - E) = (60, 70, 90, 100) \begin{pmatrix} 30 - 20 \\ 20 - 15 \\ 15 - 10 \\ 10 - 5 \end{pmatrix}$$

$$P(B - E) = (60, 70, 90, 100) \begin{pmatrix} 10 \\ 5 \\ 5 \\ 5 \end{pmatrix} = 60 \cdot 10 + 70 \cdot 5 + 90 \cdot 5 + 100 \cdot 5$$

$$P(B - E) = \$1900$$

The $1900 obtained by both methods verifies the distributive law for vector algebra.

To summarize, the commutative and associative laws of addition apply to vector addition; i.e., $A + B = B + A$ and $A + (B + C) = (A + B) + C$. The distributive law of algebra applies to vector algebra; i.e., $A(B + C) = AB + AC$. The commutative law of multiplication, however, does not apply to vector algebra; i.e., $AB \neq BA$.

3.3 Matrices

A matrix is an array of numbers. The array is said to be *ordered*, meaning that the position of each number is important. The ordered array of numbers consists of m rows and n columns that are enclosed in parentheses and designated by a capital letter. The general form of a matrix is

$$A = \begin{pmatrix} a_{11} & a_{12} & a_{13} & \cdots & a_{1n} \\ a_{21} & a_{22} & a_{23} & \cdots & a_{2n} \\ a_{31} & a_{32} & a_{33} & \cdots & a_{3n} \\ \vdots & \vdots & \vdots & & \vdots \\ a_{m1} & a_{m2} & a_{m3} & \cdots & a_{mn} \end{pmatrix} \qquad (3.2)$$

The matrix A consists of $m \cdot n$ components or elements a_{ij}. A component a_{ij} is designated by its position in the matrix, i referring to the row and j to the column. Component a_{24} thus refers to the component at the intersection of the second row and fourth column.

The matrix A is said to be of order m by n or, alternatively, is said to have dimensions of m by n, m and n again referring to the number of rows and number of columns.

Example:

$$A = \begin{pmatrix} 3 & 1 \\ 2 & 4 \\ 1 & 6 \end{pmatrix} \text{ is a 3 by 2 matrix.}$$

$B = \begin{pmatrix} 2 & 4 & 1 \\ 3 & 7 & 0 \end{pmatrix}$ is a 2 by 3 matrix.

$C = \begin{pmatrix} 3 & 2 & 6 \\ 1 & 5 & -2 \\ -1 & 6 & 1 \end{pmatrix}$ is a 3 by 3 square matrix.

$D = (3, 5, 2)$ is a 1 by 3 matrix or row vector.

$E = \begin{pmatrix} 2 \\ 1 \end{pmatrix}$ is a 2 by 1 matrix or column vector.

Matrices provide a convenient method of storing large quantities of data. As an example, consider a firm with multiple retail outlets. The inventory of the firm can be represented by the matrix G, where the columns represent the outlets and the rows represent products.

$$
\begin{array}{c}
\textit{Outlets} \\
\begin{array}{ccc}
1 & 2 & 3
\end{array}
\end{array}
$$

$$
G = \begin{pmatrix} 50 & 60 & 40 \\ 175 & 200 & 125 \\ 40 & 25 & 30 \\ 205 & 235 & 275 \end{pmatrix} \quad \begin{array}{c} \textit{Products} \\ 1 \\ 2 \\ 3 \\ 4 \end{array}
$$

Component a_{21} of matrix G is 175 units of product 2 at outlet 1.
Component a_{42} is 235 units of product 4 at outlet 2.

3.4 Matrix Algebra

Matrix algebra, also termed *linear algebra*, provides a set of rules for the addition, subtraction, multiplication, and inversion of matrices. The rules for addition and subtraction of matrices are analogous to those for ordinary algebra. The multiplication of matrices requires the introduction of a straightforward rule for matrix multiplication. Division of one matrix by another is not possible in matrix algebra. The process of matrix inversion, however, has many similarities with division in ordinary algebra. Matrix addition, subtraction, and multiplication are discussed in this section. Matrix inversion is presented in Sec. 3.5.

3.4.1 MATRIX ADDITION AND SUBTRACTION

Matrices can be added or subtracted, provided that they are conformable for addition or subtraction. Matrices are conformable for addition or subtraction if they have the same dimensions. A 2 by 3 matrix may be added to or subtracted from another 2 by 3 matrix, a 4 by 2 matrix added to or subtracted from another 4 by 2 matrix, etc. The addition or subtraction is performed by adding or subtracting corresponding components from the two matrices. The result is a new matrix of the same order or dimension as the original matrices.

Example: Let

$$A = \begin{pmatrix} 3 & 4 \\ 6 & 5 \\ 4 & 0 \end{pmatrix} \quad \text{and} \quad B = \begin{pmatrix} 2 & 5 \\ 3 & 6 \\ 4 & 2 \end{pmatrix}.$$

Determine $A + B$ and $A - B$.

$$A + B = \begin{pmatrix} 3+2 & 4+5 \\ 6+3 & 5+6 \\ 4+4 & 0+2 \end{pmatrix} = \begin{pmatrix} 5 & 9 \\ 9 & 11 \\ 8 & 2 \end{pmatrix}$$

$$A - B = \begin{pmatrix} 3-2 & 4-5 \\ 6-3 & 5-6 \\ 4-4 & 0-2 \end{pmatrix} = \begin{pmatrix} 1 & -1 \\ 3 & -1 \\ 0 & -2 \end{pmatrix}$$

Example: Let

$$C = \begin{pmatrix} 1 & 3 & 4 \\ 2 & 6 & 5 \\ 1 & -1 & 6 \end{pmatrix}, \quad D = \begin{pmatrix} 2 & 3 & 5 \\ 4 & 1 & 3 \\ 5 & 2 & 1 \end{pmatrix}, \quad E = \begin{pmatrix} 1 & -2 & 5 \\ 3 & -1 & 4 \\ 6 & 5 & 3 \end{pmatrix}$$

Show that $C + (D + E) = (C + D) + E$.

$$C + (D + E) = \begin{pmatrix} 1 & 3 & 4 \\ 2 & 6 & 5 \\ 1 & -1 & 6 \end{pmatrix} + \begin{pmatrix} 3 & 1 & 10 \\ 7 & 0 & 7 \\ 11 & 7 & 4 \end{pmatrix} = \begin{pmatrix} 4 & 4 & 14 \\ 9 & 6 & 12 \\ 12 & 6 & 10 \end{pmatrix}$$

$$(C + D) + E = \begin{pmatrix} 3 & 6 & 9 \\ 6 & 7 & 8 \\ 6 & 1 & 7 \end{pmatrix} + \begin{pmatrix} 1 & -2 & 5 \\ 3 & -1 & 4 \\ 6 & 5 & 3 \end{pmatrix} = \begin{pmatrix} 4 & 4 & 14 \\ 9 & 6 & 12 \\ 12 & 6 & 10 \end{pmatrix}$$

These examples illustrate the commutative and associative laws of addition. The commutative law states that $A + B = B + A$. The associative law states that $C + (D + E) = (C + D) + E$. The examples show that these laws apply to the addition and subtraction of matrices.

3.4.2 MATRIX MULTIPLICATION

Matrix multiplication is subdivided into two cases: (1) the multiplication of a matrix by a scalar, and (2) the multiplication of a matrix by a matrix. The The simpliest of these cases is the multiplication of a matrix by a scalar. A matrix may be multiplied by a scalar by multiplying each component in the matrix by the scalar. The resulting matrix will be of the same order as the original matrix. To illustrate, the 3 by 3 matrix

$$A = \begin{pmatrix} a_{11} & a_{12} & a_{13} \\ a_{21} & a_{22} & a_{23} \\ a_{31} & a_{32} & a_{33} \end{pmatrix}$$

can be multiplied by the scalar k to give the 3 by 3 matrix

$$kA = \begin{pmatrix} ka_{11} & ka_{12} & ka_{13} \\ ka_{21} & ka_{22} & ka_{23} \\ ka_{31} & ka_{32} & ka_{33} \end{pmatrix}$$

Similarly, if a 2 by 4 matrix were multiplied by a scalar, the resulting matrix would be a 2 by 4 matrix.

Example: Multiply the matrix $Y = \begin{pmatrix} 2 & 3 & 5 \\ 4 & 6 & -2 \end{pmatrix}$ by the scalar $k = 3$.

$$kY = 3 \begin{pmatrix} 2 & 3 & 5 \\ 4 & 6 & -2 \end{pmatrix} = \begin{pmatrix} 6 & 9 & 15 \\ 12 & 18 & -6 \end{pmatrix}$$

Example: Multiply the matrix $B = \begin{pmatrix} 2 & 1 \\ 1 & -3 \end{pmatrix}$ by the scalar $k = -2$.

$$kB = -2 \begin{pmatrix} 2 & 1 \\ 1 & -3 \end{pmatrix} = \begin{pmatrix} -4 & -2 \\ -2 & 6 \end{pmatrix}$$

Example: Multiply the matrix $A = \begin{pmatrix} 1 & 3 & -2 \\ 4 & 7 & 1 \end{pmatrix}$ by the scalar $k = 0.1$.

$$kA = 0.1 \begin{pmatrix} 1 & 3 & -2 \\ 4 & 7 & 1 \end{pmatrix} = \begin{pmatrix} 0.1 & 0.3 & -0.2 \\ 0.4 & 0.7 & 0.1 \end{pmatrix}$$

Matrix multiplication is somewhat more complicated than the multiplication of a matrix by a scalar. Two matrices may be multiplied together only if the number of columns in the first matrix equals the number of rows in the second matrix. If matrix A has dimensions a by b and matrix B has dimensions c by d, the matrix product AB is defined only if b is equal to c. Thus, a 2 by 3 matrix can be multiplied by a 3 by 4 matrix but not, for example, by a 2 by 4 matrix. The matrix resulting from the multiplication has dimensions equivalent to the number of rows in the first matrix and the number of columns in the second matrix, i.e., a by d. To summarize, matrix A with dimensions a by b may be multiplied by B with dimensions c by d to form matrix AB with dimensions a by d, provided that b equals c. If the number of columns in the first matrix equals the number of rows in the second matrix, the matrices are said to be conformable for multiplication.

To illustrate the multiplication of two matrices, multiply

$$A = \begin{pmatrix} a_{11} & a_{12} & a_{13} \\ a_{21} & a_{22} & a_{23} \end{pmatrix} \text{ by } B = \begin{pmatrix} b_{11} & b_{12} & b_{13} \\ b_{21} & b_{22} & b_{23} \\ b_{31} & b_{32} & b_{33} \end{pmatrix}$$

Since A is a 2 by 3 matrix and B is 3 by 3 matrix, the matrices are conformable for multiplication. The resulting matrix AB has dimensions 2 by 3. The multiplication is performed as follows:

$$AB = \begin{pmatrix} a_{11} & a_{12} & a_{13} \\ a_{21} & a_{22} & a_{23} \end{pmatrix} \begin{pmatrix} b_{11} & b_{12} & b_{13} \\ b_{21} & b_{22} & b_{23} \\ b_{31} & b_{32} & b_{33} \end{pmatrix}$$

$$AB = \begin{pmatrix} a_{11}b_{11} + a_{12}b_{21} + a_{13}b_{31} & a_{11}b_{12} + a_{12}b_{22} + a_{13}b_{32} \\ a_{21}b_{11} + a_{22}b_{21} + a_{23}b_{31} & a_{21}b_{12} + a_{22}b_{22} + a_{23}b_{32} \end{pmatrix}$$

$$\begin{aligned} & a_{11}b_{13} + a_{12}b_{23} + a_{13}b_{33} \\ & a_{21}b_{13} + a_{22}b_{23} + a_{23}b_{33} \end{aligned}$$

Matrix AB in the preceding example is a 2 by 3 matrix. The components of the matrix are formed by the sum of the products of the rows of the first matrix and columns of the second matrix. The sum of the products of the components in the first row of A and the first column of B give the first row, first column of AB. Similarly, the sum of the products of the components in the first row of A and second column of B gives the first row, second column component of AB. This procedure is continued until the matrix AB is formed.

Example: Determine the product of $A = \begin{pmatrix} 6 & 2 & 4 \\ 1 & 2 & 2 \end{pmatrix}$ and $B = \begin{pmatrix} 3 & 2 \\ 2 & 4 \\ 4 & 5 \end{pmatrix}$.

$$AB = \begin{pmatrix} 6 & 2 & 4 \\ 1 & 2 & 2 \end{pmatrix} \begin{pmatrix} 3 & 2 \\ 2 & 4 \\ 4 & 5 \end{pmatrix} = \begin{pmatrix} 6 \cdot 3 + 2 \cdot 2 + 4 \cdot 4 & 6 \cdot 2 + 2 \cdot 4 + 4 \cdot 5 \\ 1 \cdot 3 + 2 \cdot 2 + 2 \cdot 4 & 1 \cdot 2 + 2 \cdot 4 + 2 \cdot 5 \end{pmatrix}$$

$$AB = \begin{pmatrix} 18 + 4 + 16 & 12 + 8 + 20 \\ 3 + 4 + 8 & 2 + 8 + 10 \end{pmatrix} = \begin{pmatrix} 38 & 40 \\ 15 & 20 \end{pmatrix}$$

Example: Determine the product of $A = \begin{pmatrix} 6 & 1 \\ 3 & 7 \end{pmatrix}$ and $B = \begin{pmatrix} 2 & 4 & 7 \\ 3 & 1 & 5 \end{pmatrix}$.

$$AB = \begin{pmatrix} 6 & 1 \\ 3 & 7 \end{pmatrix} \begin{pmatrix} 2 & 4 & 7 \\ 3 & 1 & 5 \end{pmatrix}$$

$$= \begin{pmatrix} 6 \cdot 2 + 1 \cdot 3 & 6 \cdot 4 + 1 \cdot 1 & 6 \cdot 7 + 1 \cdot 5 \\ 3 \cdot 2 + 7 \cdot 3 & 3 \cdot 4 + 7 \cdot 1 & 3 \cdot 7 + 7 \cdot 5 \end{pmatrix} = \begin{pmatrix} 15 & 25 & 47 \\ 27 & 19 & 56 \end{pmatrix}$$

In the first example, a 2 by 3 matrix is multiplied by a 3 by 2 matrix. The matrices are conformable for multiplication, since the number of columns in the first matrix equals the number of rows in the second. The dimensions of the resulting matrix are equal to the number of rows of the first matrix and the number of columns of the second matrix. The second example shows that a 2 by 2 matrix and a 2 by 3 matrix are conformable for multiplication. The product of the matrices is a 2 by 3 matrix.

Example: The inventory of a firm with multiple retail outlets is given by the matrix G

$$
\begin{array}{ccc}
\textit{Outlets} & & \textit{Products}
\end{array}
$$

$$
\begin{array}{ccc}
1 & 2 & 3
\end{array}
$$

$$
G = \begin{pmatrix} 50 & 60 & 40 \\ 175 & 200 & 125 \\ 40 & 25 & 30 \\ 205 & 235 & 275 \end{pmatrix} \begin{array}{c} 1 \\ 2 \\ 3 \\ 4 \end{array}
$$

If the cost of inventory is given by the vector $C = (20, 10, 30, 40)$, determine the value of inventory at each outlet.

The value of inventory at each outlet is given by the matrix CG. Since C is a 1 by 4 matrix and G is a 4 by 3 matrix, the matrices are conformable for multiplication. The product is

$$
CG = (20, 10, 30, 40) \begin{pmatrix} 50 & 60 & 40 \\ 175 & 200 & 125 \\ 40 & 25 & 30 \\ 205 & 235 & 275 \end{pmatrix}
$$

$$
CG = (12{,}150,\ 13{,}350,\ 13{,}950)
$$

Inventory has a value of \$12,150 at outlet 1, \$13,350 at outlet 2, and \$13,950 at outlet 3.

Example: For $A = (6, 2, 3)$ and $B = \begin{pmatrix} 2 \\ 5 \\ 4 \end{pmatrix}$, find AB and BA.

The row vector A is a 1 by 3 matrix and the column vector B is a 3 by 1 matrix. The inner product AB gives the 1 by 1 matrix or scalar, $AB = 34$.

$$
AB = (6, 2, 3) \begin{pmatrix} 2 \\ 5 \\ 4 \end{pmatrix} = 12 + 10 + 12 = 34
$$

Multiplying B by A is equivalent to multiplying a 3 by 1 matrix by a 1 by 3 matrix. Since the number of columns in the first matrix equal the number of rows in the second matrix, the product is defined. The product of an n by 1 vector and a 1 by n vector is termed the *outer product* and has dimension n by n. To illustrate,

$$
BA = \begin{pmatrix} 2 \\ 5 \\ 4 \end{pmatrix} (6, 2, 3) = \begin{pmatrix} 12 & 4 & 6 \\ 30 & 10 & 15 \\ 24 & 8 & 12 \end{pmatrix}
$$

The outer product of the vectors B and A is the 3 by 3 matrix BA.

Example: For $A = \begin{pmatrix} 3 & -1 & 4 \\ 6 & 8 & 2 \\ 1 & -5 & 4 \end{pmatrix}$ and $B = \begin{pmatrix} 2 & 6 & -1 \\ 3 & -2 & 4 \\ 5 & 3 & -3 \end{pmatrix}$

find AB and BA.

$$AB = \begin{pmatrix} 3 & -1 & 4 \\ 6 & 8 & 2 \\ 1 & -5 & 4 \end{pmatrix}\begin{pmatrix} 2 & 6 & -1 \\ 3 & -2 & 4 \\ 5 & 3 & -3 \end{pmatrix} = \begin{pmatrix} 23 & 32 & -19 \\ 46 & 26 & 20 \\ 7 & 28 & -33 \end{pmatrix}$$

$$BA = \begin{pmatrix} 2 & 6 & -1 \\ 3 & -2 & 4 \\ 5 & 3 & -3 \end{pmatrix}\begin{pmatrix} 3 & -1 & 4 \\ 6 & 8 & 2 \\ 1 & -5 & 4 \end{pmatrix} = \begin{pmatrix} 41 & 51 & 16 \\ 1 & -39 & 24 \\ 30 & 34 & 14 \end{pmatrix}$$

The preceding two examples illustrate that the commutative law of multiplication does not apply to the multiplication of matrices. The commutative law of multiplication states that $a \cdot b = b \cdot a$. Since AB is not equal to BA, the commutative law of multiplication does not apply to matrix multiplication.

The associative and distributive laws of multiplication do, however, apply to matrix multiplication. Given three matrices A, B, and C that are conformable for multiplication, the associative law states that $A(BC) = (AB)C$. The distributive law, again if conformability is assumed, states that $A(B + C) = AB + AC$. These properties are illustrated by the following two examples.

Example: Show that the associative law of multiplication applies to the matrices

$$A = \begin{pmatrix} 2 & 3 \\ 4 & 5 \end{pmatrix}, \qquad B = \begin{pmatrix} 3 & 4 \\ 5 & 6 \end{pmatrix}, \qquad \text{and } C = \begin{pmatrix} 4 & 5 \\ 6 & 7 \end{pmatrix}$$

The associative law of multiplication states that $A(BC) = (AB)C$. To verify the law, we multiply A by BC.

$$BC = \begin{pmatrix} 3 & 4 \\ 5 & 6 \end{pmatrix}\begin{pmatrix} 4 & 5 \\ 6 & 7 \end{pmatrix} = \begin{pmatrix} 36 & 43 \\ 56 & 67 \end{pmatrix}$$

$$A(BC) = \begin{pmatrix} 2 & 3 \\ 4 & 5 \end{pmatrix}\begin{pmatrix} 36 & 43 \\ 56 & 67 \end{pmatrix} = \begin{pmatrix} 240 & 287 \\ 424 & 507 \end{pmatrix}$$

Next, we multiply AB by the matrix C.

$$AB = \begin{pmatrix} 2 & 3 \\ 4 & 5 \end{pmatrix}\begin{pmatrix} 3 & 4 \\ 5 & 6 \end{pmatrix} = \begin{pmatrix} 21 & 26 \\ 37 & 46 \end{pmatrix}$$

$$(AB)C = \begin{pmatrix} 21 & 26 \\ 37 & 46 \end{pmatrix}\begin{pmatrix} 4 & 5 \\ 6 & 7 \end{pmatrix} = \begin{pmatrix} 240 & 287 \\ 424 & 507 \end{pmatrix}$$

Since $A(BC)$ equals $(AB)C$, we conclude that the associative law applies in matrix multiplication.

Example: Show that the distributive law of algebra applies to the matrices

$$A = \begin{pmatrix} 2 & 3 \\ 4 & 5 \end{pmatrix}, \quad B = \begin{pmatrix} 2 & 3 & 6 \\ 3 & 5 & 1 \end{pmatrix}, \quad \text{and} \quad C = \begin{pmatrix} 1 & 4 & 5 \\ 6 & 3 & 1 \end{pmatrix}$$

The distributive law states that $A(B + C) = AB + AC$, provided that B and C are conformable for addition and that AB and AC are conformable for multiplication. The matrix $A(B + C)$ is

$$A(B + C) = \begin{pmatrix} 2 & 3 \\ 4 & 5 \end{pmatrix} \left[\begin{pmatrix} 2 & 3 & 6 \\ 3 & 5 & 1 \end{pmatrix} + \begin{pmatrix} 1 & 4 & 5 \\ 6 & 3 & 1 \end{pmatrix} \right]$$

$$A(B + C) = \begin{pmatrix} 2 & 3 \\ 4 & 5 \end{pmatrix} \begin{pmatrix} 3 & 7 & 11 \\ 9 & 8 & 2 \end{pmatrix} = \begin{pmatrix} 33 & 38 & 28 \\ 57 & 68 & 54 \end{pmatrix}$$

and the matrix $AB + AC$ is

$$AB + AC = \begin{pmatrix} 2 & 3 \\ 4 & 5 \end{pmatrix} \begin{pmatrix} 2 & 3 & 6 \\ 3 & 5 & 1 \end{pmatrix} + \begin{pmatrix} 2 & 3 \\ 4 & 5 \end{pmatrix} \begin{pmatrix} 1 & 4 & 5 \\ 6 & 3 & 1 \end{pmatrix}$$

$$AB + AC = \begin{pmatrix} 13 & 21 & 15 \\ 23 & 37 & 29 \end{pmatrix} + \begin{pmatrix} 20 & 17 & 13 \\ 34 & 31 & 25 \end{pmatrix} = \begin{pmatrix} 33 & 38 & 28 \\ 57 & 68 & 54 \end{pmatrix}$$

Since $A(B + C)$ equals $AB + AC$, the distributive law applies to matrix algebra.

3.4.3 THE IDENTITY AND NULL MATRICES

The *identity* matrix is a square matrix that has 1's on the upper left to lower right diagonal and 0's elsewhere. The 2×2 identity matrix, commonly designated by I, is

$$I = \begin{pmatrix} 1 & 0 \\ 0 & 1 \end{pmatrix}$$

Similarly, the 3 by 3 and 4 by 4 identity matrices, also designated by I, are

$$I = \begin{pmatrix} 1 & 0 & 0 \\ 0 & 1 & 0 \\ 0 & 0 & 1 \end{pmatrix} \quad \text{and} \quad I = \begin{pmatrix} 1 & 0 & 0 & 0 \\ 0 & 1 & 0 & 0 \\ 0 & 0 & 1 & 0 \\ 0 & 0 & 0 & 1 \end{pmatrix}$$

The identity matrix has properties in matrix algebra similar to those of the number 1 in ordinary algebra. Specifically, the identity matrix has the property that the product of any matrix A with a conformable identity matrix I is equal to the matrix A. Symbolically, this means that $AI = A$, or alternatively, $IA = A$. Thus, although we pointed out that the commutative law of multiplication does not apply in matrix multiplication, it does apply in the case of multiplication of a matrix by an identity matrix.

Example: For $A = \begin{pmatrix} 2 & 3 \\ 4 & 5 \end{pmatrix}$, verify that $AI = A = IA$.

$$AI = \begin{pmatrix} 2 & 3 \\ 4 & 5 \end{pmatrix} \begin{pmatrix} 1 & 0 \\ 0 & 1 \end{pmatrix} = \begin{pmatrix} 2 & 3 \\ 4 & 5 \end{pmatrix}$$

$$IA = \begin{pmatrix} 1 & 0 \\ 0 & 1 \end{pmatrix} \begin{pmatrix} 2 & 3 \\ 4 & 5 \end{pmatrix} = \begin{pmatrix} 2 & 3 \\ 4 & 5 \end{pmatrix}$$

Therefore, $AI = A = IA$.

Example: For $A = \begin{pmatrix} 2 & 3 & 4 \\ 5 & 6 & 7 \\ 8 & 9 & 10 \end{pmatrix}$, verify that $AI = A = IA$.

$$AI = \begin{pmatrix} 2 & 3 & 4 \\ 5 & 6 & 7 \\ 8 & 9 & 10 \end{pmatrix} \begin{pmatrix} 1 & 0 & 0 \\ 0 & 1 & 0 \\ 0 & 0 & 1 \end{pmatrix} = \begin{pmatrix} 2 & 3 & 4 \\ 5 & 6 & 7 \\ 8 & 9 & 10 \end{pmatrix}$$

$$IA = \begin{pmatrix} 1 & 0 & 0 \\ 0 & 1 & 0 \\ 0 & 0 & 1 \end{pmatrix} \begin{pmatrix} 2 & 3 & 4 \\ 5 & 6 & 7 \\ 8 & 9 & 10 \end{pmatrix} = \begin{pmatrix} 2 & 3 & 4 \\ 5 & 6 & 7 \\ 8 & 9 & 10 \end{pmatrix}$$

This again demonstrates that $AI = A = IA$.

The *null* matrix is a matrix in which all elements are 0. Unlike the identity matrix, the null matrix need not be a square matrix. The null matrix is commonly designated by the symbol ϕ (phi). A 2 by 3 null matrix would be

$$\phi = \begin{pmatrix} 0 & 0 & 0 \\ 0 & 0 & 0 \end{pmatrix}$$

Similarly, a 1 by 3 null matrix or null row vector is

$$\phi = (0, \quad 0, \quad 0)$$

and a 3 by 1 null matrix or null column vector is

$$\phi = \begin{pmatrix} 0 \\ 0 \\ 0 \end{pmatrix}$$

The product of any matrix with a conformable null matrix is a null matrix, i.e., $A\phi = \phi$. Thus if $A = \begin{pmatrix} 1 & 2 & 3 \\ 4 & 5 & 6 \end{pmatrix}$ and $\phi = \begin{pmatrix} 0 & 0 \\ 0 & 0 \\ 0 & 0 \end{pmatrix}$,

$$A\phi = \begin{pmatrix} 1 & 2 & 3 \\ 4 & 5 & 6 \end{pmatrix} \begin{pmatrix} 0 & 0 \\ 0 & 0 \\ 0 & 0 \end{pmatrix} = \begin{pmatrix} 0 & 0 \\ 0 & 0 \end{pmatrix}$$

3.5 Matrix Representation of Systems of Linear Equations

One of the important uses of matrices and matrix algebra is in the representation of systems of linear equations. To illustrate, consider the two linear equations

$$a_{11}x_1 + a_{12}x_2 = b_1$$
$$a_{21}x_1 + a_{22}x_2 = b_2 \tag{3.3}$$

These two equations can be represented by the matrix equation

$$AX = B \tag{3.4}$$

where A is the *coefficient* matrix $\begin{pmatrix} a_{11}a_{12} \\ a_{21}a_{22} \end{pmatrix}$, X is the *solution* vector $\begin{pmatrix} x_1 \\ x_2 \end{pmatrix}$ and B is the *right-hand-side* vector $\begin{pmatrix} b_1 \\ b_2 \end{pmatrix}$.

To show that the matrix equation $AX = B$ is an alternative way of writing the two linear equations, we write

$$AX = B$$

or alternatively,

$$\begin{pmatrix} a_{11}a_{12} \\ a_{21}a_{22} \end{pmatrix} \begin{pmatrix} x_1 \\ x_2 \end{pmatrix} = \begin{pmatrix} b_1 \\ b_2 \end{pmatrix} \tag{3.5}$$

Multiplying the coefficient matrix by the solution vector and equating the sum to the right-hand side gives

$$a_{11}x_1 + a_{12}x_2 = b_1$$
$$a_{21}x_1 + a_{22}x_2 = b_2$$

The components of the matrix equation $AX = B$ are the coefficient matrix A, the solution vector X, and the right-hand-side vector B. The coefficient matrix contains the coefficients of the linear equations. The solution vector contains the variables in the system, $x_1, x_2, \ldots, x_n$. The right-hand-side or B vector contains those elements customarily written on the right-hand side of the equals sign.†

Example: Express the following system of two equations in matrix form.

$$3x_1 + x_2 = 9$$
$$5x_1 - 3x_2 = 1$$

---

† The system of equations is termed *homogeneous* in those cases in which $B = \phi$. Systems of equations of the type illustrated in this text in which $B \neq \phi$ are, conversely, *nonhomogeneous*.

The two equations can be written as

$$\begin{pmatrix} 3 & 1 \\ 5 & -3 \end{pmatrix} \begin{pmatrix} x_1 \\ x_2 \end{pmatrix} = \begin{pmatrix} 9 \\ 1 \end{pmatrix}$$

or alternatively as

$$AX = B,$$

where

$$A = \begin{pmatrix} 3 & 1 \\ 5 & -3 \end{pmatrix}, \quad X = \begin{pmatrix} x_1 \\ x_2 \end{pmatrix}, \quad B = \begin{pmatrix} 9 \\ 1 \end{pmatrix}$$

Example: Express the system of three equations in matrix form.

$$\begin{aligned} 2x - 3y + 6z &= -18 \\ 6x + 4y - 2z &= 44 \\ 5x + 8y + 10z &= 56 \end{aligned}$$

The three equations can be written as

$$\begin{pmatrix} 2 & -3 & 6 \\ 6 & 4 & -2 \\ 5 & 8 & 10 \end{pmatrix} \begin{pmatrix} x \\ y \\ z \end{pmatrix} = \begin{pmatrix} -18 \\ 44 \\ 56 \end{pmatrix}$$

or alternatively as

$$AX = B,$$

where

$$A = \begin{pmatrix} 2 & -3 & 6 \\ 6 & 4 & -2 \\ 5 & 8 & 10 \end{pmatrix}, \quad X = \begin{pmatrix} x \\ y \\ z \end{pmatrix}, \quad \text{and} \quad B = \begin{pmatrix} -18 \\ 44 \\ 56 \end{pmatrix}$$

By applying the rules of matrix multiplication, a matrix equation is easily transformed into a system of equations. Consider the following examples.

Example: Write the matrix equation $AX = B$, where $A = \begin{pmatrix} 4 & 2 & 3 \\ 3 & 4 & -1 \\ 5 & 3 & 1 \end{pmatrix}$, $X = \begin{pmatrix} x_1 \\ x_2 \\ x_3 \end{pmatrix}$, and $B = \begin{pmatrix} 16 \\ 13 \\ 16 \end{pmatrix}$ as three equations in three variables.

The matrix equation $AX = B$ is written as

$$\begin{pmatrix} 4 & 2 & 3 \\ 3 & 4 & -1 \\ 5 & 3 & 1 \end{pmatrix} \begin{pmatrix} x_1 \\ x_2 \\ x_3 \end{pmatrix} = \begin{pmatrix} 16 \\ 13 \\ 16 \end{pmatrix}$$

Applying the rules of matrix multiplication gives

$$4x_1 + 2x_2 + 3x_3 = 16$$

$$3x_1 + 4x_2 - 1x_3 = 13$$
$$5x_1 + 3x_2 + 1x_3 = 16$$

Example: Express $\begin{pmatrix} 3 & 2 \\ 6 & -2 \end{pmatrix} \begin{pmatrix} x_1 \\ x_2 \end{pmatrix} = \begin{pmatrix} 14 \\ 4 \end{pmatrix}$ as two equations.

Multiplying the coefficient matrix by the solution vector and equating this sum to the right-hand side gives

$$3x_1 + 2x_2 = 14$$
$$6x_1 - 2x_2 = 4$$

Matrix algebra can be used in conjunction with the matrix representation of the system for solution of the system of linear equations. To illustrate, consider the preceding example of two equations with variables x_1 and x_2. The two equations were

$$3x_1 + 2x_2 = 14$$
$$6x_1 - 2x_2 = 4$$

The solution set for these two equations consists of values of x_1 and x_2 that satisfy both equalities. We shall demonstrate that these values are $x_1 = 2$ and $x_2 = 4$.

Techniques for the solution of matrix equations are based upon *row operations*. The term row operations means the application of basic algebraic operations to the rows of a matrix. Three operations are defined. These are

1. Any two rows of a matrix may be interchanged.
2. A row may be multiplied by a nonzero constant.
3. A multiple of one row may be added to another row.

The three row operations are applications of fundamental algebraic operations. To illustrate, consider the two equations

$$3x_1 + 2x_2 = 14$$
$$6x_1 - 2x_2 = 4$$

The first row operation states that the rows may be interchanged; i.e.,

$$6x_1 - 2x_2 = 4$$
$$3x_1 + 2x_2 = 14$$

The second row operation states that a row may be multiplied by a nonzero constant. For example, multiplying $6x_1 - 2x_2 = 4$ by $\frac{1}{2}$ gives

$$3x_1 - 1x_2 = 2$$

The third row operation states that a multiple of one row may be added to another row. Multiplying $3x_1 + 2x_2 = 14$ by -2 and adding the result to the equation $6x_1 - 2x_2 = 4$ gives $0x_1 - 6x_2 = -24$. To demonstrate this result, first multiply $3x_1 + 2x_2 = 14$ by -2.

$$-2(3x_1 + 2x_2 = 14) = -6x_1 - 4x_2 = -28$$

This product is then added to the equation $6x_1 - 2x_2 = 4$, i.e.,

$$
\begin{aligned}
6x_1 - 2x_2 &= 4 \\
-6x_1 - 4x_2 &= -28 \\
\hline
0x_1 - 6x_2 &= -24
\end{aligned}
$$

The resulting equation is $0x_1 - 6x_2 = -24$.

This row operation is based upon the algebraic principle that "equals may be added to equals." For the equation $a = b$ and the equation $c = d$, the principle states that $a + c = b + d$. Since $6x_1 - 2x_2 = 4$ and $-6x_1 - 4x_2 = -28$, we can add $6x_1 - 2x_2$ to $-6x_1 - 4x_2$ provided that we add -28 to 4. The result of the addition is $0x_1 - 6x_2 = -24$.

To apply row operations in the solution of matrix equations, we *augment* the coefficient matrix A with the right-hand-side vector B. The augmented matrix is designated by the symbol $A \mid B$ and is expressed in matrix form as $\begin{pmatrix} a_{11} & a_{12} & b_1 \\ a_{21} & a_{22} & b_2 \end{pmatrix}$. Referring to the preceding example, we write the augmented matrix as

$$A \mid B = \begin{pmatrix} 3 & 2 & 14 \\ 6 & -2 & 4 \end{pmatrix}$$

The solution to the system of equations is determined by applying row operations to the rows of the augmented matrix. The row operations are applied with the objective of obtaining an identity matrix in the position originally occupied by the coefficient matrix A. Once the identity matrix is obtained, the solution to the system of equations appears in the position originally occupied by the right-hand-side vector.

To demonstrate the solution technique, row operations are performed on both the augmented matrix and the system of two equations in two unknowns. Our purpose is to show that applying row operations to augmented matrices is analogous to the algebraic operations used in the solution of systems of equations and can, therefore, be used to determine the solution for systems of equations. The augmented matrix and system of equations are

$$\begin{pmatrix} 3 & 2 & 14 \\ 6 & -2 & 4 \end{pmatrix} \qquad \begin{aligned} 3x_1 + 2x_2 &= 14 \\ 6x_1 - 2x_2 &= 4 \end{aligned}$$

The first row operation is to add one times the second row or second equation to the first row or equation. This gives

$$\begin{pmatrix} 9 & 0 & 18 \\ 6 & -2 & 4 \end{pmatrix} \qquad \begin{aligned} 9x_1 + 0x_2 &= 18 \\ 6x_1 - 2x_2 &= 4 \end{aligned}$$

Next, multiply the first row or equation by $\frac{1}{9}$.

$$\begin{pmatrix} 1 & 0 & | & 2 \\ 6 & -2 & | & 4 \end{pmatrix} \qquad\qquad \begin{aligned} 1x_1 + 0x_2 &= 2 \\ 6x_1 - 2x_2 &= 4 \end{aligned}$$

Add -6 times the first row or equation to the second row or equation.

$$\begin{pmatrix} 1 & 0 & | & 2 \\ 0 & -2 & | & -8 \end{pmatrix} \qquad\qquad \begin{aligned} 1x_1 + 0x_2 &= 2 \\ 0x_1 - 2x_2 &= -8 \end{aligned}$$

The last operation is to multiply the second row or equation by $-\frac{1}{2}$.

$$\begin{pmatrix} 1 & 0 & | & 2 \\ 0 & 1 & | & 4 \end{pmatrix} \qquad\qquad \begin{aligned} 1x_1 + 0x_2 &= 2 \\ 0x_1 + 1x_2 &= 4 \end{aligned}$$

The solution to the system of equations is read directly as $x_1 = 2$ and $x_2 = 4$. The top element to the right of the vertical line is the solution value of x_1 and the bottom element the solution value of x_2. This can be demonstrated by writing the augmented matrix as the matrix equation,

$$\begin{pmatrix} 1 & 0 \\ 0 & 1 \end{pmatrix} \begin{pmatrix} x_1 \\ x_2 \end{pmatrix} = \begin{pmatrix} 2 \\ 4 \end{pmatrix}$$

and converting the matrix equation to two equations in the two unknowns,

$$1x_1 + 0x_2 = 2$$

$$0x_1 + 1x_2 = 4$$

As a second example of solving matrix equations using row operations, consider the following three equations

$$2x + 1y - 2z = -1$$
$$4x - 2y + 3z = 14$$
$$1x - 1y + 2z = 7$$

Row operations are performed on the augmented matrix and the system of equations. This will again demonstrate the parallel between the algebraic solution and the matrix solution of systems of equations. The augmented matrix and the systems of equations are

$$\begin{pmatrix} 2 & 1 & -2 & | & -1 \\ 4 & -2 & 3 & | & 14 \\ 1 & -1 & 2 & | & 7 \end{pmatrix} \qquad \begin{aligned} 2x + 1y - 2z &= -1 \\ 4x - 2y + 3z &= 14 \\ 1x - 1y + 2z &= 7 \end{aligned}$$

Add 1 times the third row to the first row.

$$\begin{pmatrix} 3 & 0 & 0 & | & 6 \\ 4 & -2 & 3 & | & 14 \\ 1 & -1 & 2 & | & 7 \end{pmatrix} \qquad \begin{aligned} 3x + 0y + 0z &= 6 \\ 4x - 2y + 3z &= 14 \\ 1x - 1y + 2z &= 7 \end{aligned}$$

Multiply the first row by $\frac{1}{3}$.

$$\begin{pmatrix} 1 & 0 & 0 & | & 2 \\ 4 & -2 & 3 & | & 14 \\ 1 & -1 & 2 & | & 7 \end{pmatrix}$$

$1x + 0y + 0z = 2$
$4x - 2y + 3z = 14$
$1x - 1y + 2z = 7$

Add -4 times the first row to the second row and -1 times the first row to the third row.

$$\begin{pmatrix} 1 & 0 & 0 & | & 2 \\ 0 & -2 & 3 & | & 6 \\ 0 & -1 & 2 & | & 5 \end{pmatrix}$$

$1x + 0y + 0z = 2$
$0x - 2y + 3z = 6$
$0x - 1y + 2z = 5$

Multiply the second row by $-\frac{1}{2}$.

$$\begin{pmatrix} 1 & 0 & 0 & | & 2 \\ 0 & 1 & -\frac{3}{2} & | & -3 \\ 0 & -1 & 2 & | & 5 \end{pmatrix}$$

$1x + 0y + 0z = 2$
$0x + 1y - \frac{3}{2}z = -3$
$0x - 1y + 2z = 5$

Add 1 times the second row to the third row.

$$\begin{pmatrix} 1 & 0 & 0 & | & 2 \\ 0 & 1 & -\frac{3}{2} & | & -3 \\ 0 & 0 & \frac{1}{2} & | & 2 \end{pmatrix}$$

$1x + 0y + 0z = 2$
$0x + 1y - \frac{3}{2}z = -3$
$0x + 0y + \frac{1}{2}z = 2$

Multiply the third row by 2.

$$\begin{pmatrix} 1 & 0 & 0 & | & 2 \\ 0 & 1 & -\frac{3}{2} & | & -3 \\ 0 & 0 & 1 & | & 4 \end{pmatrix}$$

$1x + 0y + 0z = 2$
$0x + 1y - \frac{3}{2}z = -3$
$0x + 0y + 1z = 4$

Add $\frac{3}{2}$ times the third row to the second row.

$$\begin{pmatrix} 1 & 0 & 0 & | & 2 \\ 0 & 1 & 0 & | & 3 \\ 0 & 0 & 1 & | & 4 \end{pmatrix}$$

$1x + 0y + 0z = 2$
$0x + 1y + 0z = 3$
$0x + 0y + 1z = 4$

The solution to the three equations is $x = 2$, $y = 3$, and $z = 4$.

It has been demonstrated in the preceding two examples that the row operations are applied to the augmented matrix with the objective of obtaining an identity matrix on the left side of the vertical line. Once the identity matrix is obtained, the solution vector can be read from the right side of the vertical line. This is possible since the row operations on the augmented matrix correspond to algebraic operations on the system of equations.

This method for the solution of matrix equations is termed the *Gaussian elimination* method (named for the German mathematician Karl Friedrich Gauss, 1777–1855) or simply the elimination method. To summarize, the technique involves writing the coefficient matrix and the right-hand-side vector as $A \mid B$. Row operations are applied to transform the coefficient matrix A to an identity matrix I. The solution to the system of equations is then read from the right-hand side of the vertical line.

Any combination of row operations is acceptable in obtaining the identity matrix to the left of the vertical line. The procedure many students find useful is to first obtain a 1 in the a_{11} position and to then use multiples of the first row to obtain zeros elsewhere in the first column. The student next obtains a 1 in the a_{22} position and uses multiples of the second row to obtain zeros elsewhere in the second column. This procedure is followed until the identity matrix is obtained. This method is used in the following examples. It is also used in examples later in this chapter.

Example: Determine the solution vector for the following two equations, using the Gaussian elimination method.

$$-2x_1 + 4x_2 = 4$$
$$3x_1 - 2x_2 = 10$$

Write the equations in matrix form as $A \mid B$.

$$\begin{pmatrix} -2 & 4 & | & 4 \\ 3 & -2 & | & 10 \end{pmatrix}$$

Multiply the first row by $-\frac{1}{2}$.

$$\begin{pmatrix} 1 & -2 & | & -2 \\ 3 & -2 & | & 10 \end{pmatrix}$$

Add -3 times the first row to the second row.

$$\begin{pmatrix} 1 & -2 & | & -2 \\ 0 & 4 & | & 16 \end{pmatrix}$$

Multiply the second row by $\frac{1}{4}$.

$$\begin{pmatrix} 1 & -2 & | & -2 \\ 0 & 1 & | & 4 \end{pmatrix}$$

Add 2 times the second row to the first row.

$$\begin{pmatrix} 1 & 0 & | & 6 \\ 0 & 1 & | & 4 \end{pmatrix}$$

The solution is $x_1 = 6$ and $x_2 = 4$.

Example: Determine the solution vector to the following system of equations, using the Gaussian elimination method.

$$2x + 3y - 4z = 9$$
$$3x - 2y + 3z = -15$$
$$1x + 4y - 2z = 12$$

Write the equations in matrix form as $A \mid B$.

$$\begin{pmatrix} 2 & 3 & -4 & 9 \\ 3 & -2 & 3 & -15 \\ 1 & 4 & -2 & 12 \end{pmatrix}$$

Multiply the first row by $\frac{1}{2}$.

$$\begin{pmatrix} 1 & \frac{3}{2} & -2 & \frac{9}{2} \\ 3 & -2 & 3 & -15 \\ 1 & 4 & -2 & 12 \end{pmatrix}$$

Add -3 times the first row to the second row and add -1 times the first row to the third row.

$$\begin{pmatrix} 1 & \frac{3}{2} & -2 & \frac{9}{2} \\ 0 & -\frac{13}{2} & 9 & -\frac{57}{2} \\ 0 & \frac{5}{2} & 0 & \frac{15}{2} \end{pmatrix}$$

Interchange the second and third row.

$$\begin{pmatrix} 1 & \frac{3}{2} & -2 & \frac{9}{2} \\ 0 & \frac{5}{2} & 0 & \frac{15}{2} \\ 0 & -\frac{13}{2} & 9 & -\frac{57}{2} \end{pmatrix}$$

Multiply the second row by $\frac{2}{5}$.

$$\begin{pmatrix} 1 & \frac{3}{2} & -2 & \frac{9}{2} \\ 0 & 1 & 0 & 3 \\ 0 & -\frac{13}{2} & 9 & -\frac{57}{2} \end{pmatrix}$$

Add $-\frac{3}{2}$ times the second row to the first row and add $\frac{13}{2}$ times the second row to the third row.

$$\begin{pmatrix} 1 & 0 & -2 & 0 \\ 0 & 1 & 0 & 3 \\ 0 & 0 & 9 & -9 \end{pmatrix}$$

Multiply the third row by $\frac{1}{9}$.

$$\begin{pmatrix} 1 & 0 & -2 & 0 \\ 0 & 1 & 0 & 3 \\ 0 & 0 & 1 & -1 \end{pmatrix}$$

Add 2 times the third row to the first row.

$$\begin{pmatrix} 1 & 0 & 0 & -2 \\ 0 & 1 & 0 & 3 \\ 0 & 0 & 1 & -1 \end{pmatrix}$$

The solution is $x = -2$, $y = 3$, and $z = -1$.

3.6 The Inverse: Gaussian Method

The four basic arithmetic operations of ordinary algebra are addition, subtraction, multiplication, and division. The first three operations apply in matrix algebra and have been discussed. The fourth operation, division, is not defined for matrix algebra. The inverse of a matrix, however, is used in matrix algebra in much the same manner as division in ordinary algebra.

Division is used in ordinary algebra to determine the solution for systems of equations. For instance, in the algebraic equation $ax = b$, x is determined by dividing both sides of the equation by a to give $x = b/a$. An alternative method of solving this algebraic equation is to multiply both sides of the equation $ax = b$ by the reciprocal of a. In ordinary algebra the reciprocal of a is $1/a$. The solution to the equation is thus

$$(1/a)ax = (1/a)b$$

or

$$x = (1/a)b, \qquad \text{provided } a \neq 0$$

An inverse matrix performs a function in matrix algebra similar to that of the reciprocal in ordinary algebra. The product of a number a and its reciprocal $1/a$ in ordinary algebra is 1. Similarly, in matrix algebra the product of a matrix A and its inverse, A^{-1} (read A inverse) is the identity matrix I, i.e.,

$$AA^{-1} = I \qquad (3.6)$$

The matrix equation $AX = B$ can be solved for the solution vector X by use of the inverse. The procedure is to multiply the matrix equation by the inverse,

$$A^{-1}AX = A^{-1}B \qquad (3.7)$$

Since $A^{-1}A = I$, this reduces to

$$IX = A^{-1}B, \qquad (3.8)$$

and since $IX = X$, the solution vector is

$$X = A^{-1}B \qquad (3.9)$$

In order for a matrix to have an inverse, the matrix must be square.† If A is a square matrix, the inverse of A is a square matrix with components such that the matrix product of A^{-1} and A is the identity matrix I, i.e., $AA^{-1} = I$.

As an example, consider the matrix $A = \begin{pmatrix} 3 & 2 \\ 6 & -2 \end{pmatrix}$ and the matrix $A^{-1} =$

---

† Not all square matrices have inverses. See Chapter 4, p. 125 for further explanation.

$\begin{pmatrix} \frac{1}{9} & \frac{1}{9} \\ \frac{1}{3} & -\frac{1}{6} \end{pmatrix}$. A^{-1} is the inverse of A, provided that $AA^{-1} = I$. Since

$$\begin{pmatrix} 3 & 2 \\ 6 & -2 \end{pmatrix} \begin{pmatrix} \frac{1}{9} & \frac{1}{9} \\ \frac{1}{3} & -\frac{1}{6} \end{pmatrix} = \begin{pmatrix} 1 & 0 \\ 0 & 1 \end{pmatrix}$$

we conclude that A^{-1} is the inverse of A. It also follows that the matrix $A = \begin{pmatrix} 3 & 2 \\ 6 & -2 \end{pmatrix}$ is the inverse of the matrix $A^{-1} = \begin{pmatrix} \frac{1}{9} & \frac{1}{9} \\ \frac{1}{3} & -\frac{1}{6} \end{pmatrix}$, since

$$A^{-1}A = \begin{pmatrix} \frac{1}{9} & \frac{1}{9} \\ \frac{1}{3} & -\frac{1}{6} \end{pmatrix} \begin{pmatrix} 3 & 2 \\ 6 & -2 \end{pmatrix} = \begin{pmatrix} 1 & 0 \\ 0 & 1 \end{pmatrix}.$$

In summary, for two square matrices A and A^{-1}, A^{-1} is the inverse of A if $AA^{-1} = I$. Similarly, since $A^{-1}A = I$ it follows that A is the inverse of A^{-1}. The commutative law of multiplication thus applies to inverse matrices, i.e., $AA^{-1} = I = A^{-1}A$. The relationship between a matrix and its inverse is illustrated in the following examples.

Example: Verify that $A^{-1} = \begin{pmatrix} \frac{2}{28} & \frac{3}{28} \\ \frac{6}{28} & -\frac{5}{28} \end{pmatrix}$ is the inverse matrix of $A = \begin{pmatrix} 5 & 3 \\ 6 & -2 \end{pmatrix}$.

The fact that A^{-1} is the inverse matrix of A can be verified by multiplying A by A^{-1}. If the product of the two matrices is the identity matrix, then A^{-1} is the inverse of A.

$$AA^{-1} = \begin{pmatrix} 5 & 3 \\ 6 & -2 \end{pmatrix} \begin{pmatrix} \frac{2}{28} & \frac{3}{28} \\ \frac{6}{28} & -\frac{5}{28} \end{pmatrix} = \begin{pmatrix} 1 & 0 \\ 0 & 1 \end{pmatrix}$$

Example: Verify that the commutative law of multiplication applies for inverse matrices. Use the matrices in the preceding example.

The commutative law of multiplication states that $AB = BA$. We have previously shown that this law does not hold in general in matrix multiplication. It does hold true, however, for inverse matrices. Thus,

$$AA^{-1} = \begin{pmatrix} 5 & 3 \\ 6 & -2 \end{pmatrix} \begin{pmatrix} \frac{2}{28} & \frac{3}{28} \\ \frac{6}{28} & -\frac{5}{28} \end{pmatrix} = \begin{pmatrix} 1 & 0 \\ 0 & 1 \end{pmatrix}$$

and

$$A^{-1}A = \begin{pmatrix} \frac{2}{28} & \frac{3}{28} \\ \frac{6}{28} & -\frac{5}{28} \end{pmatrix} \begin{pmatrix} 5 & 3 \\ 6 & -2 \end{pmatrix} = \begin{pmatrix} 1 & 0 \\ 0 & 1 \end{pmatrix}$$

This demonstrates that $A^{-1}A = I = AA^{-1}$.

Example: Verify that

$$A = \begin{pmatrix} 2 & 4 & -6 \\ 4 & 2 & 2 \\ 3 & -3 & 1 \end{pmatrix} \text{ and } A^{-1} = \begin{pmatrix} \frac{4}{66} & \frac{7}{66} & \frac{10}{66} \\ \frac{1}{66} & \frac{10}{66} & -\frac{14}{66} \\ -\frac{9}{66} & \frac{9}{66} & -\frac{6}{66} \end{pmatrix}$$

are inverse matrices.

The two matrices are inverse matrices if $AA^{-1} = I$.

$$AA^{-1} = \begin{pmatrix} 2 & 4 & -6 \\ 4 & 2 & 2 \\ 3 & -3 & 1 \end{pmatrix} \begin{pmatrix} \frac{4}{66} & \frac{7}{66} & \frac{10}{66} \\ \frac{1}{66} & \frac{10}{66} & -\frac{14}{66} \\ -\frac{9}{66} & \frac{9}{66} & -\frac{6}{66} \end{pmatrix} = \begin{pmatrix} 1 & 0 & 0 \\ 0 & 1 & 0 \\ 0 & 0 & 1 \end{pmatrix}$$

Since $AA^{-1} = 1$, we conclude that A and A^{-1} are inverse matrices.

3.6.1 CALCULATING THE INVERSE BY GAUSSIAN ELIMINATION

The Gaussian elimination method can be used to calculate an inverse matrix. The procedure involves augmenting the square matrix A with the identity matrix I, i.e.

$$A \mid I \tag{3.10}$$

Row operations are then employed to obtain an identity matrix on the left side of the vertical line. Concurrent with the identity matrix on the left of the vertical line, the inverse matrix is obtained on the right of the vertical line. The Gaussian elimination method involves transforming the augmented matrix $A \mid I$ to the inverse matrix $I \mid A^{-1}$ through row operations.

The justification for this procedure can be demonstrated with matrix algebra. First write the matrix A augmented with I.

$$A \mid I$$

If the inverse matrix A^{-1} were known, we could multiply the matrices on both sides of the vertical line by A^{-1}, i.e.,

$$AA^{-1} \mid IA^{-1}$$

This product would give

$$I \mid A^{-1} \tag{3.11}$$

Instead of using the inverse to obtain the identity matrix on the left side of the vertical line, we use row operations. The row operations applied to obtain the identity matrix concurrently yields the inverse matrix on the right side of the vertical line.

The Gaussian elimination method for obtaining the inverse matrix is illustrated by the following examples.

Example: Determine the inverse of $A = \begin{pmatrix} 5 & 3 \\ 6 & -2 \end{pmatrix}$.

To determine the inverse, we augment the matrix A with the identity matrix I and apply row operations to obtain an identity matrix on the left side of the vertical line. The augmented matrix is

$$A \mid I = \begin{pmatrix} 5 & 3 & 1 & 0 \\ 6 & -2 & 0 & 1 \end{pmatrix}$$

Multiply the first row by $\frac{1}{5}$.

$$\begin{pmatrix} 1 & \frac{3}{5} & \frac{1}{5} & 0 \\ 6 & -2 & 0 & 1 \end{pmatrix}$$

Add -6 times the first row to the second row.

$$\begin{pmatrix} 1 & \frac{3}{5} & \frac{1}{5} & 0 \\ 0 & -\frac{28}{5} & -\frac{6}{5} & 1 \end{pmatrix}$$

Multiply the second row by $-\frac{5}{28}$.

$$\begin{pmatrix} 1 & \frac{3}{5} & \frac{1}{5} & 0 \\ 0 & 1 & \frac{6}{28} & -\frac{5}{28} \end{pmatrix}$$

Add $-\frac{3}{5}$ times the second row to the first row.

$$\begin{pmatrix} 1 & 0 & \frac{2}{28} & \frac{3}{28} \\ 0 & 1 & \frac{6}{28} & -\frac{5}{28} \end{pmatrix}$$

This results in an identity matrix on the left side of the vertical line. From Formula (3.11) we know that the inverse of A is given on the right side of the vertical line. The inverse is

$$A^{-1} = \begin{pmatrix} \frac{2}{28} & \frac{3}{28} \\ \frac{6}{28} & -\frac{5}{28} \end{pmatrix}$$

Example: Determine the inverse of $A = \begin{pmatrix} 3 & 2 \\ 6 & -2 \end{pmatrix}$.

Augment the matrix A with the identity matrix I.

$$\begin{pmatrix} 3 & 2 & 1 & 0 \\ 6 & -2 & 0 & 1 \end{pmatrix}$$

Add 1 times the second row to the first row.

$$\begin{pmatrix} 9 & 0 & 1 & 1 \\ 6 & -2 & 0 & 1 \end{pmatrix}$$

Multiply the first row by $\frac{1}{9}$.

$$\begin{pmatrix} 1 & 0 & \frac{1}{9} & \frac{1}{9} \\ 6 & -2 & 0 & 1 \end{pmatrix}$$

Add -6 times the first row to the second row.

$$\begin{pmatrix} 1 & 0 & \frac{1}{9} & \frac{1}{9} \\ 0 & -2 & -\frac{2}{3} & \frac{1}{3} \end{pmatrix}$$

Multiply the second row by $-\frac{1}{2}$.

$$\begin{pmatrix} 1 & 0 & \frac{1}{9} & \frac{1}{9} \\ 0 & 1 & \frac{1}{3} & -\frac{1}{6} \end{pmatrix}$$

The inverse of A is given on the right side of the vertical line and is

$$A^{-1} = \begin{pmatrix} \frac{1}{9} & \frac{1}{9} \\ \frac{1}{3} & -\frac{1}{6} \end{pmatrix}$$

Example: Determine the inverse of $A = \begin{pmatrix} 2 & 3 \\ -4 & 6 \end{pmatrix}$.

Augment the matrix A with the identity matrix I.

$$\begin{pmatrix} 2 & 3 & | & 1 & 0 \\ -4 & 6 & | & 0 & 1 \end{pmatrix}$$

Add 2 times the first row to the second row.

$$\begin{pmatrix} 2 & 3 & | & 1 & 0 \\ 0 & 12 & | & 2 & 1 \end{pmatrix}$$

Multiply the first row by $\frac{1}{2}$ and the second row by $\frac{1}{12}$.

$$\begin{pmatrix} 1 & \frac{3}{2} & | & \frac{1}{2} & 0 \\ 0 & 1 & | & \frac{1}{6} & \frac{1}{12} \end{pmatrix}$$

Add $-\frac{3}{2}$ times the second row to the first row.

$$\begin{pmatrix} 1 & 0 & | & \frac{1}{4} & -\frac{3}{24} \\ 0 & 1 & | & \frac{1}{6} & \frac{1}{12} \end{pmatrix}$$

The inverse of A is

$$A^{-1} = \begin{pmatrix} \frac{1}{4} & -\frac{3}{24} \\ \frac{1}{6} & \frac{1}{12} \end{pmatrix}$$

Example: Determine the inverse of $A = \begin{pmatrix} 2 & 4 & -6 \\ 4 & 2 & 2 \\ 3 & -3 & 1 \end{pmatrix}$.

Augment the matrix A with I.

$$A \,|\, I = \begin{pmatrix} 2 & 4 & -6 & | & 1 & 0 & 0 \\ 4 & 2 & 2 & | & 0 & 1 & 0 \\ 3 & -3 & 1 & | & 0 & 0 & 1 \end{pmatrix}$$

Multiply the first row by $\frac{1}{2}$.

$$\begin{pmatrix} 1 & 2 & -3 & | & \frac{1}{2} & 0 & 0 \\ 4 & 2 & 2 & | & 0 & 1 & 0 \\ 3 & -3 & 1 & | & 0 & 0 & 1 \end{pmatrix}$$

Add -4 times the first row to the second row and add -3 times the first row to the third row.

$$\begin{pmatrix} 1 & 2 & -3 & | & \frac{1}{2} & 0 & 0 \\ 0 & -6 & 14 & | & -2 & 1 & 0 \\ 0 & -9 & 10 & | & -\frac{3}{2} & 0 & 1 \end{pmatrix}$$

Multiply the second row by $-\frac{1}{6}$.

$$\begin{pmatrix} 1 & 2 & -3 & \frac{1}{2} & 0 & 0 \\ 0 & 1 & -\frac{7}{3} & \frac{1}{3} & -\frac{1}{6} & 0 \\ 0 & -9 & 10 & -\frac{3}{2} & 0 & 1 \end{pmatrix}$$

Add -2 times the second row to the first row and add 9 times the second row to the third row. Multiply the new third row by $-\frac{1}{11}$.

$$\begin{pmatrix} 1 & 0 & \frac{5}{3} & -\frac{1}{6} & \frac{1}{3} & 0 \\ 0 & 1 & -\frac{7}{3} & \frac{1}{3} & -\frac{1}{6} & 0 \\ 0 & 0 & 1 & -\frac{3}{22} & \frac{3}{22} & -\frac{1}{11} \end{pmatrix}$$

Add $\frac{7}{3}$ times the third row to the second row and add $-\frac{5}{3}$ times the third row to the first row.

$$\begin{pmatrix} 1 & 0 & 0 & \frac{4}{66} & \frac{7}{66} & \frac{10}{66} \\ 0 & 1 & 0 & \frac{1}{66} & \frac{10}{66} & -\frac{14}{66} \\ 0 & 0 & 1 & -\frac{9}{66} & \frac{9}{66} & -\frac{6}{66} \end{pmatrix}$$

The inverse of A is

$$A^{-1} = \begin{pmatrix} \frac{4}{66} & \frac{7}{66} & \frac{10}{66} \\ \frac{1}{66} & \frac{10}{66} & -\frac{14}{66} \\ -\frac{9}{66} & \frac{9}{66} & -\frac{6}{66} \end{pmatrix}$$

3.6.2 USING THE INVERSE TO SOLVE MATRIX EQUATIONS

The solution to a matrix equation is relatively straightforward once the inverse has been determined. Given the matrix equation $AX = B$, the solution vector is $X = A^{-1}B$. This result comes from the fact that both sides of the matrix equation can be multiplied by the inverse. That is,

$$A^{-1}AX = A^{-1}B$$
$$IX = A^{-1}B$$

and

$$X = A^{-1}B$$

To illustrate, consider again the example of two simultaneous equations presented on p. 86. The two equations are written in matrix form as

$$\begin{pmatrix} 3 & 2 \\ 6 & -2 \end{pmatrix}\begin{pmatrix} x_1 \\ x_2 \end{pmatrix} = \begin{pmatrix} 14 \\ 4 \end{pmatrix}$$

The inverse of the coefficient matrix was calculated on p. 95 and is

$$A^{-1} = \begin{pmatrix} \frac{1}{9} & \frac{1}{9} \\ \frac{1}{3} & -\frac{1}{6} \end{pmatrix}$$

Multiplying A^{-1} by B gives

$$X = A^{-1}B = \begin{pmatrix} \frac{1}{9} & \frac{1}{9} \\ \frac{1}{3} & -\frac{1}{6} \end{pmatrix}\begin{pmatrix} 14 \\ 4 \end{pmatrix} = \begin{pmatrix} 2 \\ 4 \end{pmatrix}$$

The solution vector is $x_1 = 2$ and $x_2 = 4$. This is, of course, the solution that was obtained by the Gaussian elimination method.†

Example: Determine the solution vector X for the matrix equation

$$\begin{pmatrix} 5 & 3 \\ 6 & -2 \end{pmatrix}\begin{pmatrix} x_1 \\ x_2 \end{pmatrix} = \begin{pmatrix} 6 \\ 10 \end{pmatrix}$$

The inverse of the coefficient matrix was found on p. 94 to be

$$A^{-1} = \begin{pmatrix} \frac{2}{28} & \frac{3}{28} \\ \frac{6}{28} & -\frac{5}{28} \end{pmatrix}.$$

The solution vector is

$$X = A^{-1}B = \begin{pmatrix} \frac{2}{28} & \frac{3}{28} \\ \frac{6}{28} & -\frac{5}{28} \end{pmatrix}\begin{pmatrix} 6 \\ 10 \end{pmatrix} = \begin{pmatrix} \frac{3}{2} \\ -\frac{1}{2} \end{pmatrix}$$

Example: Determine the solution vector X for the matrix equation

$$\begin{pmatrix} 2 & 4 & -6 \\ 4 & 2 & 2 \\ 3 & -3 & 1 \end{pmatrix}\begin{pmatrix} x_1 \\ x_2 \\ x_3 \end{pmatrix} = \begin{pmatrix} 6 \\ 10 \\ 4 \end{pmatrix}$$

The inverse matrix was given on p. 96 as

$$A^{-1} = \begin{pmatrix} \frac{4}{66} & \frac{7}{66} & \frac{10}{66} \\ \frac{1}{66} & \frac{10}{66} & -\frac{14}{66} \\ -\frac{9}{66} & \frac{9}{66} & -\frac{6}{66} \end{pmatrix}$$

The solution vector is $X = A^{-1}B$.

$$X = \begin{pmatrix} \frac{4}{66} & \frac{7}{66} & \frac{10}{66} \\ \frac{1}{66} & \frac{10}{66} & -\frac{14}{66} \\ -\frac{9}{66} & \frac{9}{66} & -\frac{6}{66} \end{pmatrix}\begin{pmatrix} 6 \\ 10 \\ 4 \end{pmatrix} = \begin{pmatrix} \frac{67}{33} \\ \frac{25}{33} \\ \frac{6}{33} \end{pmatrix}$$

3.7 Application to Input-Output Analysis‡

One of the interesting applications of matrix algebra is in the Leontief input-output model. To illustrate the input-output model, consider the case of the Amalgamated Steel Company.

---

† The reader will note that the Gaussian elimination method can be used to determine the inverse or can be applied directly to $A|B$ to obtain the solution vector.

‡ This section can be omitted without loss of continuity.

Amalgamated Steel is one of the major producers of steel in the country. Their customers include the large automobile and truck manufacturers, the new construction industry, the metal products fabricating industry, the heating and plumbing products industry, and a multitude of other steel buyers. With minor exceptions, sales of Amalgamated Steel are made to industrial users rather than to the consuming public. Thus, the demand for Amalgamated's products is dependent upon the demand for the finished products of Amalgamated's customers.

The demand structure facing Amalgamated Steel is not at all uncommon in industry. Many firms sell to both industrial buyers and the general public. In the case of Amalgamated Steel, the majority of sales are made to industrial users of steel. The steel is used in finished products such as automobiles, office buildings, etc. The demand for Amalgamated's product is thus dependent upon the demand by the consuming public for the products of Amalgamated's customers. This type of demand is termed *derived* demand.

Products purchased directly by the consuming public represent an *autonomous* demand. In the case of an electronics manufacturer, for example, the demand for television sets is autonomous, whereas the demand for picture tubes used in the manufacture of the television sets is derived. Similarly, in the case of Amalgamated Steel, the demand for steel reinforcing bars for use in the construction of a back yard patio by the home owner is autonomous, whereas the demand for steel for use in the manufacture of an automobile is derived.

Since the majority of sales of Amalgamated Steel come from derived demand, the president of Amalgamated is interested in developing a model that describes the effect on Amalgamated of changes in the autonomous demand for Amalgamated's customer's products. Specifically, the president would like answers to questions such as "How would a 10 percent increase in new car sales affect the sales of Amalgamated Steel?" and "How would a 5 percent decrease in new car sales coupled with a 10 percent increase in new construction affect the sales of Amalgamated Steel?"

Questions such as those raised by the president of Amalgamated Steel can be answered by constructing a Leontief input-output model. The model was originally constructed for an entire economy, rather than for a segment of the economy. In applying the model to the economy, it is assumed that the economy contains a number of interacting industries, each producing products and each purchasing as intermediate products the products of other industries. To illustrate, home builders purchase lumber from the wood products industry and cement from the portland cement industry, while the wood products industry purchases steel from the steel industry and trucks from the motor vehicle industry. An industry, such as steel, can purchase some of its own product as well as purchase from an industry to which it sells (e.g., purchase trucks from the motor vehicle industry). In addition to selling its own product as an intermediate product to other industries, each industry also experiences an autonomous demand for its product from consumers, the government, or foreign governments.

The task of developing an interindustry model for the entire economy requires rather restrictive assumptions and is quite complex. Fortunately, the president of Amalgamated Steel recognized that a model that included the four major purchasers of steel would provide useful insights into the questions he had originally raised.

The first step in developing the model is the construction of an input-output matrix. Each element a_{ij} in the matrix represents the value of product i used in the manufacture of one dollar's worth of product j. The jth column thus gives the various amounts of the intermediate products used in the manufacture of one dollar's worth of product j. Conversely, the ith row describes the distribution of product i among the industries (or products) included in the matrix.

The input-output matrix for Amalgamated Steel can be used to illustrate these concepts. Amalgamated's input-output matrix was developed through cooperative studies with Amalgamated's customers and is given below.

Output (final product)

Input (intermediate product)	*Steel*	*Motor Vehicles*	*Construction*	*Metal Products*	*Iron Ore*
Steel	0.20	0.25	0.15	0.30	0.10
Motor vehicles	0.05	0.10	0.05	0	0.20
Construction	0	0.05	0.02	0.05	0.05
Metal products	0.10	0.05	0.05	0.05	0.10
Iron ore	0.25	0	0	0	0

The column for steel shows that for every dollar's worth of steel manufactured, $0.20 of steel, $0.05 of motor vehicles, $0.10 of metal products, and $0.25 of iron ore are used. Similarly, the column for motor vehicles shows that one dollar's worth of motor vehicles requires $0.25 of steel, $0.10 of motor vehicles, $0.05 of construction, and $0.05 of metal products. The columns do not sum to one dollar for two reasons. First, the matrix does not include all industries in the economy. Thus, products such as plastics that are used in motor vehicles are not included in the matrix. Second, the matrix does not include the value added by the manufacturer, i.e., the expenses incurred by the manufacturer in combining the factors of production to make a finished product.

The rows of the matrix show the distribution of the products of each of the industries. The row for steel, for instance, shows that steel is used as an input to the steel, motor vehicles, construction, metal products, and iron ore industries. The row for motor vehicles similarly shows that motor vehicles are used in the steel, motor vehicles, construction, metal products, and iron ore industries. Since the components of each row represent the value of product used in the manufacture of one dollar's worth of product shown by the column, the sum of the components of each row has no economic meaning.

The demand for each of the five products is equal to the derived demand plus the autonomous demand. Suppose that we represent the demand for each of the five products in terms of dollar sales by x_j, for $j = 1, 2, \ldots, 5$. The demand for steel is thus $x_1 = 0.20x_1 + 0.25x_2 + 0.15x_3 + 0.30x_4 + 0.10x_5 + d_1$. The autonomous demand is represented by d_1 and the derived demand by the remainder of the terms in the summation. The first term in the summation, $0.20x_1$, gives the dollar value of steel used in the manufacture of steel. The second term shows the dollar value of steel used in the manufacture of motor vehicles, etc. Similarly, the demand resulting from using steel in construction, metal products, and the mining of iron ore is given by the third, fourth, and fifth terms. The sum of the derived demands plus the autonomous demand d_1 gives the total demand for steel x_1.

The objective in the input-output model is to describe the interindustry relationships and to use these relationships to predict changes in demand caused by changes in the autonomous demand for the final products. The interindustry relationships for Amalgamated Steel are described by the following system of demand equations.

$$x_1 = 0.20x_1 + 0.25x_2 + 0.15x_3 + 0.30x_4 + 0.10x_5 + d_1$$
$$x_2 = 0.05x_1 + 0.10x_2 + 0.05x_3 + 0x_4 + 0.20x_5 + d_2$$
$$x_3 = 0x_1 + 0.05x_2 + 0.02x_3 + 0.05x_4 + 0.05x_5 + d_3$$
$$x_4 = 0.10x_1 + 0.05x_2 + 0.05x_3 + 0.05x_4 + 0.10x_5 + d_4$$
$$x_5 = 0.25x_1 + 0x_2 + 0x_3 + 0x_4 + 0x_5 + d_5$$

The system of equations can be rewritten with the variables x_j on the left of the equal sign and the autonomous demands d_j on the right of the equal sign. This gives

$$(1 - 0.20)x_1 - 0.25x_2 - 0.15x_3 - 0.30x_4 - 0.10x_5 = d_1$$
$$-0.05x_1 + (1 - 0.10)x_2 - 0.05x_3 - 0x_4 - 0.20x_5 = d_2$$
$$-0x_1 - 0.05x_2 + (1 - 0.02)x_3 - 0.05x_4 - 0.05x_5 = d_3$$
$$-0.10x_1 - 0.05x_2 - 0.05x_3 + (1 - 0.05)x_4 - 0.10x_5 = d_4$$
$$-0.25x_1 - 0x_2 - 0x_3 - 0x_4 + (1 - 0)x_5 = d_5$$

The system is alternatively represented in matrix form by

$$\begin{pmatrix} (1-0.20) & -0.25 & -0.15 & -0.30 & -0.10 \\ -0.05 & (1-0.10) & -0.05 & -0 & -0.20 \\ -0 & -0.05 & (1-0.02) & -0.05 & -0.05 \\ -0.10 & -0.05 & -0.05 & (1-0.05) & -0.10 \\ -0.25 & -0 & -0 & -0 & (1-0) \end{pmatrix} \begin{pmatrix} x_1 \\ x_2 \\ x_3 \\ x_4 \\ x_5 \end{pmatrix} = \begin{pmatrix} d_1 \\ d_2 \\ d_3 \\ d_4 \\ d_5 \end{pmatrix}$$

This system of equations describes the interindustry relationships between the five industries included in the input-output matrix. Furthermore, the demands for the products of each of the five industries can be determined for any alternative vector of autonomous demands. This means, of course, that for any vector of autonomous demands the system of equations can be solved for the variables x_j.

The solution procedure for obtaining the values of x_j involves calculating the inverse of the coefficient matrix for the system of interindustry equations. If A is the input-output matrix, I is an identity matrix, X is the demand vector, and D is the vector of autonomous demands, the system of equations is

$$(I - A)X = D$$

Providing that the inverse of $(I - A)$ exists, the demand vector X is given by

$$X = (I - A)^{-1}D$$

To illustrate the input-output model, assume that the autonomous demand for steel, motor vehicles, new construction, metal products, and iron ore is $2 billion, $20 billion, $15 billion, $10 billion, and $1 billion. The matrix $(1 - A)^{-1}$ is calculated by using the procedure explained in Sec. 3.5 and is

$$(I - A)^{-1} = \begin{pmatrix} 1.43 & 0.44 & 0.26 & 0.46 & 0.30 \\ 0.16 & 1.16 & 0.09 & 0.06 & 0.26 \\ 0.04 & 0.07 & 1.03 & 0.07 & 0.08 \\ 0.20 & 0.12 & 0.09 & 1.12 & 0.16 \\ 0.36 & 0.11 & 0.07 & 0.12 & 1.07 \end{pmatrix}$$

The solution vector is

$$X = (I - A)^{-1}D$$

$$X = \begin{pmatrix} 1.43 & 0.44 & 0.26 & 0.46 & 0.30 \\ 0.16 & 1.16 & 0.09 & 0.06 & 0.26 \\ 0.04 & 0.07 & 1.03 & 0.07 & 0.08 \\ 0.20 & 0.12 & 0.09 & 1.12 & 0.16 \\ 0.36 & 0.11 & 0.07 & 0.12 & 1.07 \end{pmatrix} \begin{pmatrix} 2 \\ 20 \\ 15 \\ 10 \\ 1 \end{pmatrix} = \begin{pmatrix} 20.46 \\ 25.73 \\ 17.71 \\ 15.51 \\ 6.24 \end{pmatrix}$$

The demand for steel, motor vehicles, new construction, metal products, and iron ore is $20.46 billion, $25.73 billion, $17.71 billion, $15.51 billion, and $6.24 billion.

After the input-output matrix A and the inverse of $(I - A)$ have been determined, it is quite simple to determine the change in demand caused by changes in the autonomous demand for the final products. For instance, suppose that the president of Amalgamated is interested in predicting the change in demand for steel caused by increases of $5 billion in motor vehicle sales, $3 billion in new construction, and a decrease of $2 billion in sales of metal products. The change in the demands for steel, represented by ΔX, caused by the change in autonomous demand, represented by ΔD, is

$$\Delta X = (I - A)^{-1}\Delta D$$

$$\Delta X = \begin{pmatrix} 1.43 & 0.44 & 0.26 & 0.46 & 0.30 \\ 0.16 & 1.16 & 0.09 & 0.06 & 0.26 \\ 0.04 & 0.07 & 1.03 & 0.07 & 0.08 \\ 0.20 & 0.12 & 0.09 & 1.12 & 0.16 \\ 0.36 & 0.11 & 0.07 & 0.12 & 1.07 \end{pmatrix} \begin{pmatrix} 0 \\ 5 \\ 3 \\ -2 \\ 0 \end{pmatrix} = \begin{pmatrix} 2.06 \\ 5.95 \\ 3.30 \\ -1.37 \\ 0.52 \end{pmatrix}$$

The total demand based upon these changes is $(X + \Delta X)$ or, alternatively, $(I - A)^{-1}(D + \Delta D)$. Thus,

$$X + \Delta X = \begin{pmatrix} 20.46 + 2.06 \\ 25.73 + 5.95 \\ 17.71 + 3.30 \\ 15.51 - 1.37 \\ 6.24 + 0.52 \end{pmatrix} = \begin{pmatrix} 22.52 \\ 31.68 \\ 21.01 \\ 14.14 \\ 6.76 \end{pmatrix}$$

Based upon these changes, the demand for steel, motor vehicles, new construction, metal products, and iron ore is, respectively, $22.52 billion, $31.68 billion, $21.01 billion, $14.14 billion, and $6.76 billion.

Example: Determine the total demand for the products of industries A, B, and C based upon the input-output matrix and autonomous demand vector shown below.

Output Industry

Input industry	A	B	C
A	0.5	0.1	0.1
B	0.2	0.6	0.2
C	0.1	0.2	0.6

$$D = \begin{pmatrix} 20 \\ 30 \\ 50 \end{pmatrix}$$

The matrix $(I - A)$ is

$$(I - A) = \begin{pmatrix} 0.5 & -0.1 & -0.1 \\ -0.2 & 0.4 & -0.2 \\ -0.1 & -0.2 & 0.4 \end{pmatrix}$$

and the inverse of $(I - A)$ is

$$(I - A)^{-1} = \frac{1}{21} \begin{pmatrix} 60 & 30 & 30 \\ 50 & 95 & 60 \\ 40 & 55 & 90 \end{pmatrix}$$

The demand vector is

$$X = (I - A)^{-1}D$$

$$X = \frac{1}{21} \begin{pmatrix} 60 & 30 & 30 \\ 50 & 95 & 60 \\ 40 & 55 & 90 \end{pmatrix} \begin{pmatrix} 20 \\ 30 \\ 50 \end{pmatrix}$$

$$X = \begin{pmatrix} 171.5 \\ 326.2 \\ 331.0 \end{pmatrix}$$

PROBLEMS

1. For the vectors $A = (10, 6, 12)$ and $B = (14, 8, 5)$, perform the operations indicated or state why the operation cannot be done.
   (a) $A - B$ (b) $B - A$
   (c) AB (d) $A + B$

2. Using the vectors $X = (6, 4, 8)$, $Y = (3, 6, -2)$, and $Z = (4, 10, 12)$, verify that the associative law of addition applies to the addition of conformable vectors.

3. For the vectors $A = \begin{pmatrix} 6 \\ 4 \\ 2 \end{pmatrix}$ and $B = \begin{pmatrix} 8 \\ -2 \\ 6 \end{pmatrix}$, determine the following.

   (a) $A + B$ (b) $2A - B$
   (c) $3A - 2B$ (d) $0.5A + 1.5B$

4. Find the inner product of $W = (5, 8, 3)$ and $V = \begin{pmatrix} 2 \\ -2 \\ 4 \end{pmatrix}$.

5. For the vectors $A = (2, 8, 6)$, $B = \begin{pmatrix} 5 \\ 4 \\ 6 \end{pmatrix}$, and $C = \begin{pmatrix} 8 \\ 10 \\ 3 \end{pmatrix}$, verify the distributive law by showing that $A(B + C)$ is equal to $AB + AC$.

6. A television dealer has an inventory of 12 black and white sets, 9 color console sets, and 8 color portable sets. The dealer's cost on the different types of sets is $80, $340, and $260, respectively. Form the inventory

and price vectors and determine the value of the inventory by finding the inner product of the two vectors.

7. For the matrices $A = \begin{pmatrix} 3 & 5 \\ 2 & 6 \\ 1 & 4 \end{pmatrix}$ and $B = \begin{pmatrix} 2 & -5 \\ 3 & 6 \\ -5 & 5 \end{pmatrix}$, determine the following.

(a) $A + B$ (b) $B - A$

(c) $3A - B$ (d) $A - 2B$

8. Find the product of $A = \begin{pmatrix} 6 & 4 & 8 \\ 3 & -3 & 5 \end{pmatrix}$ and $B = \begin{pmatrix} 8 & -3 \\ -2 & 5 \\ 6 & 4 \end{pmatrix}$.

9. Using the matrices $A = \begin{pmatrix} 5 & 2 \\ 3 & -8 \\ 2 & 6 \end{pmatrix}$, $B = \begin{pmatrix} 8 & 5 \\ 4 & 7 \end{pmatrix}$, $C = \begin{pmatrix} 4 & 7 \\ 2 & 5 \end{pmatrix}$, verify that $A(B + C) = AB + AC$.

10. For the matrices $A = \begin{pmatrix} 3 & 8 & 2 \\ 5 & -2 & 4 \\ 6 & 1 & -5 \end{pmatrix}$ and $B = \begin{pmatrix} 6 & 13 & 7 \\ 7 & 11 & 5 \\ 8 & -1 & 10 \end{pmatrix}$, find AB and BA.

11. Determine the outer product of the vectors W and V given in Problem 4.

12. The matrices A, B, and C have the following dimensions: A has 3 rows and 4 columns, B has 4 rows and 4 columns, C has 2 rows and 3 columns. Can the following operations be performed? If so, give the dimensions of the resulting matrix.

(a) AB (b) AC

(c) CA (d) $C(AB)$

13. Although the commutative law of multiplication does not in general apply to the multiplication of matrices, it does apply in the multiplication of a matrix by the identity matrix. Verify this statement, using the matrices

$$A = \begin{pmatrix} 6 & 3 & 8 \\ 4 & 2 & 5 \\ 3 & 1 & 7 \end{pmatrix} \text{ and } I = \begin{pmatrix} 1 & 0 & 0 \\ 0 & 1 & 0 \\ 0 & 0 & 1 \end{pmatrix}$$

14. Express the following systems of equations in matrix form.

(a) $2x_1 + 6x_2 = 44$
    $3x_1 + 4x_2 = 36$

(b) $14x_1 + 23x_2 = 176$
    $9x_1 + 8x_2 = 86$

(c) $3x_1 + 5x_2 + 4x_3 = 50$
    $6x_1 + 4x_2 - 3x_3 = 26$
    $2x_1 - 5x_2 + 6x_3 = 5$

(d) $2x_1 - 5x_2 + 3x_3 = 14$
    $7x_1 + 4x_2 - 6x_3 = 36$
    $5x_1 - 6x_2 + 3x_3 = 34$

15. Determine the solutions to the systems of equations in Problem 14 by the use of the Gaussian elimination method.

16. Verify that the matrices $A = \begin{pmatrix} 3 & 2 \\ 1 & 5 \end{pmatrix}$ and $A^{-1} = \begin{pmatrix} \frac{5}{13} & -\frac{2}{13} \\ -\frac{1}{13} & \frac{3}{13} \end{pmatrix}$ are inverse by determining AA^{-1}.

17. Use the Gaussian method to determine the inverses of the following matrices.

(a) $A = \begin{pmatrix} 2 & 3 \\ 4 & 7 \end{pmatrix}$

(b) $A = \begin{pmatrix} 2 & 3 \\ -5 & -2 \end{pmatrix}$

(c) $A = \begin{pmatrix} 1 & 2 & 3 \\ 1 & 3 & 5 \\ 2 & 5 & 9 \end{pmatrix}$

(d) $A = \begin{pmatrix} 5 & -6 & 7 \\ -10 & 11 & -13 \\ 1 & -1 & 1 \end{pmatrix}$

18. Use the inverses found in Problem 17 to solve the following systems of equations.

(a) $\begin{aligned} 2x_1 + 3x_2 &= 10 \\ 4x_1 + 7x_2 &= 6 \end{aligned}$

(b) $\begin{aligned} 2x_1 + 3x_2 &= 6 \\ -5x_1 - 2x_2 &= 7 \end{aligned}$

(c) $\begin{aligned} x_1 + 2x_2 + 3x_3 &= 8 \\ x_1 + 3x_2 + 5x_3 &= 5 \\ 2x_1 + 5x_2 + 9x_3 &= 10 \end{aligned}$

(d) $\begin{aligned} 5x_1 - 6x_2 + 7x_3 &= 5 \\ -10x_1 + 11x_2 - 13x_3 &= -10 \\ x_1 - x_2 + x_3 &= -12 \end{aligned}$

19. Determine the total demand for the products of industries A, B, C, and D based on the input-output matrix and the autonomous demand vector shown below.

Output Industry

Input industry	A	B	C	D		
A	0.4	0.2	0	0.1		30
B	0.1	0.3	0.2	0.1	$D =$	40
C	0.2	0.3	0.1	0.1		60
D	0.1	0.3	0.2	0.2		40

20. Determine the change in total demand for each of the industries in Problem 19 caused by the following change in autonomous demand: industry A, an increase of 3; industry B, a decrease of 5; industry C, an increase of 8; industry D, no change.

SUGGESTED REFERENCES

AYRES, FRANK JR., *Theory and Problems of Matrices*, Schaum's Outline Series (New York, N.Y.: McGraw-Hill Book Company, Inc., 1962).

DORFMAN, R., et al., *Linear Programming and Economic Analysis* (New York, N.Y.: McGraw-Hill Book Company, Inc., 1958).

FULLER, LEONARD E., *Basic Matrix Theory* (Englewood Cliffs, N.J.: Prentice-Hall, Inc., 1962).

GRAYBILL, FRANKLIN A., *Introduction to Matrices with Applications in Statistics* (Belmont, Ca.: Wadsworth Publishing Company, Inc., 1969).

HADLEY, G., *Linear Algebra* (Reading, Mass.: Addison-Wesley Publishing Company, Inc., 1961).

KEMENY, JOHN G., et al., *Finite Mathematical Structures* (Englewood Cliffs, N.J.: Prentice-Hall, Inc., 1958), 4.

LIPSCHUTZ, SEYMOUR, *Linear Algebra*, Schaum's Outline Series (New York, N.Y.: McGraw-Hill Book Company, Inc., 1968).

STEIN, F. MAX, *An Introduction to Matrices and Determinants* (Belmont, Ca.: Wadsworth Publishing Company, Inc., 1967).

YAMANE, TARO, *Mathematics for Economists*, 2nd ed. (Englewood Cliffs, N.J.: Prentice-Hall, Inc., 1962), 10–12.

Chapter 4

Additional Elements
of
Matrix Algebra

The preceding chapter introduced the concepts of vectors and matrices. This chapter continues the discussion of matrices and matrix algebra by introducing the transpose matrix and matrix partitioning. This is followed by a discussion of the determinant of a matrix, the use of the determinant in establishing inverse matrices, Cramer's rule, and the significance of the rank of a matrix to the solution set of systems of equations.

4.1 Transpose

The *transpose* of the matrix A is the matrix A^t (read A transpose) that is formed by writing the rows of matrix A as columns in matrix A^t. In forming the transpose matrix, the first row in A is the same as the first column in A^t, the second row in A is the same as the second column in A^t, and, in general, the ith row in A is the same as the ith column in A^t. To illustrate for the 2 by 3 matrix A, the 3 by 2 matrix A^t is formed as follows.

$$A = \begin{pmatrix} a_{11} & a_{12} & a_{13} \\ a_{21} & a_{22} & a_{23} \end{pmatrix}$$

Interchanging the rows and columns gives

$$A^t = \begin{pmatrix} a_{11} & a_{21} \\ a_{12} & a_{22} \\ a_{13} & a_{23} \end{pmatrix}$$

The matrix A with m rows and n columns is thus transposed to form the matrix A^t with n rows and m columns. The transposition of a matrix is illustrated by the following examples.

Example: Form the transpose of $A = \begin{pmatrix} 2 & 1 & 3 \\ 4 & 3 & 6 \\ 1 & 5 & 4 \end{pmatrix}$. Interchanging the rows and columns gives

$$A^t = \begin{pmatrix} 2 & 4 & 1 \\ 1 & 3 & 5 \\ 3 & 6 & 4 \end{pmatrix}$$

Example: Form the transpose of $A = \begin{pmatrix} 1 & 4 & 3 & 7 \\ 2 & 5 & 1 & 3 \end{pmatrix}$. Interchanging the rows and columns gives

$$A^t = \begin{pmatrix} 1 & 2 \\ 4 & 5 \\ 3 & 1 \\ 7 & 3 \end{pmatrix}$$

Example: Form the transpose of the row vector $A = (6, 3, 1, 7)$. The transpose of the 1 by 4 row vector is the 4 by 1 column vector

$$A^t = \begin{pmatrix} 6 \\ 3 \\ 1 \\ 7 \end{pmatrix}$$

4.2 Submatrices and Partitioning

In applications of matrix algebra discussed later in this chapter, it is necessary to delete selected rows or columns from a matrix. The components that remain after the rows or columns are deleted form a *submatrix*. If, for instance, the second row and third column are deleted from the matrix

$$A = \begin{pmatrix} 1 & 3 & 4 & 6 \\ 2 & 1 & 3 & 7 \\ 0 & 4 & 7 & 1 \end{pmatrix}$$

the submatrix $B = \begin{pmatrix} 1 & 3 & 6 \\ 0 & 4 & 1 \end{pmatrix}$ remains.

Example: Determine the submatrix B formed by the deletion of the third row and first column of the matrix A.

$$A = \begin{pmatrix} 1 & 3 & 6 \\ 2 & 4 & 7 \\ 5 & 5 & 8 \\ 4 & 6 & 9 \end{pmatrix} \qquad B = \begin{pmatrix} 3 & 6 \\ 4 & 7 \\ 6 & 9 \end{pmatrix}$$

Example: Determine the submatrices B_1, B_2, and B_3 formed by the deletion of the second and third columns, the first and third columns, and the first and second columns, respectively, of the matrix

$$A = \begin{pmatrix} 1 & 2 & 4 \\ 2 & 3 & 5 \\ 3 & 4 & 6 \end{pmatrix}$$

The submatrices are

$$B_1 = \begin{pmatrix} 1 \\ 2 \\ 3 \end{pmatrix}, \qquad B_2 = \begin{pmatrix} 2 \\ 3 \\ 4 \end{pmatrix}, \quad \text{and} \quad B_3 = \begin{pmatrix} 4 \\ 5 \\ 6 \end{pmatrix}$$

In certain applications it is useful to consider matrices in which the components of the matrix are submatrices. For instance, the matrix $A = \begin{pmatrix} B_1 & B_2 \\ B_3 & B_4 \end{pmatrix}$ with components $B_1 = \begin{pmatrix} 1 & 2 \\ 3 & 4 \end{pmatrix}$, $B_2 = \begin{pmatrix} 6 \\ 7 \end{pmatrix}$, $B_3 = \begin{pmatrix} 4 & 3 \\ 2 & 1 \end{pmatrix}$, and $B_4 = \begin{pmatrix} 4 \\ 5 \end{pmatrix}$ is a matrix whose components are submatrices. The matrix A can be written as

$$A = \begin{pmatrix} 1 & 2 & 6 \\ 3 & 4 & 7 \\ 4 & 3 & 4 \\ 2 & 1 & 5 \end{pmatrix}$$

Matrices such as A whose components are submatrices are said to have been *partitioned*.

One interesting use of partitioning of matrices is in the multiplication of large matrices. Given two matrices A and B that are partitioned into $A = \begin{pmatrix} A_1 & A_2 \\ A_3 & A_4 \end{pmatrix}$ and $B = \begin{pmatrix} B_1 & B_2 \\ B_3 & B_4 \end{pmatrix}$, the product AB (where the matrices are conformable for multiplication) is

$$AB = \begin{pmatrix} A_1 & A_2 \\ A_3 & A_4 \end{pmatrix} \begin{pmatrix} B_1 & B_2 \\ B_3 & B_4 \end{pmatrix} = \begin{pmatrix} A_1B_1 + A_2B_3 & A_1B_2 + A_2B_4 \\ A_3B_1 + A_4B_3 & A_3B_2 + A_4B_4 \end{pmatrix}$$

The only requirement for the multiplication is that all matrices must be conformable for multiplication. Matrices A and B must be conformable and the matrices must be partitioned such that the submatrices are conformable.

Example: Determine the product of A and B, where

$$A = \begin{pmatrix} 2 & 1 & 0 & 4 \\ -1 & 3 & 2 & 1 \\ -3 & 0 & 2 & -4 \\ 0 & -2 & 3 & 1 \end{pmatrix} \quad \text{and} \quad B = \begin{pmatrix} 3 & 0 \\ -2 & 2 \\ 1 & 3 \\ 3 & 4 \end{pmatrix}$$

The matrices are partitioned such that the submatrices are conformable.

$$AB = \begin{pmatrix} A_1 & A_2 \\ A_3 & A_4 \end{pmatrix} \begin{pmatrix} B_1 \\ B_2 \end{pmatrix} = \begin{pmatrix} A_1B_1 + A_2B_2 \\ A_3B_1 + A_4B_2 \end{pmatrix}$$

The components can be calculated by matrix multiplication and addition to give

$$A_1B_1 + A_2B_2 = \begin{pmatrix} 4 & 2 \\ -9 & 6 \end{pmatrix} + \begin{pmatrix} 12 & 16 \\ 5 & 10 \end{pmatrix} = \begin{pmatrix} 16 & 18 \\ -4 & 16 \end{pmatrix}$$

$$A_3B_1 + A_4B_2 = \begin{pmatrix} -9 & 0 \\ 4 & -4 \end{pmatrix} + \begin{pmatrix} -10 & -10 \\ 6 & 13 \end{pmatrix} = \begin{pmatrix} -19 & -10 \\ 10 & 9 \end{pmatrix}$$

The matrix resulting from the product of A and B is

$$AB = \begin{pmatrix} 16 & 18 \\ -4 & 16 \\ -19 & -10 \\ 10 & 9 \end{pmatrix}$$

The student should verify that this result is the same as obtained by multiplication without partitioning.

4.3 Determinant

One of the important concepts in matrix algebra is that of the *determinant* of a matrix. The determinant of a matrix is a number that is associated with a square matrix. This number can be positive, negative, or zero. The number is important in that it is used in determining the solutions to systems of simultaneous equations.

The determinant is used in two alternative solution techniques for calculating the solution set to a system of simultaneous equations. One approach uses the determinant to calculate the inverse of a matrix. The inverse of the matrix is then used to determine the solution to the system of equations. The second approach uses the determinant in a solution technique termed *Cramer's rule*. This technique is discussed in Sec. 4.5.

4.3.1 DETERMINANTS OF 2 BY 2 AND
3 BY 3 MATRICES

The determinant of the matrix A is commonly represented by one of two symbols, det A or $|A|$. These symbols are used interchangeably in many texts. We will, however, use the symbol $|A|$ exclusively in this text to represent the determinant.

The determinants of 2 by 2 and 3 by 3 matrices are easily calculated. The determinant of the 2 by 2 matrix A is given by

$$|A| = \begin{vmatrix} a_{11} & a_{12} \\ a_{21} & a_{22} \end{vmatrix} = a_{11} \cdot a_{22} - a_{21} \cdot a_{12} \tag{4.1}$$

To illustrate the formula, assume that $A = \begin{pmatrix} 1 & 3 \\ -2 & 4 \end{pmatrix}$.

$$|A| = \begin{vmatrix} 1 & 3 \\ -2 & 4 \end{vmatrix} = 1 \cdot 4 - (-2) \cdot 3 = 10$$

The determinant of A is the scalar 10.

Example: Find the determinant of $C = \begin{pmatrix} -2 & 3 \\ 6 & 4 \end{pmatrix}$.

$$|C| = \begin{vmatrix} -2 & 3 \\ 6 & 4 \end{vmatrix} = -8 - 18 = -26$$

Example: Find the determinant of $D = \begin{pmatrix} 1 & 3 \\ 1 & 3 \end{pmatrix}$.

$$|D| = \begin{vmatrix} 1 & 3 \\ 1 & 3 \end{vmatrix} = 3 - 3 = 0$$

The determinant of a 3 by 3 matrix is calculated by a similar formula. The determinant of the 3 by 3 matrix A is given by

$$|A| = \begin{vmatrix} a_{11} & a_{12} & a_{13} \\ a_{21} & a_{22} & a_{23} \\ a_{31} & a_{32} & a_{33} \end{vmatrix} = \begin{matrix} a_{11}a_{22}a_{33} + a_{12}a_{23}a_{31} + a_{13}a_{21}a_{32} \\ - a_{31}a_{22}a_{13} - a_{32}a_{23}a_{11} - a_{33}a_{21}a_{12} \end{matrix} \tag{4.2}$$

Some students find the formula easy to remember by employing the following scheme. First, repeat the first two columns of the matrix. Next, draw diagonals through the components as shown below. The products of the components joined by

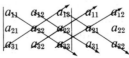

each diagonal are then determined, e.g., $a_{11} \cdot a_{22} \cdot a_{33}$. The products of the components on the diagonals from upper left to lower right are summed, and the products of the components on the diagonals from lower left to upper right are subtracted from this sum. This results in Eq. (4.2) given above, i.e.,

$$|A| = a_{11}a_{22}a_{33} + a_{12}a_{23}a_{31} + a_{13}a_{21}a_{32} - a_{31}a_{22}a_{13} - a_{32}a_{23}a_{11} - a_{33}a_{21}a_{12}$$

Example: Find the determinant of $A = \begin{pmatrix} 1 & 3 & -4 \\ 2 & -1 & 6 \\ 3 & 0 & -2 \end{pmatrix}$. Use the method of repeating the first two columns.

$$|A| = 1(-1)(-2) + 3 \cdot 6 \cdot 3 + (-4)2 \cdot 0 - 3(-1)(-4) - (0)6 \cdot 1 - (-2)2 \cdot 3$$
$$|A| = 2 + 54 + 0 - 12 - 0 + 12 = 56$$

Example: Find the determinant of $A = \begin{pmatrix} 1 & 5 & 4 \\ -3 & 2 & 3 \\ 6 & -1 & 4 \end{pmatrix}$.

$$|A| = 1 \cdot 2 \cdot 4 + 5 \cdot 3 \cdot 6 + 4(-3)(-1) - 6 \cdot 2 \cdot 4 - (-1)3 \cdot 1 - 4(-3)5$$
$$|A| = 8 + 90 + 12 - 48 + 3 + 60 = 125$$

Example: Find the determinant of $A = \begin{pmatrix} 2 & -1 & 3 \\ 3 & -2 & 6 \\ 2 & 1 & 4 \end{pmatrix}$.

$$|A| = 2(-2)4 + (-1)6 \cdot 2 + 3 \cdot 3 \cdot 1 - 2(-2)3 - 1 \cdot 6 \cdot 2 - 4 \cdot 3(-1)$$
$$|A| = -16 - 12 + 9 + 12 - 12 + 12 = -7$$

4.3.2 MINOR, COFACTOR, AND COFACTOR EXPANSION

The method discussed in Sec. 4.3.1 applies for calculating the determinant of a 2 by 2 or 3 by 3 matrix. It does not, however, apply to matrices of higher dimensions. This section introduces a procedure for calculating a determinant that applies to any square matrix. This procedure is termed the method of *cofactor expansion.*

Before discussing the method of cofactor expansion, we must define two terms. There are *minor* and *cofactor.* Both of these terms are defined with reference to an element a_{ij} of a square matrix A. We thus speak of the minor of the a_{ij} element or the cofactor of the a_{ij} element.

With reference again to the a_{ij} element, the minor of the a_{ij} element is defined as the determinant of the submatrix formed by deleting the ith row and the jth column of the matrix. The minor of an element is thus a determinant. It is, again, the determinant of the submatrix resulting from deleting the ith row and jth column. For the 3 by 3 matrix A shown below, the minor of a_{11}

$$A = \begin{pmatrix} a_{11} & a_{12} & a_{13} \\ a_{21} & a_{22} & a_{23} \\ a_{31} & a_{32} & a_{33} \end{pmatrix}$$

is the determinant of the submatrix formed by deleting the first row and first column of A, i.e.,

$$\text{minor } a_{11} = \begin{vmatrix} a_{11} & a_{12} & a_{13} \\ a_{21} & a_{22} & a_{23} \\ a_{31} & a_{32} & a_{33} \end{vmatrix} = \begin{vmatrix} a_{22} & a_{23} \\ a_{32} & a_{33} \end{vmatrix}$$

Similarly, the minor of the element a_{21} is $\begin{vmatrix} a_{12} & a_{13} \\ a_{32} & a_{33} \end{vmatrix}$ and the minor of a_{22} is $\begin{vmatrix} a_{11} & a_{13} \\ a_{31} & a_{33} \end{vmatrix}$.

A method of determining the elements included in the submatrix is to line out the row and column to which the a_{ij} element belongs. The minor of a_{32} is thus

$$\text{minor } a_{32} = \begin{vmatrix} a_{11} & a_{12} & a_{13} \\ a_{21} & a_{22} & a_{23} \\ a_{31} & a_{32} & a_{33} \end{vmatrix} = \begin{vmatrix} a_{11} & a_{13} \\ a_{21} & a_{23} \end{vmatrix}$$

Example: Find the minors of the elements of the first row and of the third column of the matrix $A = \begin{pmatrix} 2 & 3 & 1 \\ -1 & 3 & 2 \\ 4 & 2 & 3 \end{pmatrix}$. Since the minor of the a_{ij} element is the determinant of the submatrix formed by eliminating the ith row and jth column, the minors of a_{11}, a_{12}, a_{13}, a_{23}, and a_{33} are

$$\text{minor } a_{11} = \begin{vmatrix} 3 & 2 \\ 2 & 3 \end{vmatrix} = 9 - 4 = 5$$

$$\text{minor } a_{12} = \begin{vmatrix} -1 & 2 \\ 4 & 3 \end{vmatrix} = -3 - 8 = -11$$

$$\text{minor } a_{13} = \begin{vmatrix} -1 & 3 \\ 4 & 2 \end{vmatrix} = -2 - 12 = -14$$

$$\text{minor } a_{23} = \begin{vmatrix} 2 & 3 \\ 4 & 2 \end{vmatrix} = 4 - 12 = -8$$

$$\text{minor } a_{33} = \begin{vmatrix} 2 & 3 \\ -1 & 3 \end{vmatrix} = 6 - (-3) = 9$$

Example: Find the minors of the second row of the matrix

$$A = \begin{pmatrix} 3 & 1 & 6 & 4 \\ 2 & 3 & -1 & 7 \\ 2 & 0 & 3 & 2 \\ 4 & 2 & -3 & 1 \end{pmatrix}$$

$$\text{minor } a_{21} = \begin{vmatrix} 1 & 6 & 4 \\ 0 & 3 & 2 \\ 2 & -3 & 1 \end{vmatrix} = 3 + 24 + 0 - 24 + 6 + 0 = 9$$

$$\text{minor } a_{22} = \begin{vmatrix} 3 & 6 & 4 \\ 2 & 3 & 2 \\ 4 & -3 & 1 \end{vmatrix} = 9 + 48 - 24 - 48 + 18 - 12 = -9$$

$$\text{minor } a_{23} = \begin{vmatrix} 3 & 1 & 4 \\ 2 & 0 & 2 \\ 4 & 2 & 1 \end{vmatrix} = 0 + 8 + 16 - 0 - 12 - 2 = 10$$

$$\text{minor } a_{24} = \begin{vmatrix} 3 & 1 & 6 \\ 2 & 0 & 3 \\ 4 & 2 & -3 \end{vmatrix} = 0 + 12 + 24 - 0 - 18 + 6 = 24$$

We stated earlier that both cofactor and minor were defined with reference to an element a_{ij} of a square matrix. The *cofactor* is the product of the minor of the a_{ij} element and $(-1)^{i+j}$. The cofactor is thus simply the minor multiplied by either $+1$ or -1, the sign depending upon the location of the a_{ij} element. For instance, the cofactor of the a_{12} element would be the product of the minor of this element and $(-1)^{1+2}$, or -1. The cofactor of the a_{22} element would be the product of the minor of the a_{22} element and $(-1)^{2+2}$, or $+1$. The cofactor of the a_{ij} element is designated by the symbol A_{ij}. The general formula is

$$A_{ij} = (-1)^{i+j} \text{ minor } a_{ij} \tag{4.3}$$

Example: Find the cofactors of the elements of the first row and of the third column of the matrix $A = \begin{pmatrix} 2 & 3 & 1 \\ -1 & 3 & 2 \\ 4 & 2 & 3 \end{pmatrix}$. The minors of this matrix were calculated on p. 114. Since the cofactor A_{ij} is the product of $(-1)^{i+j}$ and the minor of a_{ij}, the cofactors are

$$A_{11} = (-1)^2 \cdot \text{minor } a_{11} = +1(5) = 5$$
$$A_{12} = (-1)^3 \cdot \text{minor } a_{12} = -1(-11) = 11$$
$$A_{13} = (-1)^4 \cdot \text{minor } a_{13} = +1(-14) = -14$$
$$A_{23} = (-1)^5 \cdot \text{minor } a_{23} = -1(-8) = 8$$
$$A_{33} = (-1)^6 \cdot \text{minor } a_{33} = +1(9) = 9$$

Example: Determine the cofactors of the elements of the second row of the matrix

$$A = \begin{pmatrix} 3 & 1 & 6 & 4 \\ 2 & 3 & -1 & 7 \\ 2 & 0 & 3 & 2 \\ 4 & 2 & -3 & 1 \end{pmatrix}$$

The minors of the elements of the second row were determined on p. 115. The cofactors of the elements of the second row are

$$A_{21} = (-1)^3 \cdot \text{minor } a_{21} = -1(9) = -9$$
$$A_{22} = (-1)^4 \cdot \text{minor } a_{22} = +1(-9) = -9$$
$$A_{23} = (-1)^5 \cdot \text{minor } a_{23} = -1(10) = -10$$
$$A_{24} = (-1)^6 \cdot \text{minor } a_{24} = +1(24) = 24$$

We can now introduce a method for calculating a determinant that applies to any square matrix. To calculate the determinant we multiply the cofactors for any row or column by the elements of that row or column. The sum of these products is the determinant of the matrix. To illustrate, the determinant of the 3 by 3 matrix

$$A = \begin{pmatrix} a_{11} & a_{12} & a_{13} \\ a_{21} & a_{22} & a_{23} \\ a_{31} & a_{32} & a_{33} \end{pmatrix}$$

is given by multiplying the elements in the first row by the cofactors of that row,

$$|A| = a_{11}A_{11} + a_{12}A_{12} + a_{13}A_{13}$$

or alternatively, by multiplying the elements in the third column by the co-factors of the column,

$$|A| = a_{13}A_{13} + a_{23}A_{23} + a_{33}A_{33}$$

or by the sum of the products of the cofactors and the elements of any other row or column. This method of calculating determinants is termed *cofactor expansion.*

Example: Calculate the determinant by cofactor expansion of the matrix $A = \begin{pmatrix} 2 & 3 & 1 \\ -1 & 3 & 2 \\ 4 & 2 & 3 \end{pmatrix}$. Verify that expansion about the first row or the third column yields the same value of the determinant.

The cofactors of the first row and of the third column of the matrix were calculated on p. 115. Cofactor expansion about the first row gives

$$|A| = 2A_{11} + 3A_{12} + 1A_{13}$$

$$|A| = 2(5) + 3(11) + 1(-14) = 29$$

Cofactor expansion about the third column gives

$$|A| = 1A_{13} + 2A_{23} + 3A_{33}$$

$$|A| = 1(-14) + 2(8) + 3(9) = 29$$

The reader can verify that the cofactor expansion about any other row or column gives the same value for the determinant.

Example: Find the determinant of the matrix

$$A = \begin{pmatrix} 3 & 1 & 6 & 4 \\ 2 & 3 & -1 & 7 \\ 2 & 0 & 3 & 2 \\ 4 & 2 & -3 & 1 \end{pmatrix}$$

The cofactors of the elements of the second row were calculated on p. 116. The determinant of A can be easily determined by cofactor expansion about the second row.

$$|A| = 2A_{21} + 3A_{22} - 1A_{23} + 7A_{24}$$

$$|A| = 2(-9) + 3(-9) - 1(-10) + 7(24)$$

$$|A| = 133$$

These examples show that the determinant of a 4 by 4 matrix can be obtained by cofactor expansion. To illustrate for a larger matrix, consider the 5 by 5 matrix

$$A = \begin{pmatrix} a_{11} & a_{12} & a_{13} & a_{14} & a_{15} \\ a_{21} & a_{22} & a_{23} & a_{24} & a_{25} \\ a_{31} & a_{32} & a_{33} & a_{34} & a_{35} \\ a_{41} & a_{42} & a_{43} & a_{44} & a_{45} \\ a_{51} & a_{52} & a_{53} & a_{54} & a_{55} \end{pmatrix}$$

The determinant of A may be calculated by cofactor expansion about any row or column. If it is assumed that the first row is selected for expansion, the determinant of A is

$$|A| = a_{11}A_{11} + a_{12}A_{12} + a_{13}A_{13} + a_{14}A_{14} + a_{15}A_{15}$$

where A_{1j} are cofactors of the j submatrices. Since the cofactor is the product of the minor a_{ij} and $(-1)^{i+j}$, we must calculate the minors of the elements of the first row. In this instance, the minors must be calculated for 4 by 4 submatrices. The calculation of the determinant of a 4 by 4 matrix by cofactor expansion was illustrated by the preceding example.

4.3.3 PROPERTIES OF DETERMINANTS

The reader has undoubtedly recognized that the calculation of determinants of higher-order matrices requires a considerable amount of arithmetic. The arithmetic involved in the calculation of a determinant by cofactor expansion is reduced if the row or column selected for cofactor expansion contains mainly zeros. For instance, the determinant of a 5 by 5 matrix was given as

$$|A| = a_{11}A_{11} + a_{12}A_{12} + a_{13}A_{13} + a_{14}A_{14} + a_{15}A_{15}$$

The arithmetic calculations required to determine the determinant are reduced in proportion to the number of elements a_{ij} that have values of zero.

Certain properties of determinants are useful in obtaining zeros as elements in rows or columns. Many of these properties will also prove useful in more advanced studies of matrix applications. These properties of determinants are offered without proof as follows.

1. Interchanging any two rows or any two columns of a matrix changes the sign of the determinant of the matrix but not the absolute value of the determinant. Thus if

$$A = \begin{pmatrix} 1 & 3 & 2 \\ 3 & 1 & -3 \\ 2 & 0 & 4 \end{pmatrix} \quad \text{and}$$

$$B = \begin{pmatrix} 2 & 3 & 1 \\ -3 & 1 & 3 \\ 4 & 0 & 2 \end{pmatrix}, \quad \text{then} \quad |A| = -|B|$$

2. The determinant of a matrix and of its transpose are equal, i.e.,

$$|A| = |A^t|$$

3. If all of the elements in any row or column are zero, the determinant of the matrix is zero.

4. If the elements in any two rows or two columns are equal or proportional, the determinant of the matrix is zero.

5. If the elements of any row or column are multiplied by a constant k, the determinant of the matrix is multiplied by the constant k, i.e.,

$$A = \begin{pmatrix} 1 & 3 & 2 \\ 3 & 1 & -3 \\ 2 & 0 & 4 \end{pmatrix} \quad \text{and}$$

$$B = \begin{pmatrix} 1k & 3k & 2k \\ 3 & 1 & -3 \\ 2 & 0 & 4 \end{pmatrix}, \quad \text{then} \quad |B| = k|A|$$

6. Any nonzero multiple of one row may be added to another row or any nonzero multiple of one column may be added to another column without changing the value of the determinant.
7. The determinant of any identity matrix is 1.

These properties can be applied in many ingenious ways to obtain the determinant of a matrix. To illustrate, assume that we must calculate the determinant of the 5 by 5 matrix

$$A = \begin{pmatrix} 3 & 1 & 0 & -2 & 4 \\ 2 & 3 & 1 & 2 & -3 \\ 4 & -5 & 0 & 6 & 2 \\ 0 & 6 & 1 & 3 & -1 \\ 5 & 2 & 3 & 1 & 1 \end{pmatrix}$$

Using the method of cofactor expansion, we examine the rows and columns and select the third column for expansion. This column is selected because it has more zeros than any other column or row.

To reduce the computation required in the expansion, subtract the second row from the fourth row and three times the second row from the fifth row. This gives

$$|A| = \begin{vmatrix} 3 & 1 & 0 & -2 & 4 \\ 2 & 3 & 1 & 2 & -3 \\ 4 & -5 & 0 & 6 & 2 \\ -2 & 3 & 0 & 1 & 2 \\ -1 & -7 & 0 & -5 & 10 \end{vmatrix}.$$

Cofactor expansion about the third column gives

$$|A| = a_{13}A_{13} + a_{23}A_{23} + a_{33}A_{33} + a_{43}A_{43} + a_{53}A_{53}$$

and since all elements in the column with the exception of a_{23} are zero, the expression becomes

$$|A| = a_{23}A_{23} = 1(-1)^5 \cdot \text{minor } a_{23} = -1 \begin{vmatrix} 3 & 1 & -2 & 4 \\ 4 & -5 & 6 & 2 \\ -2 & 3 & 1 & 2 \\ -1 & -7 & -5 & 10 \end{vmatrix}$$

The determinant of A equals -1 times the minor of a_{23}. This minor can be evaluated by applying the properties of determinants. Subtracting twice the fourth row from the third row and adding four times the fourth row to the second row and three times the fourth row to the first row gives

$$|A| = -1 \begin{vmatrix} 0 & -20 & -17 & 34 \\ 0 & -33 & -14 & 42 \\ 0 & 17 & 11 & -18 \\ -1 & -7 & -5 & 10 \end{vmatrix}$$

Multiplying the fourth row by -1 gives

$$|A| = +1 \begin{vmatrix} 0 & -20 & -17 & 34 \\ 0 & -33 & -14 & 42 \\ 0 & 17 & 11 & -18 \\ 1 & 7 & 5 & -10 \end{vmatrix}$$

Expanding about the first column gives

$$|A| = 1(-1)^{4+1} \text{ minor } a_{41} = -1 \begin{vmatrix} -20 & -17 & 34 \\ -33 & -14 & 42 \\ 17 & 11 & -18 \end{vmatrix}$$

The determinant of the 3 by 3 matrix can be calculated directly by the method presented in Sec. 4.3.1.

$$|A| = -1\{(-20)(-14)(-18) + (-17)(42)(17) + (34)(-33)(11)$$
$$- (17)(-14)(34) - (11)(42)(-20) - (-18)(-33)(-17)\}$$
$$|A| = -1\{-5040 - 12138 - 12342 + 8092 + 9240 + 10098\}$$
$$|A| = -1(-2090) = 2090$$

Example: Verify that expansion about the first row of the 5 by 5 matrix A results in the same determinant as expansion about the third column.

$$|A| = \begin{vmatrix} 3 & 1 & 0 & -2 & 4 \\ 2 & 3 & 1 & 2 & -3 \\ 4 & -5 & 0 & 6 & 2 \\ 0 & 6 & 1 & 3 & -1 \\ 5 & 2 & 3 & 1 & 1 \end{vmatrix}$$

Cofactor expansion about the third column in the preceding example resulted in $|A| = 2090$. To expand about the first row, we subtract three times the second column from the first column, add two times the second column to the fourth column, and subtract four times the second column from the fifth column. This gives

$$|A| = \begin{vmatrix} 0 & 1 & 0 & 0 & 0 \\ -7 & 3 & 1 & 8 & -15 \\ 19 & -5 & 0 & -4 & 22 \\ -18 & 6 & 1 & 15 & -25 \\ -1 & 2 & 3 & 5 & -7 \end{vmatrix}$$

The cofactor expansion about the first row is

$$|A| = a_{11}A_{11} + a_{12}A_{12} + a_{13}A_{13} + a_{14}A_{14} + a_{15}A_{15}$$

which reduces to

$$|A| = a_{12}A_{12} = 1(-1)^3 \text{ minor } a_{12} = -1 \begin{vmatrix} -7 & 1 & 8 & -15 \\ 19 & 0 & -4 & 22 \\ -18 & 1 & 15 & -25 \\ -1 & 3 & 5 & -7 \end{vmatrix}$$

To reduce the number of arithmetic operations, subtract the first row from the third row and three times the first row from the fourth row. This gives

$$|A| = -1 \begin{vmatrix} -7 & 1 & 8 & -15 \\ 19 & 0 & -4 & 22 \\ -11 & 0 & 7 & -10 \\ 20 & 0 & -19 & 38 \end{vmatrix}$$

Expanding about the second column gives

$$|A| = -1(a_{12}A_{12} + a_{22}A_{22} + a_{32}A_{32} + a_{42}A_{42})$$

or alternatively,

$$|A| = -1(1 \cdot (-1)^3 \text{ minor } a_{12}) = +1 \begin{vmatrix} 19 & -4 & 22 \\ -11 & 7 & -10 \\ 20 & -19 & 38 \end{vmatrix}$$

The determinant can now be calculated directly by the method presented in Sec. 4.3.1.

$$|A| = 19 \cdot 7 \cdot 38 + (-4)(-10)(20) + 22(-11)(-19) - 20 \cdot 7 \cdot 22$$
$$\quad - (-19)(-10)(19) - 38(-11)(-4)$$
$$|A| = 5054 + 800 + 4598 - 3080 - 3610 - 1672$$
$$|A| = 2090$$

The determinants of A calculated by expansion about the third column and about the first row are equal.

Example: One of the properties of determinants is that the value of the determinant is zero if any two rows or two columns are equal. Verify this property for the 3 by 3 matrix $A = \begin{pmatrix} 1 & 2 & 1 \\ 2 & 1 & 2 \\ 3 & 4 & 3 \end{pmatrix}$.

The determinant is

$$|A| = 1 \cdot 1 \cdot 3 + 2 \cdot 2 \cdot 3 + 1 \cdot 2 \cdot 4 - 3 \cdot 1 \cdot 1 - 4 \cdot 2 \cdot 1 - 3 \cdot 2 \cdot 2$$
$$|A| = 3 + 12 + 8 - 3 - 8 - 12 = 0$$

Example: Verify that the determinant of any matrix in which two rows or two columns are proportional is zero.

Given the matrix $A = \begin{pmatrix} 1 & 2 & 3 \\ 2 & 4 & 6 \\ 3 & 1 & 7 \end{pmatrix}$, it can be seen that the second row is twice the first row. Subtracting twice the first row from the second row gives

$$|A| = \begin{vmatrix} 1 & 2 & 3 \\ 0 & 0 & 0 \\ 3 & 1 & 7 \end{vmatrix}$$

Since all elements in the second row are zero, the determinant of A is zero.

Example: One of the properties of a determinant is that if the elements of any row or column are multiplied by a constant k, the determinant of the matrix is multiplied by the constant. This property often proves useful in calculating a determinant. For instance, if $A = \begin{pmatrix} 3 & 2 & 8 \\ 1 & 4 & -6 \\ 6 & 6 & 12 \end{pmatrix}$ the determinant of A is

$$|A| = \begin{vmatrix} 3 & 2 & 8 \\ 1 & 4 & -6 \\ 6 & 6 & 12 \end{vmatrix}$$

The second column can be factored to give

$$|A| = 2 \begin{vmatrix} 3 & 1 & 8 \\ 1 & 2 & -6 \\ 6 & 3 & 12 \end{vmatrix}$$

Factoring a 3 from the third row gives

$$|A| = 2 \cdot 3 \begin{vmatrix} 3 & 1 & 8 \\ 1 & 2 & -6 \\ 2 & 1 & 4 \end{vmatrix}$$

The third column is factored to give

$$|A| = 2 \cdot 3 \cdot 4 \begin{vmatrix} 3 & 1 & 2 \\ 1 & 2 & -1.5 \\ 2 & 1 & 1 \end{vmatrix}$$

The determinant is

$$|A| = 24(3 \cdot 2 \cdot 1 + 1(-1.5)(2) + 2 \cdot 1 \cdot 1 - 2 \cdot 2 \cdot 2 - 1(-1.5)(3) - 1 \cdot 1 \cdot 1)$$

$$|A| = 24(6 - 3 + 2 - 8 + 4.5 - 1) = 12$$

4.3.4 SINGULAR AND NONSINGULAR MATRICES

The terms *singular* and *nonsingular* are used with reference to the value of the determinant of a square matrix. A square matrix A is said to be singular if the determinant of A is zero. Conversely, matrices whose determinants are nonzero are termed nonsingular.

Example: Determine if the matrix A is singular or nonsingular.

$$A = \begin{pmatrix} 2 & 3 \\ 4 & 3 \end{pmatrix}$$

Since $|A| = -6$ is nonzero, the matrix is nonsingular.

Example: Determine if the matrix A is singular or nonsingular.

$$A = \begin{pmatrix} 4 & 6 \\ 2 & 3 \end{pmatrix}$$

Since $|A| = 0$ is zero, the matrix is singular.

4.4 The Inverse: Adjoint Method

The inverse of the matrix A was defined in Chapter 3 as the matrix A^{-1} such that

$$AA^{-1} = I$$

where A and A^{-1} are both square matrices and I is the identity matrix. A method of calculating the inverse, termed the Gaussian elimination method, was presented in that chapter.

An alternative method of calculating the inverse is given in this section. Before giving the procedure for this technique, we must define the cofactor and adjoint matrices.

4.4.1 COFACTOR MATRIX

The cofactor of an element a_{ij} was defined in Sec. 4.3.2. The cofactor of the element a_{ij} of matrix A was designated by the symbol A_{ij}, where i refers to the row location and j the column location. The *cofactor matrix*, designated cof A, is the matrix of cofactors. That is,

$$\text{cof } A = \begin{pmatrix} A_{11} & A_{12} & \cdots & A_{1n} \\ A_{21} & A_{22} & \cdots & A_{2n} \\ \vdots & \vdots & & \vdots \\ A_{m1} & A_{m2} & \cdots & A_{mn} \end{pmatrix}, \tag{4.4}$$

where the component in row i and column j is the cofactor of the element a_{ij} of the square matrix A.

Example: Determine the cofactor matrix for the matrix

$$A = \begin{pmatrix} 3 & -2 & 4 \\ 6 & 4 & 5 \\ 3 & -1 & 6 \end{pmatrix}$$

The minors of the element a_{ij} are

minor $a_{11} = \quad 29$	minor $a_{12} = \quad 21$	minor $a_{13} = \ -18$
minor $a_{21} = \ -8$	minor $a_{22} = \quad 6$	minor $a_{23} = \quad 3$
minor $a_{31} = \ -26$	minor $a_{32} = \ -9$	minor $a_{33} = \quad 24$

The cofactor of each element is given by the product of the minor and $(-1)^{i+j}$. The cofactor matrix is

$$\text{cof } A = \begin{pmatrix} 29 & -21 & -18 \\ 8 & 6 & -3 \\ -26 & 9 & 24 \end{pmatrix}$$

Example: Determine the cofactor matrix for the matrix

$$A = \begin{pmatrix} 6 & -2 & -3 \\ -1 & 8 & -7 \\ 4 & -3 & 6 \end{pmatrix}$$

The minors of the elements a_{ij} are

minor $a_{11} =$ 27	minor $a_{12} =$ 22	minor $a_{13} = -29$
minor $a_{21} = -21$	minor $a_{22} =$ 48	minor $a_{23} = -10$
minor $a_{31} =$ 38	minor $a_{32} = -45$	minor $a_{33} =$ 46

The cofactor matrix is

$$\text{cof } A = \begin{pmatrix} 27 & -22 & -29 \\ 21 & 48 & 10 \\ 38 & 45 & 46 \end{pmatrix}$$

4.4.2 ADJOINT MATRIX

The adjoint matrix is the transpose of the cofactor matrix. That is,

$$\text{adj } A = (\text{cof } A)^t \tag{4.5}$$

where adj A designates the adjoint matrix of A. This definition is illustrated by the following two examples.

Example: Determine the adjoint matrix for

$$A = \begin{pmatrix} 3 & -2 & 4 \\ 6 & 4 & 5 \\ 3 & -1 & 6 \end{pmatrix}$$

The cofactor matrix was calculated on p. 123 and is

$$\text{cof } A = \begin{pmatrix} 29 & -21 & -18 \\ 8 & 6 & -3 \\ -26 & 9 & 24 \end{pmatrix}$$

The adjoint matrix is the transpose of the cofactor matrix, i.e.,

$$\text{adj } A = \begin{pmatrix} 29 & 8 & -26 \\ -21 & 6 & 9 \\ -18 & -3 & 24 \end{pmatrix}$$

Example: Determine the adjoint matrix for

$$A = \begin{pmatrix} 6 & -2 & -3 \\ -1 & 8 & -7 \\ 4 & -3 & 6 \end{pmatrix}$$

The cofactor matrix was calculated on p. 124. The adjoint matrix is the transpose of the cofactor matrix, i.e.,

$$\text{adj } A = \begin{pmatrix} 27 & 21 & 38 \\ -22 & 48 & 45 \\ -29 & 10 & 46 \end{pmatrix}$$

The technique for calculating the inverse of the square matrix A can now be given. The inverse matrix, A^{-1}, is given by

$$A^{-1} = \left(\frac{1}{|A|}\right) \text{adj } A \tag{4.6}$$

provided A is nonsingular. This definition states that the inverse of A exists provided that the determinant of A is nonsingular, i.e., nonzero. The inverse is given by the product of the scalar $1/|A|$ and the adjoint matrix. This method of calculating an inverse matrix is illustrated by the following examples.

Example: Calculate the inverse of

$$A = \begin{pmatrix} 3 & -2 & 4 \\ 6 & 4 & 5 \\ 3 & -1 & 6 \end{pmatrix}$$

Applying the properties of determinants, we obtain

$$|A| = 3 \begin{vmatrix} 1 & -2 & 4 \\ 2 & 4 & 5 \\ 1 & -1 & 6 \end{vmatrix} = 3 \cdot 2 \begin{vmatrix} 1 & -2 & 4 \\ 1 & 2 & 2.5 \\ 1 & -1 & 6 \end{vmatrix}$$

$$|A| = 3 \cdot 2 \begin{vmatrix} 1 & -2 & 4 \\ 0 & 4 & -1.5 \\ 0 & 1 & 2 \end{vmatrix} = 3 \cdot 2[(1)(8 + 1.5)]$$

$$|A| = 57$$

The adjoint matrix was calculated on p. 124 and is

$$\text{adj } A = \begin{pmatrix} 29 & 8 & -26 \\ -21 & 6 & 9 \\ -18 & -3 & 24 \end{pmatrix}$$

The inverse is

$$A^{-1} = \frac{1}{57} \begin{pmatrix} 29 & 8 & -26 \\ -21 & 6 & 9 \\ -18 & -3 & 24 \end{pmatrix} = \begin{pmatrix} \frac{29}{57} & \frac{8}{57} & -\frac{26}{57} \\ -\frac{21}{57} & \frac{6}{57} & \frac{9}{57} \\ -\frac{18}{57} & -\frac{3}{57} & \frac{24}{57} \end{pmatrix}$$

Example: Determine the inverse of

$$A = \begin{pmatrix} 6 & -2 & -3 \\ -1 & 8 & -7 \\ 4 & -3 & 6 \end{pmatrix}$$

The determinant of A is

$$|A| = 6 \cdot 8 \cdot 6 + (-2)(-7)(4) + (-3)(-1)(-3) - 4 \cdot 8(-3)$$
$$-(-3)(-7)(6) - 6(-1)(-2)$$
$$|A| = 288 + 56 - 9 + 96 - 126 - 12$$
$$|A| = 293$$

The adjoint matrix was calculated on p. 125 as

$$\text{adj } A = \begin{pmatrix} 27 & 21 & 38 \\ -22 & 48 & 45 \\ -29 & 10 & 46 \end{pmatrix}$$

The inverse matrix is given by $A^{-1} = (1/|A|)\,\text{adj } A$.

$$A^{-1} = \begin{pmatrix} 27/293 & 21/293 & 38/293 \\ -22/293 & 48/293 & 45/293 \\ -29/293 & 10/293 & 46/293 \end{pmatrix}$$

Example: Solve the following system of linear equations by determining the inverse.

$$3x_1 + 3x_2 + 4x_3 = 20$$
$$-2x_1 + 4x_2 - 2x_3 = -6$$
$$4x_1 - 2x_2 + 3x_3 = 16$$

In matrix form, the system of equations is $AX = B$.

$$\begin{pmatrix} 3 & 3 & 4 \\ -2 & 4 & -2 \\ 4 & -2 & 3 \end{pmatrix} \begin{pmatrix} x_1 \\ x_2 \\ x_3 \end{pmatrix} = \begin{pmatrix} 20 \\ -6 \\ 16 \end{pmatrix}$$

The inverse of A is calculated by Formula (4.6),

$$A^{-1} = (1/|A|)\,\text{adj } A \qquad (4.6)$$

where

$$|A| = \begin{vmatrix} 3 & 3 & 4 \\ -2 & 4 & -2 \\ 4 & -2 & 3 \end{vmatrix}$$

$$|A| = 3 \cdot 4 \cdot 3 + 3(-2)4 + 4(-2)(-2) - 4 \cdot 4 \cdot 4 - (-2)(-2)3 - 3(-2)(3),$$
$$|A| = -30$$

The cofactors of A are

$$A_{11} = (-1)^2 \begin{vmatrix} 4 & -2 \\ -2 & 3 \end{vmatrix} = +1(12 - 4) = 8$$

$$A_{12} = (-1)^3 \begin{vmatrix} -2 & -2 \\ 4 & 3 \end{vmatrix} = -1(-6 + 8) = -2$$

$$A_{13} = (-1)^4 \begin{vmatrix} -2 & 4 \\ 4 & -2 \end{vmatrix} = +1(4 - 16) = -12$$

$$A_{21} = (-1)^3 \begin{vmatrix} 3 & 4 \\ -2 & 3 \end{vmatrix} = -1(9 + 8) = -17$$

$$A_{22} = (-1)^4 \begin{vmatrix} 3 & 4 \\ 4 & 3 \end{vmatrix} = +1(9 - 16) = -7$$

$$A_{23} = (-1)^5 \begin{vmatrix} 3 & 3 \\ 4 & -2 \end{vmatrix} = -1(-6 - 12) = 18$$

$$A_{31} = (-1)^4 \begin{vmatrix} 3 & 4 \\ 4 & -2 \end{vmatrix} = +1(-6 - 16) = -22$$

$$A_{32} = (-1)^5 \begin{vmatrix} 3 & 4 \\ -2 & -2 \end{vmatrix} = -1(-6 + 8) = -2$$

$$A_{33} = (-1)^6 \begin{vmatrix} 3 & 3 \\ -2 & 4 \end{vmatrix} = +1(12 + 6) = 18$$

The cofactor matrix is

$$\operatorname{cof} A = \begin{pmatrix} 8 & -2 & -12 \\ -17 & -7 & 18 \\ -22 & -2 & 18 \end{pmatrix}$$

The adjoint matrix is given by the transpose of cof A,

$$\operatorname{adj} A = \begin{pmatrix} 8 & -17 & -22 \\ -2 & -7 & -2 \\ -12 & 18 & 18 \end{pmatrix}$$

The inverse matrix is

$$A^{-1} = -\tfrac{1}{30} \begin{pmatrix} 8 & -17 & -22 \\ -2 & -7 & -2 \\ -12 & 18 & 18 \end{pmatrix}$$

The solution vector for the three simultaneous equations is

$$X = A^{-1}B = -\tfrac{1}{30} \begin{pmatrix} 8 & -17 & -22 \\ -2 & -7 & -2 \\ -12 & 18 & 18 \end{pmatrix} \begin{pmatrix} 20 \\ -6 \\ 16 \end{pmatrix}$$

$$X = -\tfrac{1}{30} \begin{pmatrix} -90 \\ -30 \\ -30 \end{pmatrix} = \begin{pmatrix} 3 \\ 1 \\ 2 \end{pmatrix}$$

4.5 Cramer's Rule

A technique called Cramer's rule is available for solving a system of n equations with n variables. This technique bypasses the calculation of the inverse of the coefficient matrix or the reduction of the augmented coefficient matrix through row operations. Instead, Cramer's rule requires the calculation of determinants for $n + 1$ matrices. The rationale underlying Cramer's rule is shown for a 2 by 2 system of simultaneous equations and extended to the general case of n by n simultaneous equations.

Given the 2 by 2 system of equations,

(1) $ax_1 + cx_2 = e$

(2) $bx_1 + dx_2 = f$

Cramer's rule is developed as follows.
Multiply Eq. (1) by $(1/a)$.

(1′) $x_1 + \left(\dfrac{c}{a}\right) x_2 = \dfrac{e}{a}$

Subtract b times Eq. (1′) from Eq. (2).

(2′) $\left(d - \dfrac{bc}{a}\right) x_2 = f - \dfrac{be}{a}$

Solve for x_2.

$$x_2 = \dfrac{\dfrac{af - be}{a}}{\dfrac{ad - bc}{a}} = \dfrac{af - be}{ad - bc}$$

This can also be expressed as

$$x_2 = \dfrac{\begin{vmatrix} a & e \\ b & f \end{vmatrix}}{\begin{vmatrix} a & c \\ b & d \end{vmatrix}}$$

Solve Eq. (1′) for x_1.

$$x_1 = \dfrac{e}{a} - \left(\dfrac{c}{a}\right) x_2$$

$$x_1 = \dfrac{e}{a} - \left(\dfrac{c}{a}\right) \left(\dfrac{af - be}{ad - bc}\right)$$

$$x_1 = \dfrac{e}{a} \left(\dfrac{ad - bc}{ad - bc}\right) - \dfrac{c}{a} \left(\dfrac{af - be}{ad - bc}\right)$$

$$x_1 = \frac{ead - ebc}{a(ad - bc)} - \frac{caf + ebc}{a(ad - bc)}$$

$$x_1 = \frac{ed - cf}{ad - bc}$$

This can also be expressed as

$$x_1 = \frac{\begin{vmatrix} e & c \\ f & d \end{vmatrix}}{\begin{vmatrix} a & c \\ b & d \end{vmatrix}}$$

For the 2 by 2 system of equations, the solution is

$$x_1 = \frac{\begin{vmatrix} e & c \\ f & d \end{vmatrix}}{\begin{vmatrix} a & c \\ b & d \end{vmatrix}} \quad \text{and} \quad x_2 = \frac{\begin{vmatrix} a & e \\ b & f \end{vmatrix}}{\begin{vmatrix} a & c \\ b & d \end{vmatrix}}$$

Cramer's rule can be applied to a system of n by n simultaneous equations. For the general case, define A_i as the matrix formed by interchanging the ith *column* of the coefficient matrix A with the right-hand-side vector B. Provided that A is nonsingular (i.e., $|A| \neq 0$), the solution value for the ith variable x_i is given by

$$x_i = \frac{|A_i|}{|A|} \tag{4.7}$$

To illustrate, consider the 3 by 3 system of equations,

$$a_{11}x_1 + a_{12}x_2 + a_{13}x_3 = b_1$$
$$a_{21}x_1 + a_{22}x_2 + a_{23}x_3 = b_2$$
$$a_{31}x_1 + a_{32}x_2 + a_{33}x_3 = b_3$$

The value of x_1 is given by

$$x_1 = \frac{|A_1|}{|A|} = \frac{\begin{vmatrix} b_1 & a_{12} & a_{13} \\ b_2 & a_{22} & a_{23} \\ b_3 & a_{32} & a_{33} \end{vmatrix}}{\begin{vmatrix} a_{11} & a_{12} & a_{13} \\ a_{21} & a_{22} & a_{23} \\ a_{31} & a_{32} & a_{33} \end{vmatrix}}$$

Similarly, the values of x_2 and x_3 are given by

$$x_2 = \frac{|A_2|}{|A|} = \frac{\begin{vmatrix} a_{11} & b_1 & a_{13} \\ a_{21} & b_2 & a_{23} \\ a_{31} & b_3 & a_{33} \end{vmatrix}}{\begin{vmatrix} a_{11} & a_{12} & a_{13} \\ a_{21} & a_{22} & a_{23} \\ a_{31} & a_{32} & a_{33} \end{vmatrix}}$$

and

$$x_3 = \frac{|A_3|}{|A|} = \frac{\begin{vmatrix} a_{11} & a_{12} & b_1 \\ a_{21} & a_{22} & b_2 \\ a_{31} & a_{32} & b_3 \end{vmatrix}}{\begin{vmatrix} a_{11} & a_{12} & a_{13} \\ a_{21} & a_{22} & a_{23} \\ a_{31} & a_{32} & a_{33} \end{vmatrix}}$$

Example: Solve the 2 by 2 system of equations, using Cramer's rule.

$$3x_1 + 6x_2 = 6$$
$$2x_1 - 2x_2 = 16$$

$$x_1 = \frac{\begin{vmatrix} 6 & 6 \\ 16 & -2 \end{vmatrix}}{\begin{vmatrix} 3 & 6 \\ 2 & -2 \end{vmatrix}} = \frac{-108}{-18} = 6$$

$$x_2 = \frac{\begin{vmatrix} 3 & 6 \\ 2 & 16 \end{vmatrix}}{\begin{vmatrix} 3 & 6 \\ 2 & -2 \end{vmatrix}} = \frac{36}{-18} = -2$$

Example: Solve the 3 by 3 system of equations, using Cramer's rule.

$$3x_1 - 4x_2 + 8x_3 = 26$$
$$6x_1 + 3x_2 - 5x_3 = 1$$
$$-2x_1 + 1x_2 + 3x_3 = 11$$

$$x_1 = \frac{\begin{vmatrix} 26 & -4 & 8 \\ 1 & 3 & -5 \\ 11 & 1 & 3 \end{vmatrix}}{\begin{vmatrix} 3 & -4 & 8 \\ 6 & 3 & -5 \\ -2 & 1 & 3 \end{vmatrix}} = \frac{340}{170} = 2$$

$$x_2 = \frac{\begin{vmatrix} 3 & 26 & 8 \\ 6 & 1 & -5 \\ -2 & 11 & 3 \end{vmatrix}}{\begin{vmatrix} 3 & -4 & 8 \\ 6 & 3 & -5 \\ -2 & 1 & 3 \end{vmatrix}} = \frac{510}{170} = 3$$

$$x_3 = \frac{\begin{vmatrix} 3 & -4 & 26 \\ 6 & 3 & 1 \\ -2 & 1 & 11 \end{vmatrix}}{\begin{vmatrix} 3 & -4 & 8 \\ 6 & 3 & -5 \\ -2 & 1 & 3 \end{vmatrix}} = \frac{680}{170} = 4$$

4.6 Rank

Section 2.4 first introduced the concept of a system of equations. One of the important points made in that section was that a system of equations may be consistent or inconsistent and that a consistent system of equations may have a unique or an infinite number of solutions. We can now introduce a method for determining if a system of equations is consistent, has a unique solution, or has an infinite number of solutions. The method is based upon the concept of the rank of a matrix.

4.6.1 RANK OF A MATRIX

The *rank* of a matrix is the number of linearly independent rows (or columns) in the matrix. The rank of a matrix can be determined in two ways. First, row operations can be applied to the matrix with the objective of obtaining zeros in as many rows as possible. The number of rows in the matrix that cannot be reduced to zero by row operations gives the number of linearly independent rows. This number is the rank of the matrix.

A second method of determining the rank of a matrix is to determine the order of the largest square submatrix whose determinant is not zero. If, for instance, the matrix A is square and has a nonzero determinant, the rank of A is equal to the order (i.e., number of rows or columns) of A. If, however, the matrix A is not square or is square but has a zero determinant, the rank of A is equal to the order of the largest square submatrix in A whose determinant is nonzero. If there is no square submatrix in A with a nonzero determinant, the rank of A is zero.

To illustrate these methods, consider the matrix

$$A = \begin{pmatrix} 2 & 3 & -2 \\ 3 & -2 & 1 \\ 1 & -5 & 3 \end{pmatrix}$$

The rank of the matrix is determined by applying row operations with the objective of reducing the elements in the rows to zero. Multiplying the first row by $\frac{1}{2}$ gives

$$\begin{pmatrix} 1 & \frac{3}{2} & -1 \\ 3 & -2 & 1 \\ 1 & -5 & 3 \end{pmatrix}$$

Adding -3 times the new first row to the second row and -1 times the new first row to the third row gives

$$\begin{pmatrix} 1 & \frac{3}{2} & -1 \\ 0 & -\frac{13}{2} & 4 \\ 0 & -\frac{13}{2} & 4 \end{pmatrix}$$

Adding -1 times the second row to the third row gives

$$\begin{pmatrix} 1 & \frac{3}{2} & -1 \\ 0 & -\frac{13}{2} & 4 \\ 0 & 0 & 0 \end{pmatrix}$$

Since no further combinations of row operations can reduce all components of either the first or second row to zero, the rank of the matrix is 2.

To verify that the order of the largest square submatrix of A whose determinant is nonzero is also 2, the determinant of A is calculated.

$$|A| = \begin{vmatrix} 2 & 3 & -2 \\ 3 & -2 & 1 \\ 1 & -5 & 3 \end{vmatrix}$$

$$|A| = 2(-2)(3) + 3 \cdot 1 \cdot 1 + (-2)(3)(-5) - 1(-2)(-2)$$
$$\qquad -(-5)(1)(2) - 3 \cdot 3 \cdot 3$$

$$|A| = -12 + 3 + 30 - 4 + 10 - 27$$

$$|A| = 0$$

Since the determinant of A is zero, we know that the rank of A is not 3. Next, we must determine if one of the square submatrices of order 2 has a nonzero determinant. The determinant of the submatrix $\begin{pmatrix} 2 & 3 \\ 3 & -2 \end{pmatrix}$ is -13. Since A has at least one square submatrix of order 2, we again conclude that the rank of A is 2.

Example: Determine the rank of the matrix A by determining the number of independent rows.

$$A = \begin{pmatrix} 1 & 4 & 6 & 5 \\ 3 & 2 & 4 & 4 \\ 2 & -2 & 3 & -10 \end{pmatrix}$$

To reduce the elements to zero, add -3 times the first row to the second row and -2 times the first row to the third row. This gives

$$\begin{pmatrix} 1 & 4 & 6 & 5 \\ 0 & -10 & -14 & -11 \\ 0 & -10 & -9 & -20 \end{pmatrix}$$

Next, add -1 times the second row to the third row.

$$\begin{pmatrix} 1 & 4 & 6 & 5 \\ 0 & -10 & -14 & -11 \\ 0 & 0 & 5 & -9 \end{pmatrix}$$

Since there is no further combination of row operations that leads to all zeros in one or more of the rows, we conclude that the rank of the matrix is 3.

Example: Determine the rank of the matrix A in the preceding example by determining the order of the largest square submatrix of A whose determinant is nonzero.

The largest square submatrices of A are matrices with dimensions 3 by 3. To find if one or more of these submatrices has a nonzero determinant, we begin by calculating $|S_1|$.

$$|S_1| = \begin{vmatrix} 1 & 4 & 6 \\ 3 & 2 & 4 \\ 2 & -2 & 3 \end{vmatrix}$$

$$|S_1| = 1 \cdot 2 \cdot 3 + 4 \cdot 4 \cdot 2 + 6 \cdot 3(-2) - 2 \cdot 2 \cdot 6 - (-2)(4)(1) - 3 \cdot 3 \cdot 4$$

$$|S_1| = -50$$

The determinant of a 3 by 3 submatrix of A is nonzero; therefore, the rank of A is 3.

4.6.2 RANK OF A SYSTEM OF EQUATIONS

The concept of rank can be extended to a system of equations. Remembering that the rank of a matrix was defined as the number of linearly independent rows in a matrix, we similarly define the rank of a system of equations as the number of linearly independent equations in the system.

To determine the rank of the system of equations $AX = B$, the coefficient matrix A is augmented with the right-hand-side vector B to form the augmented matrix $A \mid B$. The rank of the system of equations is given by the rank of this augmented matrix. To calculate the rank, the augmented matrix can be reduced by row operations and the number of nonzero rows counted. Alternatively, the rank can be determined by finding the order of the largest square submatrix of $A \mid B$ whose determinant is nonzero.

To illustrate, consider the system of equations

$$6x_1 - 2x_2 - 4x_3 = 16$$
$$2x_1 - 2x_2 - x_3 = 2$$
$$x_1 + x_2 - x_3 = 6$$

Augmenting the coefficient matrix with the right-hand-side vector gives

$$A \mid B = \begin{pmatrix} 6 & -2 & -4 & | & 16 \\ 2 & -2 & -1 & | & 2 \\ 1 & 1 & -1 & | & 6 \end{pmatrix}$$

To reduce the augmented matrix, add -2 times the third row to the second row and -6 times the third row to the first row.

$$A \mid B = \begin{pmatrix} 0 & -8 & 2 & -20 \\ 0 & -4 & 1 & -10 \\ 1 & 1 & -1 & 6 \end{pmatrix}$$

Next, add -2 times the second row to the first row.

$$A \mid B = \begin{pmatrix} 0 & 0 & 0 & 0 \\ 0 & -4 & 1 & -10 \\ 1 & 1 & -1 & 6 \end{pmatrix}$$

Since no further combination of row operations leads to a zero row, the rank of the augmented matrix is 2.

The rank of the system of equations can also by found by calculating the determinants of the square submatrices of $A \mid B$. The determinants of the four submatrices whose dimensions are 3 by 3 are zero, i.e.,

$$|S_1| = \begin{vmatrix} 6 & -2 & -4 \\ 2 & -2 & -1 \\ 1 & 1 & -1 \end{vmatrix} = 0$$

$$|S_2| = \begin{vmatrix} -2 & -4 & 16 \\ -2 & -1 & 2 \\ 1 & -1 & 6 \end{vmatrix} = 0$$

$$|S_3| = \begin{vmatrix} 6 & -4 & 16 \\ 2 & -1 & 2 \\ 1 & -1 & 6 \end{vmatrix} = 0$$

$$|S_4| = \begin{vmatrix} 6 & -2 & 16 \\ 2 & -2 & 2 \\ 1 & 1 & 6 \end{vmatrix} = 0$$

The determinant of at least one of the square submatrices of order 2 is, however, nonzero. Consequently, the rank of the augmented matrix and therefore the system of equations is 2.

Example: Determine the rank of the following system of equations,

$$2x_1 + 3x_2 = 6$$

$$6x_1 - 2x_2 = 3$$

The rank of the system of equations is found by augmenting the coefficient matrix with the right-hand-side vector, i.e.,

$$A \mid B = \begin{pmatrix} 2 & 3 & 6 \\ 6 & -2 & 3 \end{pmatrix}$$

The determinant of at least one of the three submatrices whose dimensions are 2 by 2 is nonzero. Therefore, the rank of the system of equations is 2.

4.6.3 RANK, CONSISTENCY, AND UNIQUE SOLUTIONS

At the beginning of this section, we stated that the concept of rank is used to determine if a system of equations is consistent or inconsistent. We also stated that the rank is used to distinguish between consistent systems of equations with a unique solution and consistent systems with an infinite number of solutions. This relationship among rank, consistency, and the number of solutions is given by the following two rules.

Rule 1. A system of equations is consistent if and only if the rank of the augmented matrix is equal to the rank of the coefficient matrix. This rank is termed the rank of the system.

Rule 2. A system of linear equations has a unique solution if and only if the rank of the system equals the number of variables.

To illustrate these rules, consider the following system of equations.

$$2x + 3y - 2z = 40$$
$$3x - 2y + 1z = 50$$
$$1x - 5y + 3z = 10$$

It was shown in Chapter 2, p. 61, that this system is consistent with an infinite number of solutions. This can be verified by augmenting the coefficient matrix with the right-hand-side vector, i.e.,

$$A \mid B = \begin{pmatrix} 2 & 3 & -2 & | & 40 \\ 3 & -2 & 1 & | & 50 \\ 1 & -5 & 3 & | & 10 \end{pmatrix}$$

This system is reduced by adding 1 times the first row to the third row. This gives

$$A \mid B = \begin{pmatrix} 2 & 3 & -2 & | & 40 \\ 3 & -2 & 1 & | & 50 \\ 3 & -2 & 1 & | & 50 \end{pmatrix}$$

Adding -1 times the second row to the third row gives

$$A \mid B = \begin{pmatrix} 2 & 3 & -2 & | & 40 \\ 3 & -2 & 1 & | & 50 \\ 0 & 0 & 0 & | & 0 \end{pmatrix}$$

The example shows that the rank of both the augmented matrix $A \mid B$ and the coefficient matrix A is 2. Based upon Rule 1 the system of equations is consistent. Since the rank of the system is less than the number of variables, we conclude from Rule 2 that the system has an infinite number of solutions.

Example: Verify that the following system of equations is consistent with a unique solution.

$$2x + y = 9$$
$$x - y = 3$$
$$x + 2y = 6$$

The rank of $A \mid B$ and A is found by reducing the augmented matrix by row operations or by calculating the determinants of the square submatrices of A and $A \mid B$. The determinant of $A \mid B$ is zero, i.e.,

$$|A \mid B| = \begin{vmatrix} 2 & 1 & 9 \\ 1 & -1 & 3 \\ 1 & 2 & 6 \end{vmatrix}$$

$$|A \mid B| = 2(-1)(6) + 1 \cdot 3 \cdot 1 + 9 \cdot 1 \cdot 2 - 1(-1)(9) - 2 \cdot 3 \cdot 2 - 6 \cdot 1 \cdot 1$$

$$|A \mid B| = 0$$

The determinant of at least one of the submatrices of dimensions 2 by 2 in the augmented matrix $A \mid B$ is nonzero. Similarly, the determinant of at least one of the square submatrices of dimension 2 by 2 in the coefficient matrix A is nonzero. Consequently, the rank of both the augmented matrix and the coefficient matrix is 2. On the basis of Rule 1 we therefore conclude that the system of equations is consistent. Furthermore, since the rank of the system of equation is the same as the number of variables, we also conclude from Rule 2 that the system has a unique solution.

Example: Verify that the following system of equations is inconsistent.

$$x_1 + 3x_2 + x_3 = 2$$
$$x_1 - 2x_2 - 2x_3 = 3$$
$$2x_1 + x_2 - x_3 = 6$$

To determine if the system of equations is consistent we augment the coefficient matrix with the right-hand-side vector and reduce the resulting matrix using row operations

$$A \mid B = \begin{pmatrix} 1 & 3 & 1 & | & 2 \\ 1 & -2 & -2 & | & 3 \\ 2 & 1 & -1 & | & 6 \end{pmatrix}$$

Add the second row to the first row.

$$\begin{pmatrix} 2 & 1 & -1 & | & 5 \\ 1 & -2 & -2 & | & 3 \\ 2 & 1 & -1 & | & 6 \end{pmatrix}$$

Add -1 times the third row to the first row.

$$\begin{pmatrix} 0 & 0 & 0 & | & -1 \\ 1 & -2 & -2 & | & 3 \\ 2 & 1 & -1 & | & 6 \end{pmatrix}$$

No further combination of row operations results in a row composed entirely of zeros. Consequently, the rank of the augmented matrix $A \mid B$ is 3. The rank of the coefficient matrix A, however, is 2. Since the rank of the coefficient matrix is less than the rank of the augmented matrix, the system of equations is inconsistent.

PROBLEMS

1. Form the transpose of each of the following matrices.

(a) $A = \begin{pmatrix} 3 & 1 \\ 4 & 3 \end{pmatrix}$

(b) $A = (1, 5, 6)$

(c) $A = \begin{pmatrix} 2 & 1 & -3 \\ 3 & -2 & 4 \end{pmatrix}$

(d) $A = \begin{pmatrix} -2 & 1 \\ 1 & 3 \\ 3 & 2 \end{pmatrix}$

2. Determine the submatrix B formed by the deletion of the second row and third column of the following matrices.

(a) $A = \begin{pmatrix} 3 & 2 & -4 \\ 2 & -3 & 3 \\ 5 & 4 & -1 \end{pmatrix}$

(b) $A = \begin{pmatrix} 6 & 4 & 5 & 2 \\ -5 & 3 & 1 & 6 \\ 7 & 2 & 3 & 7 \end{pmatrix}$

3. Given two matrices A and B that are partitioned into $A = \begin{pmatrix} A_1 & A_2 \\ A_3 & A_4 \end{pmatrix}$ and $B = \begin{pmatrix} B_1 \\ B_2 \end{pmatrix}$, determine the product AB by multiplying and adding the submatrices. The submatrices are

$$A_1 = \begin{pmatrix} 3 & 2 \\ 6 & 3 \end{pmatrix}, A_2 = \begin{pmatrix} 4 & 1 \\ 2 & 5 \end{pmatrix}, A_3 = \begin{pmatrix} 3 & 1 \\ 2 & 3 \end{pmatrix}, A_4 = \begin{pmatrix} 5 & 6 \\ 4 & 8 \end{pmatrix},$$

$$B_1 = \begin{pmatrix} 3 & 5 \\ 6 & 3 \end{pmatrix}, \text{ and } B_2 = \begin{pmatrix} 2 & 7 \\ 1 & 5 \end{pmatrix}.$$

4. Find the determinants of the following matrices.

(a) $A = \begin{pmatrix} 6 & 4 \\ 3 & 8 \end{pmatrix}$

(b) $A = \begin{pmatrix} 5 & -3 \\ 2 & 4 \end{pmatrix}$

(c) $A = \begin{pmatrix} 3 & -2 & 2 \\ 1 & 4 & 3 \\ 2 & 5 & -4 \end{pmatrix}$

(d) $A = \begin{pmatrix} 3 & 5 & 3 \\ 6 & 4 & -5 \\ 5 & 8 & 2 \end{pmatrix}$

5. Find the minors of the elements in the matrix A.

$$A = \begin{pmatrix} 6 & 3 & 4 \\ 3 & -5 & 2 \\ 4 & 3 & -3 \end{pmatrix}$$

6. Find the cofactors of each of the elements in matrix A of Problem 5.

7. Use the method of cofactor expansion to find the determinant of the matrix in Problem 5. Verify the solution by expanding about both the first row and the first column.

8. In each of the following matrices, how can we know that the determinant is zero without actually calculating the determinant?

(a) $A = \begin{pmatrix} 3 & 4 & 6 \\ 6 & 8 & 5 \\ 12 & 16 & 2 \end{pmatrix}$ (b) $A = \begin{pmatrix} 6 & 4 & 7 \\ 0 & 0 & 0 \\ 5 & 9 & 4 \end{pmatrix}$

(c) $A = \begin{pmatrix} 2 & 5 & 21 \\ 3 & 6 & 9 \\ 6 & 15 & 63 \end{pmatrix}$ (d) $A = \begin{pmatrix} 1 & 3 & 1 \\ 1 & 1 & 3 \\ 1 & 3 & 1 \end{pmatrix}$

9. Find the determinants of the following matrices by the method of cofactor expansion.

(a) $A = \begin{pmatrix} 3 & 7 & 1 & 5 \\ 6 & 2 & 0 & 4 \\ 1 & 6 & 1 & 3 \\ 2 & 5 & 0 & 2 \end{pmatrix}$ (b) $A = \begin{pmatrix} 7 & -2 & 3 & 1 \\ 2 & 3 & -1 & 0 \\ 5 & 7 & 3 & 1 \\ 4 & -6 & 6 & 2 \end{pmatrix}$

10. Determine the cofactor matrix for the matrix A in Problem 5.

11. Determine the adjoint matrix for the matrix A in Problem 5.

12. Determine the inverse of the matrix A in Problem 5.

13. Use the adjoint method to calculate the inverses of the following matrices.

(a) $A = \begin{pmatrix} 3 & 1 & 1 \\ 2 & -2 & 2 \\ 1 & 0 & 1 \end{pmatrix}$ (b) $A = \begin{pmatrix} 1 & 2 & 3 \\ 1 & 3 & 5 \\ 2 & 5 & 9 \end{pmatrix}$

(c) $A = \begin{pmatrix} 2 & 6 & 4 \\ 6 & 6 & 4 \\ 3 & 2 & 4 \end{pmatrix}$ (d) $A = \begin{pmatrix} 2 & 8 & -11 \\ -1 & -5 & 7 \\ 1 & 2 & -3 \end{pmatrix}$

14. Solve the following systems of equations. (*Hint:* Use the results of Problem 13 to simplify the calculations.)

(a) $3x_1 + x_2 + x_3 = 16$ (b) $x_1 + 2x_2 + 3x_3 = 16$
    $2x_1 - 2x_2 + 2x_3 = 24$ $x_1 + 3x_2 + 5x_3 = 20$
    $x_1 + x_3 = 8$ $2x_1 + 5x_2 + 9x_3 = 10$

(c) $2x_1 + 6x_2 + 4x_3 = 320$
$6x_1 + 6x_2 + 4x_3 = 480$
$3x_1 + 2x_2 + 4x_3 = 192$

(d) $2x_1 + 8x_2 - 11x_3 = 25$
$-x_1 - 5x_2 + 7x_3 = 10$
$x_1 + 2x_2 + -3x_3 = 15$

15. Resolve Problem 14, using Cramer's rule.

16. Determine the solutions to the following systems of equations, using Cramer's rule.

(a) $7x_1 + 13x_2 = 91$
$13x_1 + 8x_2 = 104$

(b) $14x_1 + 23x_2 = 176$
$9x_1 + 8x_2 = 86$

(c) $3x_1 + 5x_2 + 4x_3 = 42$
$6x_1 + 4x_2 - 3x_3 = 26$
$2x_1 - 5x_2 + 6x_3 = 5$

(d) $2x_1 - 5x_2 + 3x_3 = 14$
$7x_1 + 4x_2 - 6x_3 = 36$
$5x_1 - 6x_2 + 3x_3 = 34$

17. Determine the ranks of the following matrices.

(a) $A = \begin{pmatrix} 6 & 4 \\ 4 & 6 \end{pmatrix}$

(b) $A = \begin{pmatrix} 3 & 9 \\ 2 & 6 \end{pmatrix}$

(c) $A = \begin{pmatrix} 1 & 2 & 0 \\ 4 & 8 & 5 \end{pmatrix}$

(d) $I = \begin{pmatrix} 1 & 0 & 0 \\ 0 & 1 & 0 \\ 0 & 0 & 1 \end{pmatrix}$

(e) $A = (3 \quad 4 \quad 1)$

(f) $A = (6)$

18. Use the concept of rank to determine if the following systems of equations are consistent or inconsistent. For those systems that are consistent, state whether the system of equations has a unique solution or has an infinite number of solutions.

(a) $3x_1 + 2x_2 = 8$
$x_1 + 4x_2 = 5$

(b) $2x_1 + 5x_2 = 8$
$4x_1 + 10x_2 = 4$

(c) $x_1 + 2x_2 = 4$
$2x_1 + 4x_2 = 8$

(d) $2x_1 + 4x_2 = 6$
$x_1 + 2x_2 = 8$

(e) $2x_1 + x_2 + 3x_3 = 2$
$x_1 + 2x_2 + x_3 = 4$
$2x_1 + 4x_2 + 2x_3 = 10$

(f) $x_1 + 2x_2 + x_3 = 2$
$2x_1 + x_2 + 3x_3 = 1$
$x_1 + 2x_2 + x_3 = 2$

(g) $4x_1 + 2x_2 + 3x_3 + x_4 = 18$
$2x_1 + 3x_2 - x_3 + 2x_4 = 10$

SUGGESTED REFERENCES

The references for this chapter are listed in Chapter 3.

Chapter 5

Linear Programming

Linear programming is one of the most frequently and successfully applied mathematical approaches to managerial decisions. The objective in using linear programming is to develop a model to aid the decision maker in determining the optimal allocation of the firm's resources among the alternative products of the firm. Since the resources employed in a firm have an economic value and the products of the firm lead to profits and costs, the linear programming problem becomes that of allocating the scarce resources to the products in a manner such that profits are a maximum, or alternatively, costs are a minimum.

The popularity of linear programming stems directly from its usefulness to the business firm. One of the fundamental tasks of management involves the allocation of resources. For instance, the petroleum firm must allocate sources of crude oil for the production of various petroleum products. Similarly, the wood products firm must allocate timber resources for the production of different lumber and paper products. In both examples, the allocation must be made subject to constraints on the availability of resources and the demand for the products. The objective in allocating the resources to the products is to maximize the profit of the firm.

Our objective in this and the following two chapters is twofold. First, we shall describe the linear programming problem and the solution techniques available for this problem. Our approach will be to introduce the important characteristics of linear programming through the use of very simple two-product linear programming problems and to solve these problems using a graphical solution technique. This, in turn, will lead to a discussion in Chapter 6 of a more general solution technique. This technique, termed the *simplex algorithm*, relies largely upon the use of the elementary row operations of matrix algebra to obtain the solution of systems of simultaneous equations. A second, and equally important objective, will be to explain and offer examples of important applications of linear programming. These examples are in sufficient detail for the student to understand industrial applications of linear programming and, hopefully, to use the technique in his job.

The importance of the computer in the solution of linear programming problems should be emphasized. Although "linear programming" and "computer programming" are not related, the computer is extremely valuable in obtaining the solution to linear programming problems. Many industrial applications of linear programming require more than one hundred equations and variables. Problems of this size can be solved only through the use of modern electronic computers.

5.1 The Linear Programming Problem

The linear programming problem consists of allocating scarce resources among competing products or activities. Two factors give rise to the allocation problem. First, resources available to management have a cost and are limited in supply. Therefore, management must determine how these limited resources will be used. Second, the allocation of these resources must be made in accordance with some overall objective. In the business firm, this objective is normally the maximization of profit or the minimization of cost.

As an illustration of the allocation of resources using linear programming, consider the following problem concerning the scheduling of production in a machine shop. Production must be scheduled for two types of machines, machine 1 and machine 2. One hundred twenty hours of time can be scheduled for machine 1, and 80 hours can be scheduled for machine 2. Production during the scheduling period is limited to two products, A and B. Each unit of product A requires 2 hours of process time on each machine. Each unit of product B requires 3 hours on machine 1 and 1.5 hours on machine 2. The contribution margin is $4.00 per unit of product A and $5.00 per unit of product B.† Both types of products can be readily marketed; consequently, production should be scheduled with the objective of maximizing profit. The

---

† Contribution margin is defined as contribution to fixed expenses and profit.

linear programming problem can be formulated in the following manner. Let:

x_1 = the number of units of product A scheduled for production
x_2 = the number of units of product B scheduled for production
P = contribution to fixed expenses and profit

The linear programming problem is

Maximize: $P = 4x_1 + 5x_2$ (profit)
Subject to: $2x_1 + 3x_2 \leq 120$ (machine 1 resource)
                   $2x_1 + 1.5x_2 \leq 80$ (machine 2 resource)
                   $x_1, x_2 \geq 0$ (nonnegativity)

The mathematical formulation states that production of products A and B should be scheduled with the objective of maximizing profit. Production is limited, however, by the availability of resources, i.e., 120 hours of machine 1 time and 80 hours of machine 2 time. An additional limitation, shown by $x_1, x_2 \geq 0$, is that negative production is not allowed, i.e., the number of units of either product must be zero or positive.

The solution in terms of the production quantities of products A and B that maximizes profit can be determined in this problem by either a graphical approach or the simplex method. The graphical approach for solving two variable problems provides important insights into the nature of linear programming problems. This approach is shown in Sec. 5.2. The simplex method is presented in Chapter 6.

To illustrate a linear programming problem in which costs are minimized, consider the problem facing a speciality metals manufacturer. The firm produces an alloy that is made from steel and scrap metal. The cost per ton of steel is $50 and the cost per ton of scrap is $20. The technological requirements for the alloy are (1) a minimum of one ton of steel is required for every two tons of scrap; (2) one hour of processing time is required for each ton of steel, and four hours of processing time are required for each ton of scrap; (3) the steel and scrap combine linearly to make the alloy. The process loss from the steel is 10 percent and the loss from scrap is 20 percent. Although production may exceed demand, a minimum of 40 tons of the alloy must be manufactured. To maintain efficient plant operation, a minimum of 80 hours of processing time must be used. The supply of both scrap and steel is adequate for production of the alloy.

The objective of the manufacturer is to produce the alloy at a minimum cost. The problem can be formulated as follows.

Let: x_1 = the number of tons of steel
              x_2 = the number of tons of scrap
              C = the cost of the alloy

The linear programming problem is

Minimize: $C = 50x_1 + 20x_2$ (cost)
Subject to: $2x_1 - 1x_2 \geq 0$ (tech. req'm't.)
             $1x_1 + 4x_2 \geq 80$ (process)
             $0.90x_1 + 0.80x_2 \geq 40$ (demand)
             $x_1, x_2 \geq 0$ (nonnegativity)

The problem is to minimize the cost of producing the alloy. The cost function is equal to $50 per ton of steel multiplied by the number of tons of steel plus $20 per ton of scrap multiplied by the number of tons of scrap. The first constraint states that a minimum of one ton of steel must be used for every two tons of scrap. The second constraint concerns process time, and the third constraint shows that the steel and scrap combine linearly to make the alloy with a 10 percent and 20 percent loss, respectively. The solution to the problem is determined using the graphical method in Sec. 5.2.

5.2 Graphical Solution

The graphical solution technique provides a convenient method for solving simple two-variable linear programming problems. It is also quite useful in illustrating basic concepts in linear programming. For problems in which more than two variables are required, the simplex algorithm is used to determine the solution.

5.2.1 MAXIMIZATION

The problem concerning the scheduling of production in a machine shop was expressed mathematically as

Maximize: $P = 4x_1 + 5x_2$
Subject to: $2x_1 + 3x_2 \leq 120$
            $2x_1 + 1.5x_2 \leq 80$
            $x_1, x_2 \geq 0$

The function to be optimized, $P = 4x_1 + 5x_2$, is termed the *objective function*. If x_1 and x_2 were unrestricted, there would be no limit on the value of the objective function. There are, however, *constraints* upon the values of the variables.† The constraints are expressed as inequalities and are

$$2x_1 + 3x_2 \leq 120$$
$$2x_1 + 1.5x_2 \leq 80$$
$$x_1, x_2 \geq 0$$

---

† Constraints can take the form $ax_1 + bx_2 \leq r$, $ax_1 + bx_2 = r$, or $ax_1 + bx_2 \geq r$.

The first two constraints come from the technological requirements specified in the problem. The remaining constraints, $x_1 \geq 0$ and $x_2 \geq 0$, are termed *nonnegativity* constraints. These constraints are necessary to assure that the variables take on only zero or positive values.

The constraints are graphed in Fig. 5.1 by plotting the equations

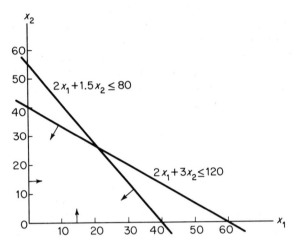

Figure 5.1

Since there are four constraints in the problem, four equations must be plotted. These are

$$2x_1 + 3x_2 = 120$$

$$2x_1 + 1.5x_2 = 80$$

$$x_1 = 0$$

$$x_2 = 0$$

Each of the equations is linear; consequently, each equation can be plotted from two points. The intercept points ($x_1 = 0$, $x_2 = 40$) and ($x_1 = 60$, $x_2 = 0$) are used to plot the first equation. Similarly, the second equation is plotted by using ($x_1 = 0$, $x_2 = 53.3$) and ($x_1 = 40$, $x_2 = 0$). The remaining equations, $x_1 = 0$ and $x_2 = 0$, represent the vertical and horizontal axis of the graph.

The constraints in this problem are expressed as inequalities. It was shown in Chapter 2 that the solution to an inequality consisted of the set $\{(x_1, x_2) \mid ax_1 + bx_2 \leq r\}$. The direction of each inequality is shown in Fig. 5.1 by an arrow. For instance, the inequality $2x_1 + 1.5x_2 \leq 80$ can be thought of as consisting of all ordered pairs for which $2x_1 + 1.5x_2 = 80$ and for which $2x_1 + 1.5x_2 < 80$. The line shows the ordered pairs for which $2x_1 + 1.5x_2 =$

80, and the arrow pointing downward and to the left shows the ordered pairs for which $2x_1 + 1.5x_2 < 80$.

The linear programming problem consists of maximizing the objective function, $P = 4x_1 + 5x_2$, subject to the constraints. To define feasible values of (x_1, x_2), we denote the four constraints by c_1, c_2, c_3, and c_4.

$$c_1: \quad 2x_1 + 3x_2 \leq 120$$

$$c_2: \quad 2x_1 + 1.5x_2 \leq 80$$

$$c_3: \quad x_1 \geq 0$$

$$c_4: \quad x_2 \geq 0$$

The values of x_1 and x_2 are limited to the set of ordered pairs that are solutions to the constraints. Using set notation, we express this set of ordered pairs as

$$F = \{(x_1, x_2) \mid c_1 \cap c_2 \cap c_3 \cap c_4\}$$

The set F contains all ordered pairs that are *feasible* solutions to the linear programming problem. These ordered pairs are shown in Fig. 5.2 by the

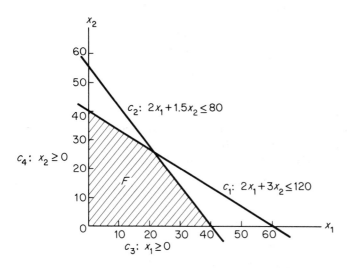

Figure 5.2

shaded area. This area is termed the *feasible region*. The ordered pairs defined by F are members of the *solution set* of the linear programming problem.

The optimal solution to the linear programming problem is the member of F that maximizes the objective function. Even in this simplified example, it is obvious that there are an infinite number of ordered pairs in the feasible region (i.e., shaded area). Since each of these ordered pairs is a member of the solution set, it would not be practical to investigate each ordered pair for optimality. Instead, we can make use of one of the important theorems in linear programming. This theorem, termed the *extreme point theorem*, states that the *optimal value of the objective function occurs at one of the extreme points of the feasible region.*† An extreme point in two-dimensional space is defined by the intersection of two equations. By referring to Fig. 5.2 we see that there are four extreme points. The values of x_1 and x_2 for each of these extreme points are determined by solving the appropriate pairs of intersecting equations. The solutions are

$$E = \{(0, 0), (40, 0), (20, 26.67), (0, 40)\}$$

According to the theorem, the objective function will be a maximum at one of these four ordered pairs. To determine which ordered pair maximizes the objective function, the objective function is evaluated for each extreme point. Table 5.1 shows that the objective function is a maximum when $x_1 = 20$ and

Table 5.1

(x_1, x_2)	P
(0, 0)	$0
(40, 0)	160.00
(20, 26.67)	213.33
(0, 40)	200.00

$x_2 = 26.67$.

The validity of the extreme point theorem can be demonstrated for the two-variable linear programming problem by superimposing the objective function on a graph of the feasible region. The objective function is

$$P = 4x_1 + 5x_2$$

which can also be written as

$$x_2 = 0.2P - 0.8x_1$$

The objective function, superimposed on the feasible region in Fig. 5.3, is shown as a series of parallel lines. Each line shows the combination of ordered pairs (x_1, x_2) that gives a specific value of P. For instance, the set of ordered pairs that are solutions to the equation $x_2 = 0.2(\$100) - 0.8x_1$ represents all combinations of (x_1, x_2) that result in profit of $100. The objective function is also plotted for $P = \$150$, $P = \$200$, and $P = \$213.33$.

Profit increases as the objective function moves upward and to the right

---

† This theorem is true provided that a unique, finite, optimal solution exists for the linear programming problem. There are three exceptions to this theorem. These are discussed in Sec. 5.3.3.

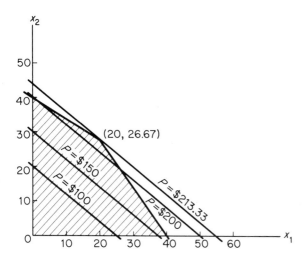

Figure 5.3

in the direction perpendicular to the slope of the function. Figure 5.3 shows that profit will be a maximum when the objective function passes through an extreme point of the feasible region. In this particular problem the extreme point is (20, 26.67), and profit is \$213.33.

Example: Use the graphical solution technique to determine the solution to the following linear programming problem.

$$\text{Maximize:} \quad P = 4x_1 + 3x_2$$
$$\text{Subject to:} \quad 21x_1 + 16x_2 \leq 336$$
$$13x_1 + 25x_2 \leq 325$$
$$15x_1 + 18x_2 \leq 270$$
$$x_1, x_2 \geq 0$$

The feasible region is defined by the intersection of the five linear inequalities. The inequalities $x_1 \geq 0$ and $x_2 \geq 0$ limit the solution to the first quadrant, i.e., nonnegative values of x_1 and x_2. The remaining three inequalities act as an upper bound on the values of x_1 and x_2. The feasible region is shown as the shaded area of Fig. 5.4.

The objective function has an optimal value at an extreme point of the feasible region. If the ordered pair (0, 0) is included as an extreme point, a total of five extreme points must be evaluated. In two-dimensional space, an extreme point is defined by the intersection of two equations. The five extreme points can thus be determined by specifying the five pairs of intersecting equations and solving each pair of equations simultaneously. The sets of ordered pairs specified by the five pairs of intersecting equations are

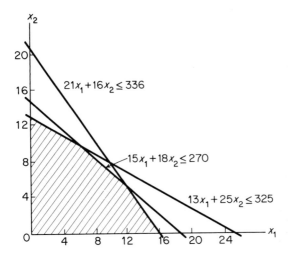

Figure 5.4

1. $\{(x_1, x_2) \mid x_1 = 0 \cap x_2 = 0\}$
2. $\{(x_1, x_2) \mid x_1 = 0 \cap 13x_1 + 25x_2 = 325\}$
3. $\{(x_1, x_2) \mid 15x_1 + 18x_2 = 270 \cap 13x_1 + 25x_2 = 325\}$
4. $\{(x_1, x_2) \mid 21x_1 + 16x_2 = 336 \cap 15x_1 + 18x_2 = 270\}$
5. $\{(x_1, x_2) \mid 21x_1 + 16x_2 = 336 \cap x_2 = 0\}$

Solving each pair of equations simultaneously gives the set of extreme points, $E = \{(0, 0), (0, 13), (6.38, 9.68), (12.45, 4.57), (16,0)\}$.

Not all pairs of intersecting equations in a linear programming problem specify an extreme point. For instance, the equations $21x_1 + 16x_2 = 336$ and $13x_1 + 25x_2 = 325$ intersect at $(9.7, 8.2)$. This point is not in the feasible region. Consequently, the ordered pair $(9.7, 8.2)$ is *nonfeasible* and is not considered in evaluating the objective function. Table 5.2 shows that the

Table 5.2

(x_1, x_2)	P
(0, 0)	\$0
(0, 13)	39.00
(6.38, 9.68)	54.56
(12.45, 4.57)	63.51
(16, 0)	64.00

objective function is a maximum when $x_1 = 16$ and $x_2 = 0$.

5.2.2 MINIMIZATION

The graphical solution for a minimization problem is found by using the technique introduced in the preceding section. The difference between the minimization and maximization problems is that the extreme point that results in a minimum value of the objective function is the solution to the minimization problem rather than that which results in a maximum value of the function. To illustrate, consider the problem of the speciality metals manufacturer previously described in Sec. 5.1. The mathematical formulation of the problem was

$$\text{Minimize:} \quad C = \quad 50x_1 + 20x_2$$
$$\text{Subject to:} \quad 2x_1 - \quad 1x_2 \geq 0$$
$$1x_1 + \quad 4x_2 \geq 80$$
$$0.90x_1 + 0.80x_2 \geq 40$$
$$x_1, x_2 \geq 0$$

All ordered pairs (x_1, x_2) that satisfy the five inequalities are feasible solutions. These are shown by the shaded area in Fig. 5.5.

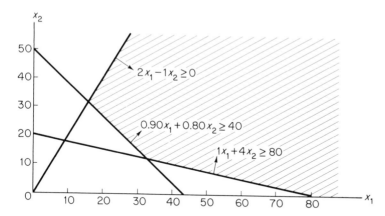

Figure 5.5

The objective function is minimized at an extreme point of the feasible region. Figure 5.5 shows that three extreme points must be evaluated. These points are described by the sets

1. $\{(x_1, x_2) \mid 2x_1 - 1x_2 = 0 \cap 0.90x_1 + 0.80x_2 = 40\}$
2. $\{(x_1, x_2) \mid 0.90x_1 + 0.80x_2 = 40 \cap 1x_1 + 4x_2 = 80\}$
3. $\{(x_1, x_2) \mid 1x_1 + 4x_2 = 80 \cap x_2 = 0\}$

The solution of each of the three pairs of simultaneous equations gives the set of extreme points, $E = \{(16, 32), (34.32, 11.42), (80, 0)\}$. The value of the objective function at each extreme point is shown in Table 5.3. The objective

Table 5.3

(x_1, x_2)	C
(16, 32)	$1440
(34.32, 11.42)	$1944.40
(80, 0)	$4000

function is minimized for $x_1 = 16$ and $x_2 = 32$.

Example: Use the graphical solution technique to determine the solution to the following linear programming problem.

$$
\begin{aligned}
\text{Minimize:} \quad & C = 3x_1 + 2x_2 \\
\text{Subject to:} \quad & 2x_1 + x_2 \geq 5 \\
& x_1 + 3x_2 \geq 6 \\
& x_1 + x_2 \geq 4 \\
& x_1, x_2 \geq 0
\end{aligned}
$$

The feasible region for the linear programming problem is shown in **Fig. 5.6.** From the extreme point theorem, we know that the objective function

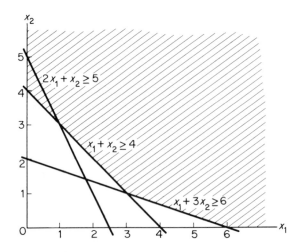

Figure 5.6

is a minimum at one of the extreme points. The extreme points are determined by selecting the pairs of equations that define the extreme points and solving each pair of equations simultaneously. This results in the set of extreme points

$$E = \{(0, 5), (1, 3), (3, 1), (6, 0)\}$$

By evaluating the objective function at each of these extreme points, we find that the objective function has a minimum value of $C = 9$ at the extreme point $x_1 = 1$, $x_2 = 3$.

5.2.3 CHARACTERISTICS ILLUSTRATED BY THE GRAPHIC SOLUTION

The graphical solution illustrates several important characteristics of the linear programming problem. As implied by the name "linear programming," one of these characteristics is that the inequalities and the objective function are linear. This assumption of linearity is made for all linear programming problems, regardless of the number of inequalities or variables.

A second characteristic illustrated by the graphical solution is that an optimal solution occurs at an extreme point of the feasible region. Feasible solutions to the linear programming problem are defined by sets of linear inequalities or equations. The feasible region, as illustrated in Fig. 5.2, contains an infinite number of ordered pairs that are solutions to the system of inequalities. At least one of these solutions will be optimal. The graphical solution shows that this optimal solution occurs at one of the finite number of extreme points rather than at one of the infinite number of ordered pairs that are solutions to the system of inequalities.

A third characteristic important in the linear programming problem is that the extreme points are defined by the intersection of linear equations. With two variables, two equations are required to define an extreme point. Three variables would require three equations to define an extreme point. Similarly, with n variables, n equations are required to define an extreme point. Since an optimal solution occurs at an extreme point, the number of candidates for the optimal solution is limited to the finite number of extreme points.

The simplex solution procedure for the general linear programming problem is presented in Chapter 6. This algorithm provides a series of rules that permits a systematic evaluation of selected extreme points of the feasible region. One important difference between the simplex algorithm and the graphical solution technique is that the algorithm provides a method of reducing the number of extreme points that must be evaluated to a relatively small subset of the total set of extreme points. Since there are often more than one hundred thousand extreme points in a linear programming problem, the importance of the simplex algorithm in reducing the number of extreme points that must be investigated to a manageable number is obvious.

5.3 Assumptions, Terminology, and Special Cases

Certain assumptions and terms are required in the linear programming model. Several of these have been introduced. The purpose of this section is to introduce the additional assumptions of linearity, continuity, and convexity, to show that linear programming implies constant returns to scale, and to illustrate exceptions to the extreme point theorem of linear programming.

5.3.1 LINEARITY, CONTINUITY, AND RETURNS TO SCALE

One of the assumptions in linear programming is that the objective function and constraints are linear. The problem of scheduling production in a machine shop illustrates this assumption. The problem was

$$
\begin{aligned}
\text{Maximize:} \quad & P = 4x_1 + 5x_2 \\
\text{Subject to:} \quad & 2x_1 + 3x_2 \leq 120 \\
& 2x_1 + 1.5x_2 \leq 80 \\
& x_1, x_2 \geq 0
\end{aligned}
$$

The assumption of linearity imposes an important restriction on the structure of the problem. This restriction is that both profits and resource usage are a linear combination of the number of units of each product. In this problem profit is given by $P = 4x_1 + 5x_2$. The linear form of the objective function thus prohibits the competitive and substitutive effects that sometimes exist between products or activities. In certain industries, for instance, the manufacture of one product enables the manufacture of a second product at lesser cost. Such "second-order effects" are not included in the linear programming model.†

The result of the assumption of linearity involves what economists term "constant returns to scale." In the example, two units of both resource one and resource two are required to manufacture one unit of product one. This requirement remains regardless of the number of units of product one manufactured. The assumption of linearity, therefore, implies that the productivity of the resource is independent of the number of units of the product manu-

---

† These types of effects can, however, be included in more advanced programming models. See Harvey M. Wagner, *Principles of Operations Research* (Prentice-Hall, Inc., Englewood Cliffs, New Jersey, 1969), Chapter 14.

factured; i.e., the returns from the manufacturing process are constant regardless of the scale of production.

An additional restriction in the linear programming model comes from the assumption of continuity. The solution to the linear programming problem occurs at an extreme point. Since the equations that define extreme points are continuous, the solutions need not have integer values. This implies that fractional units of products or activities are possible. For instance, the optimal solution to the problem of scheduling production in the machine shop was $x_1 = 20$ and $x_2 = 26.67$.

The restrictions on the linear programming model due to the assumptions of linearity and continuity are not as severe as one might expect. Profit (or cost) and resource usage are often described quite satisfactorily by linear relationships, especially in the relevant range of production. Fractional solutions may present no problem. For those cases in which fractional solutions are impossible, judicial rounding sometimes provides a practical suboptimal solution. If rounding is unsatisfactory, *integer programming* can be employed.†

5.3.2 CONVEX SETS

The solution set or feasible region in a linear programming problem is sometimes referred to as a *convex set*. By definition, a set is *convex* if each point on a straight line that joins two points in the set is also in the set. To illustrate this definition, consider the two sets in Fig. 5.7. The set in Fig. 5.7(a) represents a typical solution set for a two-variable linear programming problem and is convex. It is impossible to select two points in this set such that all points on a straight line that joins the two points are not also in the set. This convex solution set was formed by the intersection of a system of linear inequalities.

The set in Fig. 5.7(b) does not have the property of convexity. This is shown by the fact that points on the line connecting the points $(1, 4)$ and $(3, 3)$ are not in the set.

It can be seen that the set in Fig. 5.7(b) is not formed by the intersection of a system of linear inequalities. If the lines that form the boundaries of the set are extended, we see that there is some area on both sides of at least one line. This means, of course, that at least one of the inequalities that forms the set in Fig. 5.7(b) is not satisfied.

Since the intersection of a system of linear inequalities defines the solution set for a linear programming problem and the set of points generated by this

---

† A good discussion of integer programming is provided by Donald R. Plane and Claude McMillan, Jr., *Discrete Optimization* (Prentice-Hall, Inc., Englewood Cliffs, N.J., 1971).

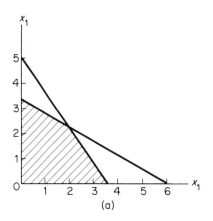
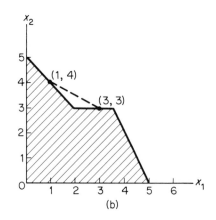

(a)   (b)

Figure 5.7

intersection has the property of convexity, one may ask the reason for introducing the concept of convex sets. The answer, aside from the theoretical importance of convexity, is that the analyst need be aware that the inequalities in a linear programming problem must be satisfied for all possible values of the variables rather than only a selected subset of values.

5.3.3 NONFEASIBLE, MULTIPLE OPTIMAL, AND UNBOUNDED SOLUTIONS

The linear programming problems discussed in the preceding sections have had a single optimal solution. Although problems of this type are common, other possibilities exist. These include: (1) no feasible solution; (2) multiple optimal solutions; and (3) unbounded optimal solutions. These are the exceptions referred to in the footnote on p. 146 to the extreme point theorem.

No FEASIBLE SOLUTION. The solution set for a linear programming problem is formed by the intersection of the system of constraining inequalities. If this intersection is empty, no feasible solution exists for the problem. As an example, assume that the machine shop problem is modified by adding the constraint, $x_1 + x_2 \geq 50$. The problem becomes

$$
\begin{aligned}
\text{Maximize:} \quad & P = 4x_1 + 5x_2 \\
\text{Subject to:} \quad & 2x_1 + 3x_2 \leq 120 \\
& 2x_1 + 1.5x_2 \leq 80 \\
& x_1 + x_2 \geq 50 \\
& x_1, x_2 \geq 0
\end{aligned}
$$

The constraints are shown in Fig. 5.8. It can be seen from this figure that

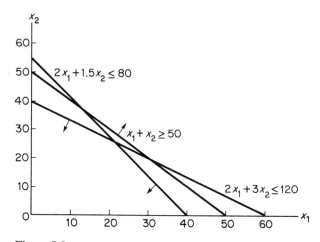

Figure 5.8

there is no set of points that simultaneously satisfies all the inequalities. Consequently, there is no feasible solution to the linear programming problem.

MULTIPLE OPTIMAL SOLUTIONS. We have stressed that an optimal solution occurs at an extreme point of the feasible region. If the objective function is parallel to one of the constraints, however, there will be multiple optimal solutions rather than a unique optimal solution. To illustrate, assume that the objective function for the machine shop problem is changed to $P = 4x_1 + 6x_2$. The problem becomes

$$\text{Maximize:} \quad P = 4x_1 + 6x_2$$
$$\text{Subject to:} \quad 2x_1 + 3x_2 \leq 120$$
$$2x_1 + 1.5x_2 \leq 80$$
$$x_1, x_2 \geq 0$$

The constraints and the objective function are shown in Fig. 5.9.

The figure illustrates that the objective function is parallel to the constraint $2x_1 + 3x_2 \leq 120$. Increases in profit are shown by moving the objective function upward and to the right. Profit is a maximum when the objective function and the constraint are coincident. All ordered pairs described by the constraint between the extreme points $(0, 40)$ and $(20, 26.67)$ are optimal solutions. Profit is $240 at either extreme point, or at any linear combination of these extreme points.

UNBOUNDED OPTIMAL SOLUTION. In problems with multiple optimal solutions, the objective function has the same value for all optimal solutions. In

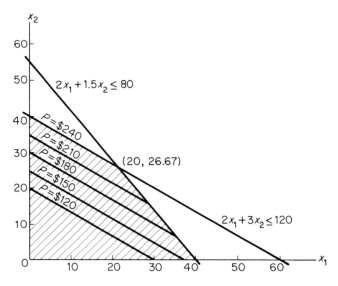

Figure 5.9

the preceding example this value was $P = \$240$. Based upon the constraints, the upper *bound* on the objective function was $240.

Certain linear programming problems do not have a bound on the value of the objective function. In these cases the objective function is *unbounded*. To illustrate, consider the problem

$$
\begin{aligned}
\text{Maximize:} \quad P = \quad & x_1 + 2x_2 \\
\text{Subject to:} \quad & -x_1 + x_2 \le 2 \\
& x_1 + x_2 \ge 4 \\
& x_1, x_2 \ge 0
\end{aligned}
$$

The problem is shown in Fig. 5.10.

The feasible region for this problem is convex. The extreme points of the feasible region are $(4, 0)$ and $(1, 3)$. Neither of these extreme points, however, provides an optimal solution to the problem. This is shown in Fig. 5.10 by the fact that the objective function can increase indefinitely without reaching an upper bound. In fact, given any value of the objective function, there is always a solution that gives a greater value of the objective function. Consequently, the objective function for this problem is unbounded.

These examples illustrate the exception given in the footnote on p. 146 to the extreme point theorem of linear programming. The theorem stated that the optimal value of the objective function occurs at an extreme point of the feasible region. The footnote added that the theorem is true provided that a unique finite optimal solution exists. In the first example, the solution set was

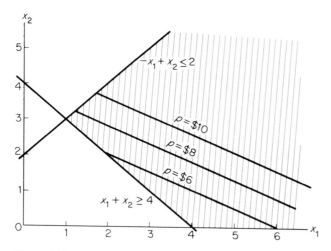

Figure 5.10

empty. There was no feasible solution for the problem. The second example illustrated the case of multiple optimal solutions. Instead of a unique optimal solution, there were an infinite number of optimal solutions, each having the same value. In the final example, the linear programming problem had no finite optimal solution. For any value of the objective function, it was possible to find a solution for which the objective function had a greater value. This third and final exception to the extreme point theorem occurred when the objective function was unbounded.

5.3.4 CHANGING THE SENSE OF OPTIMIZATION

Certain computer programs for solving linear programming problems are written to accept either a maximization problem or a minimization problem. This means that it is sometimes necessary to change the sense of optimization in a linear programming problem, i.e., change a maximization problem to a minimization problem or vice versa.

Changing the sense of optimization in a linear programming problem is straightforward. Any linear maximization problem can be solved as a linear minimization problem by changing the signs of the objective function coefficients and minimizing the objective function subject to the original constraints. Similarly, any linear minimization problem can be solved as a maximization problem by changing the signs of the objective function coefficients and maximizing the objective function subject to the constraints. To illustrate, consider the problem

$$\begin{aligned}
\text{Minimize:} \quad C &= x_1 + x_2 \\
\text{Subject to:} \quad 3x_1 + 8x_2 &\geq 24 \\
3x_1 + 2x_2 &\geq 12 \\
x_1 &\geq 1 \\
x_2 &\geq 0
\end{aligned}$$

The minimization problem can be solved as the following maximization problem simply by changing the signs of the objective function coefficients.

$$\begin{aligned}
\text{Maximize:} \quad -C &= -x_1 - x_2 \\
\text{Subject to:} \quad 3x_1 + 8x_2 &\geq 24 \\
3x_1 + 2x_2 &\geq 12 \\
x_1 &\geq 1 \\
x_2 &\geq 0
\end{aligned}$$

The graphical solution is shown in Fig. 5.11. The extreme points are $E =$

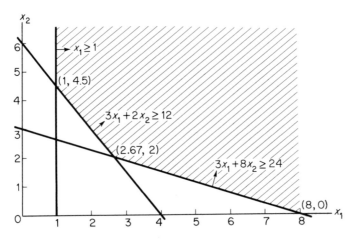

Figure 5.11

$\{(1, 4.5), (2.67, 2), (8, 0)\}$. Table 5.4 gives the value of $C = x_1 + x_2$ and $-C = -x_1 - x_2$.

Table 5.4

(x_1, x_2)	C	$-C$
(1, 4.5)	5.5	−5.5
(2.67, 2)	4.67	−4.67
(8, 0)	8.0	−8.0

The minimum value of C is 4.67 and the maximum value of $-C$ is -4.67.† The solution obtained by minimizing C is thus equivalent to that obtained by maximizing $-C$.

5.4 Applications of Linear Programming

One of the objectives of this chapter is to provide examples of important applications of linear programming. On the basis of these examples, the reader can begin to express problems in the framework of the linear programming model. These examples involve problems of allocation, such as product mix, feed mix, fluid blending, portfolio selection, and multiperiod scheduling. The analytical techniques needed to solve these problems are presented in Chapter 6.‡

5.4.1 PRODUCT MIX

In product mix problems management must determine the quantities of products to be manufactured during a specific time period. To illustrate, consider the case of the Hall Manufacturing Company. Hall manufactures four types of electronic subassemblies for use in aircraft avionic equipment. Each subassembly requires assembly labor, test labor, resistors, and capacitors. The requirements, availability of parts and labor, and the profit per subassembly is projected for the month of September in Table 5.5. Determine the quanti-

Table 5.5

Resource	Electronic Subassembly				Resource Availabilities
	Type 1	Type 2	Type 3	Type 4	
Assembly labor	6.0	5.0	3.5	4.0	600 hours
Test labor	1.0	1.5	1.2	1.2	120 hours
Resistors	4	3	3	3	400 units
Capacitors	2	2	2	3	300 units
Profit	$4.25	$6.25	$5.00	$4.50	

---

† Note that $-4.67 > -5.5$ and $-4.67 > -8.0$. Thus, -4.67 is the maximum.

‡ These applications, while representative, by no means illustrate all uses of linear programming. Additional examples are given in the references at the end of the chapter. Problems that utilize the transportation and assignment algorithms of linear programming are discussed in Chapter 8.

ties of the four products that should be manufactured during September in order to maximize profit. Assume that Hall is contractually required to supply 40 type 1 subassemblies.

To formulate the linear programming problem we let x_j, for $j = 1, 2, 3, 4$, represent the quantity of product j manufactured. The problem can be expressed as

$$\text{Maximize:} \quad P = 4.25x_1 + 6.25x_2 + 5.00x_3 + 4.50x_4$$
$$\text{Subject to:} \quad 6.0x_1 + 5.0x_2 + 3.5x_3 + 4.0x_4 \leq 600$$
$$1.0x_1 + 1.5x_2 + 1.2x_3 + 1.2x_4 \leq 120$$
$$4x_1 + 3x_2 + 3x_3 + 3x_4 \leq 400$$
$$2x_1 + 2x_2 + 2x_3 + 3x_4 \leq 300$$
$$x_1 \geq 40$$
$$x_2, x_3, x_4 \geq 0$$

Because of the number of variables, the problem cannot be solved graphically. The solution is instead obtained by using the procedure discussed in Chapter 6. The optimal solution is $x_1 = 76.9$ type 1 subassemblies, $x_2 = 20.5$ type 2 subassemblies, $x_3 = 10.3$ type 3 subassemblies, and $x_1 = 0$ type 4 subassemblies. Profit for the month is \$506.40.

The following two examples provide additional illustrations of the product mix problem.

Example: A manufacturer can produce three different products during the month of October. Each of these products requires casting, grinding, assembly, and testing. The maximum available hours of capacity of each of the processes during the month, the requirements for each unit of product, and the profit per unit of product are given in Table 5.6. Formulate the linear

Table 5.6

	Product A	B	C	Available Resource Hours
Casting	1.0	1.5	2.0	200
Grinding	0.6	1.0	0.8	120
Assembly	1.5	2.0	1.4	240
Testing	0.3	0.2	0.2	40
Profit	\$12.00	\$18.00	\$16.00	

programming problem. Assume that all products produced during the month can be sold.

Let x_j, for $j = 1, 2, 3$, represent quantities of products A, B, and C. The linear programming problem is

$$\begin{aligned}
\text{Maximize:} \quad P &= 12x_1 + 18x_2 + 16x_3 \\
\text{Subject to:} \quad 1.0x_1 &+ 1.5x_2 + 2.0x_3 \leq 200 \\
0.6x_1 &+ 1.0x_2 + 0.8x_3 \leq 120 \\
1.5x_1 &+ 2.0x_2 + 1.4x_3 \leq 240 \\
0.3x_1 &+ 0.2x_2 + 0.2x_3 \leq 40 \\
x_j &\geq 0 \quad \text{for } j = 1, 2, 3
\end{aligned}$$

Example: McGraw Chemical Company uses nitrates, phosphates, potash, and an inert filler material in the manufacture of chemical fertilizers. The firm mixes these ingredients to make three basic fertilizers: 5-10-5, 5-8-8, and 8-12-12 (numbers represent percent by weight of nitrates, phosphates, and potash in each ton of fertilizer). McGraw receives $70, $75, and $80, respectively, for each ton of the 5-10-5, 5-8-8, and 8-12-12 fertilizer. The cost of the ingredients is $200 per ton of nitrates, $60 per ton of phosphates, $100 per ton of potash, and $20 per ton of the inert filler material. The direct costs of mixing, packing, and selling the fertilizer are $25 per ton.

McGraw has 1200 tons of nitrates, 2000 tons of phosphates, 1500 tons of potash, and an unlimited supply of the inert filler on hand. They will not receive additional chemicals until next month. McGraw has a delivery contract for 8000 tons of 5-8-8 during the month. They believe that they can sell or store at negligible cost all fertilizer produced during the month. Formulate the linear programming problem to maximize profits.

Let x_j, for $j = 1, 2, 3$, represent tons of 5-10-5, 5-8-8, and 8-12-12 fertilizers. The coefficients of the objective function are calculated by subtracting the cost of the ingredients and the $25 mixing cost from the sales price. To illustrate, consider the objective function coefficient of x_1 (i.e., 5-10-5 fertilizer). The cost of the nitrates in one ton of 5-10-5 fertilizer is 0.05($200) or $10. The cost of the phosphates is 0.10($60) or $6. Similarly, the cost of the potash is 0.05($100) = $5, and the cost of the inert filler is 0.80($20) = $16. Subtracting these costs plus the cost of mixing, packing, and selling the 5-10-5 fertilizer from the sales price gives an objective function coefficient of

$$c_1 = \$70 - 0.05(\$200) - 0.10(\$60) - 0.05(\$100)$$
$$- 0.80(\$20) - \$25 = \$8.00$$

The remaining objective function coefficients are calculated in the same manner.

The constraints shown below describe the resource limitations and the contractual requirements. The linear programming problem is

$$\text{Maximize:} \quad P = 8.00x_1 + 11.40x_2 + 6.20x_3$$
$$\text{Subject to:} \quad 0.05x_1 + 0.05x_2 + 0.08x_3 \leq 1200$$
$$0.10x_1 + 0.08x_2 + 0.12x_3 \leq 2000$$
$$0.05x_1 + 0.08x_2 + 0.12x_3 \leq 1500$$
$$x_2 \geq 8000$$
$$x_1, x_3 \geq 0.$$

5.4.2 FEED MIX

The feed mix problem is another form of the allocation problem. This problem involves determining the proportion of several ingredients that, when mixed, satisfy certain criteria and give a minimum cost. To illustrate, consider the problem of mixing feed for a dairy herd. The dairyman may purchase and mix quantities of three types of grain, each containing differing amounts of four nutritional elements. The dairyman has developed nutritional requirements for the feed mixture. His objective is to minimize the cost of the mixture while meeting the nutritional requirements. The data are given in Table 5.7.

Table 5.7

Nutritional Elements	Units of Nutritional Elements Contained in One Pound of			Minimum No. of Units Required
	Grain 1	Grain 2	Grain 3	
1	4	6	8	2500
2	2	3	1	750
3	4	3	1	900
4	0.7	0.3	0.8	250
Cost per pound	$0.025	$0.020	$0.017	
Available	Unlimited	1000 lbs.	Unlimited	

Let x_j, for $j = 1, 2, 3$, represent pounds of grain j in the feed mix. The linear programming problem to minimize cost is

$$\text{Minimize:} \quad C = 0.025x_1 + 0.020x_2 + 0.017x_3$$
$$\text{Subject to:} \quad 4x_1 + 6x_2 + 8x_3 \geq 2500$$
$$2x_1 + 3x_2 + 1x_3 \geq 750$$
$$4x_1 + 3x_2 + 1x_3 \geq 900$$

$$0.7x_1 + \quad 0.3x_2 + \quad 0.8x_3 \geq 250$$
$$x_2 \qquad\qquad \leq 1000$$
$$x_1, x_2, x_3 \geq 0$$

The optimal solution is found by using the simplex algorithm. The solution is $x_1 = 75.0$ pounds or grain 1, $x_2 = 134.4$ pounds of grain 2, and $x_3 = 196.4$ pounds of grain 3. The cost of this mix is $7.90.

Example: American Foods, Inc., is developing a low-calorie, high-protein diet supplement called Hi-Pro. The specifications for Hi-Pro have been established by a panel of medical experts. These specifications, along with the calorie, protein, and vitamin content of three basic foods, are given in Table 5.8.

Table 5.8

Nutritional Elements	Units of Nutritional Elements per 8-Ounce Serving of Basic Foods			
	Basic Foods			Hi-Pro Specifications
	No. 1	*No. 2*	*No. 3*	
Calories	350	250	200	≤ 300
Protein	250	300	150	≥ 200
Vitamin A	100	150	75	≥ 100
Vitamin C	75	125	150	≥ 100
Cost per serving	$0.15	$0.20	$0.12	

Formulate the linear programming model to minimize cost.

Let x_j, for $j = 1, 2, 3$, represent the proportion of basic foods 1, 2, and 3 in an 8-ounce serving of Hi-Pro. The linear programming problem is

Minimize: $C = 0.15x_1 + 0.20x_2 + 0.12x_3$

Subject to:
$$350x_1 + 250x_2 + 200x_3 \leq 300$$
$$250x_1 + 300x_2 + 150x_3 \geq 200$$
$$100x_1 + 150x_2 + 75x_3 \geq 100$$
$$75x_1 + 125x_2 + 150x_3 \geq 100$$
$$x_1 + x_2 + x_3 = 1$$
$$x_j \geq 0 \text{ for } j = 1, 2, 3$$

Example: Chico Candy Company, Inc., mixes three types of candies to obtain a one-pound box of candy. The box of candy sells for $0.85 per pound, and the three ingredient candies sell for $1.00, $0.50, and $0.25 per pound, respectively. The mixture must contain at least 0.3 pounds of the first type of candy, and the weight of the first two candies must at least equal the weight of the third. Formulate the linear programming problem to maximize profit.

Let x_j, for $j = 1, 2, 3$, represent the quantity in pounds of each of the three ingredient candies. The objective function coefficients are found by subtracting the cost of each type of candy from the sales price of the box. For instance, the coefficient of x_1 is $0.85 - \$1.00 = -\$0.15$. The linear programming problem is

$$\text{Maximize:} \quad P = -0.15x_1 + 0.35x_2 + 0.60x_3$$

$$
\begin{aligned}
\text{Subject to:} \quad & x_1 & & & & & \geq 0.3 \\
& x_1 + & & x_2 - & & x_3 & \geq 0 \\
& x_1 + & & x_2 + & & x_3 & = 1 \\
& & & x_2, x_3 & & & \geq 0
\end{aligned}
$$

5.4.3 FLUID BLENDING

Another variation of the allocation problem occurs when fluids, such as chemicals, plastics, molten metals, and oils, must be blended to form a product. Each of the resources included in the blend has certain properties and costs. The fluid blending problem involves blending the different resources to form a final product that meets certain criteria and has a minimum cost or a maximum profit. To illustrate, consider a simplified example of gasoline blending.

The Sigma Oil Company markets three brands of gasoline: Sigma Supreme, Sigma Plus, and Sigma. The three gasolines are made by blending two grades of gasoline, each with a different octane rating, and a special high-octane lead additive. In the gasoline blends, the octane rating of the blend is a linear combination of the octane ratings of the component gasolines and additives. The relevant data are shown in Table 5.9.

Table 5.9

Brands	Minimum Octane	Sales Price to Retailer (per gallon)
Sigma Supreme	100	$0.20
Sigma Plus	94	0.16
Sigma	86	0.14

Blending Component	Octane	Cost (per gallon)	Supply (gallons)
1	110	$0.10	20,000
2	80	0.07	12,000
3	1500	1.00	500

It is assumed that all gasoline blended during the period can be sold or stored at negligible cost. An alternative assumption, not included in this problem, is that forecasts of demand for the three brands of gasoline are available and that gasoline unsold at the end of the period is stored at a cost and sold during the following period.

To formulate the linear programming model, let x_{ij}, for $i = 1, 2, 3$ and $j = 1, 2, 3$, represent the gallons of blending component i used in gasoline blend j. The problem can be expressed as

Maximize: $P = 0.10x_{11} + 0.13x_{21} - 0.80x_{31} + 0.06x_{12} + 0.09x_{22}$
$- 0.84x_{32} + 0.04x_{13} + 0.07x_{23} - 0.86x_{33}$

Subject to:
$$110x_{11} + 80x_{21} + 1500x_{31} \geq 100(x_{11} + x_{21} + x_{31})$$
$$110x_{12} + 80x_{22} + 1500x_{32} \geq 94(x_{12} + x_{22} + x_{32})$$
$$110x_{13} + 80x_{23} + 1500x_{33} \geq 86(x_{13} + x_{23} + x_{33})$$
$$x_{11} + x_{12} + x_{13} \leq 20{,}000$$
$$x_{21} + x_{22} + x_{23} \leq 12{,}000$$
$$x_{31} + x_{32} + x_{33} \leq 500$$
$$x_{ij} \geq 0 \quad \text{for } i = 1, 2, 3 \text{ and } j = 1, 2, 3$$

The objective function shows that there is a $0.10 profit for each gallon of blending component 1 used in Sigma Supreme, a $0.13 profit for each gallon of blending component 2 used in Sigma Supreme, an $0.80 loss for each gallon of blending component 3 used in Sigma Supreme, etc. These numbers are calculated by subtracting the cost of the blending component from the sales price of the gasoline. For instance, each gallon of blending component 1 used in Sigma Supreme costs $0.10 and is sold for $0.20, giving a profit of $0.10.

The first three constraints specify that the blends must have minimum octane ratings. The next three constraints restrict the quantity of each component used for blending to the supply available during the period. The solution to the problem is given in Table 5.10. Profit for the period is $3238.

Table 5.10   Gallons of Blend i Used in Gasoline j

Blend	Supreme	Plus	Sigma	Total
1	20,000	0	0	20,000
2	12,000	0	0	12,000
3	403	0	0	403
Total	32,403	0	0	32,403

Example: American whiskeys are made by blending different types of bourbon whiskeys. Assume that Kentucky Whiskeys, Inc. mixes three bourbon whiskeys, A, B, and C, to obtain two blends, Kentucky Premium and Kentucky Smooth. The recipes used by Kentucky for their two blends are given in Table 5.11.

Table 5.11

Blend	Ingredients	Price (per gallon)
Kentucky Premium	Not less than 60% A Not more than 20% C	$7.00
Kentucky Smooth	Not less than 30% A Not more than 60% C	$5.50

Supplies of the basic bourbon whiskeys and their costs are shown in Table 5.12. Using this data, formulate the linear programming problem to maximize profits.

Table 5.12

Bourbon	Quantity Available (gallons)	Cost (per gallon)
A	1500	$6.75
B	2000	$6.00
C	1000	$4.00

Let x_{ij}, for $i = 1, 2, 3$ and $j = 1, 2$ represent the quantity of bourbon i used in one gallon of blend j. The linear programming problem is

Maximize: $P = 0.25x_{11} - 1.25x_{12} + 1.00x_{21} - 0.50x_{22}$
$$+ 3.00x_{31} + 1.50x_{32}$$

Subject to:

$$x_{11} \geq 0.60(x_{11} + x_{21} + x_{31})$$
$$x_{31} \leq 0.20(x_{11} + x_{21} + x_{31})$$
$$x_{12} \geq 0.30(x_{12} + x_{22} + x_{32})$$
$$x_{32} \leq 0.60(x_{12} + x_{22} + x_{32})$$

$$x_{11} + x_{12} \leq 1500$$
$$x_{21} + x_{22} \leq 2000$$
$$x_{31} + x_{32} \leq 1000$$
$$x_{ij} \geq 0$$

The objective function shows that there is a $0.25 profit on each gallon of bourbon A used in Kentucky Premium. This is determined by subtracting the cost per gallon of bourbon A from the price per gallon of Kentucky Premium. Similarly, there is a loss of $1.25 per gallon for each gallon of bourbon A used in Kentucky Smooth.

The first four constraints specify the maximum and minimum quantities of bourbons used in the blended whiskeys. For example, the first constraint shows that the proportion of blend A in Kentucky Premium must be not less than 60 percent. Similarly, the second constraint shows that the proportion of blend C in Kentucky Premium must not exceed 20 percent. The last three constraints limit the supplies of the bourbon whiskeys available for blending during the period.

Example: Plains Coffee Company mixes Brazilian, Colombian, and Mexican coffees to make two brands of coffee, Plains X and Plains XX. The characteristics used in blending the coffees include strength, acidity, and caffeine. Test results of the available supplies of Brazilian, Colombian, and Mexican coffees are shown in Table 5.13.

Table 5.13

Imported Coffee	Price per Pound	Strength Index	Acidity Index	Percent Caffeine	Supply Available
Brazilian	$0.30	6	4.0	2.0	40,000 lb
Colombian	0.40	8	3.0	2.5	20,000 lb
Mexican	0.35	5	3.5	1.5	15,000 lb

The requirements for Plains X and Plains XX coffees are given in Table 5.14. Assume that 35,000 lb of Plains X and 25,000 lb of Plains XX are to be sold. Formulate the linear programming problem to maximize profits.

Table 5.14

Plains Coffee	Price per Pound	Minimum Strength	Maximum Acidity	Maximum Percent Caffeine	Quantity Demanded
X	$0.45	6.5	3.8	2.2	35,000 lb
XX	0.55	6.0	3.5	2.0	25,000 lb

Let x_{ij}, for $i = 1, 2, 3$ and $j = 1, 2$ represent the pounds of imported coffees used in Plains coffees. The linear programming problem is

Maximize: $P = 0.15x_{11} + 0.25x_{12} + 0.05x_{21} + 0.15x_{22}$
$+ 0.10x_{31} + 0.20x_{32}$

Subject to: $6x_{11} + \quad 8x_{21} + \quad 5x_{31} \geq \quad 6.5(x_{11} + x_{21} + x_{31})$
$6x_{12} + \quad 8x_{22} + \quad 5x_{32} \geq \quad 6.0(x_{12} + x_{22} + x_{32})$

$4.0x_{11} + \quad 3.0x_{21} + \quad 3.5x_{31} \leq \quad 3.8(x_{11} + x_{21} + x_{31})$
$4.0x_{12} + \quad 3.0x_{22} + \quad 3.5x_{32} \leq \quad 3.5(x_{12} + x_{22} + x_{32})$

$2.0x_{11} + \quad 2.5x_{21} + \quad 1.5x_{31} \leq \quad 2.2(x_{11} + x_{21} + x_{31})$
$2.0x_{12} + \quad 2.5x_{22} + \quad 1.5x_{32} \leq \quad 2.0(x_{12} + x_{22} + x_{32})$

$x_{11} + x_{12} \leq 40{,}000$
$x_{21} + x_{22} \leq 20{,}000$
$x_{31} + x_{32} \leq 15{,}000$

$x_{11} + x_{21} + x_{31} = 35{,}000$
$x_{12} + x_{22} + x_{32} = 25{,}000$
$x_{ij} \geq 0$

The coefficients of the objective function represent profits from imported coffee i in mix j. The first six constraints establish restrictions on strength, acidity, and caffeine. Limitations on the supply of imported coffees are given by the next three constraints and the quantity of Plains X and Plains XX coffees demanded during the period is described by the final two constraints.

5.4.4 PORTFOLIO AND MEDIA SELECTION

Another example of the allocation problem occurs in portfolio selection. Various financial institutions such as pension funds, insurance companies, and mutual funds, along with private individuals, are frequently faced with the problem of investing funds among the alternatives available. These alternatives vary in terms of safety, liquidity, income, growth, and other factors. Linear programming can be used to suggest the allocation of funds among the competing alternatives.

To illustrate the use of linear programming in portfolio selection, consider the case of the ABC Mutual Fund. ABC is an income-oriented mutual fund. To aid in its investment decisions, ABC has developed the investment alternatives given in Table 5.15. The return on investment is expressed as an annual rate of return on the invested capital. The risk is a subjective estimate on a scale from 0 to 10 made by the portfolio manager of the safety of the investment. The term of investment is the average length of time required to realize the return on investment indicated in Table 5.15.

ABC's objective is to maximize the return on investment. The guide lines for selecting the portfolio are (1) the average risk should not exceed 2.5; (2)

Table 5.15

Investment Alternative	Yearly Return on Investment	Risk	Term of Investment (yrs)
1. Blue chip stock	12%	2	4
2. Bonds	10	1	8
3. Growth stock	15	3	2
4. Speculation	25	4	10
5. Cash	0	0	0

the average term of investment should not exceed six years; and (3) at least 15 percent of the funds should be retained in the form of cash. To maximize the return on investment, let x_j, for $j = 1, 2, 3, 4, 5$ represent the proportion of funds to be invested in the jth alternative. The objective function and constraints are

Maximize: $P = 12x_1 + 10x_2 + 15x_3 + 25x_4 + 0x_5$

Subject to:
$$x_1 + x_2 + x_3 + x_4 + x_5 = 1$$
$$2x_1 + 1x_2 + 3x_3 + 4x_4 \leq 2.5$$
$$4x_1 + 8x_2 + 2x_3 + 10x_4 \leq 6$$
$$x_5 \geq 0.15$$
$$x_j \geq 0 \quad \text{for } j = 1, 2, 3, 4, 5$$

The first constraint states that the proportions of funds invested in the various alternatives must total 1. The remainder of the constraints describe the risk, term of investment, and cash requirements. The solution to the problem is $x_1 = 40.7$ percent, $x_2 = 2.85$ percent, $x_3 = 0$, $x_4 = 41.4$ percent, and and $x_5 = 15.0$ percent. The overall rate of return on the portfolio is 15.5 percent.

The portfolio selection model is quite similar to the media selection model. Media selection involves the allocation of the advertising budget of a firm among various communication media in an effort to reach appropriate audiences. The objective is to maximize total effective exposures, subject to constraints on the advertising budget and the number of exposures per media.

To illustrate a simplified media selection model, consider the case of the Edwards Advertising Agency. Edwards is formulating an advertising campaign for one of its customers. A careful analysis of the available media has narrowed the allocation problem to that of determining the number of messages to appear in each of three magazines. The cost of advertisements and other relevant data are given in Table 5.16.

The cost per advertisement, characteristics of the media, and the audience size are available from industry sources. Edwards has established the maximum and minimum number of advertisements per medium and the relative

Table 5.16

	Media		
	1	*2*	*3*
Cost per advertisement	$500	$750	$1,000
Maximum number of ads	24	52	24
Minimum number of ads	6	12	0
Reader characteristics			
Age: 21–40	50%	60%	65%
Income: $10,000 or more	40%	50%	80%
Education: college	40%	40%	60%
Audience size	500,000	800,000	1,200,000

importance of the characteristics. The relative importance of the characteristics are: age, 0.5; income, 0.3; education, 0.2.

The linear programming problem is to maximize total effective exposures. The coefficients of the objective function are the product of the audience size multiplied by the "effectiveness coefficient" for each medium. The effectiveness coefficient is defined as equalling the sum of the products of the characteristics and their relative importance. The effectiveness coefficient for media 1 is

$$\text{effectiveness coefficient} = 0.50(0.5) + 0.40(0.3) + 0.40(0.2) = 0.450$$

Similarly, the effectiveness coefficients for media 2 and 3 are 0.530 and 0.685, respectively. The product of the effectiveness coefficients and the audience size gives the coefficients of the objective function. To illustrate, the coefficient of the objective function for media 1 is 0.450(500,000) or 225,000.

If it is assumed that an advertising budget of $35,000 has been established, the linear problem to maximize total effective exposure is

Maximize: $E = 225{,}000x_1 + 424{,}000x_2 + 822{,}000x_3$

Subject to:
$$500x_1 + 750x_2 + 1000x_3 \leq 35{,}000$$
$$x_1 \geq 6$$
$$x_1 \leq 24$$
$$x_2 \geq 12$$
$$x_2 \leq 52$$
$$x_3 \leq 24$$
$$x_3 \geq 0$$

where x_j, for $j = 1, 2, 3$, represents the number of advertisements in the jth media.

The solution to the problem is $x_1 = 6$, $x_2 = 12$, and $x_3 = 23$. The total number of effective exposures is 25,344,000.

5.4.5 MULTIPERIOD SCHEDULING

Linear programming can often be applied to establish production schedules that extend over several time periods. For instance, many manufacturers must plan for seasonal variations in demand for their products. In these cases the costs of overtime must be balanced against the cost of carrying inventories from period to period. The objective is to establish a multiperiod schedule that minimizes all relevant costs of production of the product.

To illustrate multiperiod scheduling, consider the case of Kim Manufacturing, Inc. On October 1, Kim received a contract to supply 6000 units of a specialized product. The terms of the contract require that 1000 units be shipped in October, 3000 in November, and 2000 in December. Kim can manufacture 1500 units per month on regular time and 750 units per month on overtime. The manufacturing cost per item produced during a regular hour is \$3, and the cost per item produced during overtime is \$5. The monthly storage cost is \$1. Formulate the linear programming problem to minimize total cost of production.

Let x_{ijk}, for $i = 1, 2, 3, j = 1, 2$, and $k = 1, 2, 3$, represent the number of units manufactured in month i using shift j and shipped in month k. With this notation, x_{121} represents the number of units manufactured in October using overtime and shipped in October. Similarly, x_{213} represents the number of units manufactured during November using regular time and shipped in December. The linear programming problem is

Minimize: $C = 3x_{111} + 5x_{121} + 4x_{112} + 6x_{122} + 5x_{113} + 7x_{123}$
$$+ 3x_{212} + 5x_{222} + 4x_{213} + 6x_{223} + 3x_{313} + 5x_{323}$$

Subject to:
$$x_{111} + x_{112} + x_{113} \le 1500$$
$$x_{212} + x_{213} \le 1500$$
$$x_{313} \le 1500$$

$$x_{121} + x_{122} + x_{123} \le 750$$
$$x_{222} + x_{223} \le 750$$
$$x_{323} \le 750$$

$$x_{111} + x_{121} = 1000$$
$$x_{112} + x_{122} + x_{212} + x_{222} = 3000$$
$$x_{113} + x_{123} + x_{213} + x_{223} + x_{313} + x_{323} = 2000$$
$$x_{ijk} \ge 0$$

The coefficients of the objective function represent manufacturing, overtime, and storage cost. The first three constraints limit production during regular hours to 1500 units per month. The next three constraints limit production during overtime to 750 units per month. The final three constraints assure that the shipping schedule of 1000 units during October, 3000 units during November, and 2000 units during December is met. The solution is $x_{111} = 750$, $x_{121} = 250$, $x_{112} = 750$, $x_{122} = 0$, $x_{113} = 0$, $x_{123} = 0$, $x_{212} = 1500$, $x_{222} = 750$, $x_{213} = 0$, $x_{223} = 0$, $x_{313} = 1500$, and $x_{323} = 500$. The cost of this schedule is $21,750.

The approach to multiperiod scheduling suggested in the preceding example can easily be adapted to scheduling capital expenditures. To illustrate, assume that the investment alternatives described in Table 5.17 are available

Table 5.17

In- vestment	Amount of Investment	Commitment of Funds	Reinvestment Possible	Total Return on Investment
1	Unlimited	one year*	yes	10%
2	$30,000 maximum	two years*	no	30
3	$20,000 maximum	two years†	no	50
4	$20,000 maximum	one year‡	no	40

* Investment available at the beginning of the first year.
† Investment available at the beginning of the second year.
‡ Investment available at the beginning of the third year.

to a firm. The firm has $50,000 available to invest at the beginning of the scheduling period. The objective is to schedule the investments during the three years in order to maximize the return on the investment.

Let x_{ij}, for $i = 1, 2, 3, 4$ and $j = 1, 2, 3$, represent the investment in alternative i at the beginning of period j. The linear programming problem is to

$$\text{Maximize:} \quad R = 0.10x_{11} + 0.10x_{12} + 0.10x_{13} + 0.30x_{21} + 0.50x_{32} + 0.40x_{43}$$

The constraint on the available funds at the beginning of the first year is

$$x_{11} + x_{21} \leq 50,000$$

The amount available for investing at the beginning of the second year is $50,000 - x_{21} + 0.10x_{11}$. The constraint on the investments at the beginning of the second year is

$$x_{12} + x_{32} \leq 50,000 - x_{21} + 0.10x_{11}$$

The amount available for investing at the beginning of the third year is

$$50,000 + 0.30x_{21} + 0.10x_{11} + 0.10x_{12} - x_{32}$$

The constraint on the investments at the beginning of the third year is

$$x_{13} + x_{43} \leq 50,000 + 0.30x_{21} + 0.10x_{11} + 0.10x_{12} - x_{32}$$

The constraints on the maximum investment in each of the alternatives are

$$x_{21} \leq 30,000$$
$$x_{32} \leq 20,000$$
$$x_{43} \leq 20,000$$

The linear programming formulation for the multiperiod investment schedule is expressed as

Maximize: $R = 0.10x_{11} + 0.10x_{12} + 0.10x_{13} + 0.30x_{21} + 0.50x_{32} + 0.40x_{43}$

Subject to:

$$x_{11} + \quad x_{21} \leq 50,000$$
$$x_{12} + \quad x_{32} + \quad x_{21} - 0.10x_{11} \leq 50,000$$
$$x_{13} + \quad x_{43} - 0.30x_{21} - 0.10x_{11} - 0.10x_{12}$$
$$+ \quad x_{32} \leq 50,000$$
$$x_{21} \leq 30,000$$
$$x_{32} \leq 20,000$$
$$x_{43} \leq 20,000$$
$$x_{ij} \geq 0$$

The solution is $x_{11} = \$20,000$, $x_{12} = \$2000$, $x_{13} = \$21,200$, $x_{21} = \$30,000$, $x_{32} = \$20,000$, $x_{43} = \$20,000$. The total return on the investment is \$31,320.

PROBLEMS

1. Determine the solutions to the following linear programming problems by graphing the linear inequalities and evaluating the objective function at each extreme point of the feasible region.

(a) Maximize: $Z = 5x_1 + 6x_2$

Subject to:
$$3x_1 + 2x_2 \leq 32$$
$$1x_1 + 4x_2 \leq 34$$
$$x_1, x_2 \geq 0$$

(b) Maximize: $Z = 3.0x_1 + 1.0x_2$

Subject to:
$$6.0x_1 + 4.0x_2 \leq 48$$
$$3.5x_1 + 6.0x_2 \leq 42$$
$$x_1, x_2 \geq 0$$

(c) Minimize: $C = 5x_1 + 7x_2$

Subject to:
$$10x_1 + 3x_2 \geq 90$$
$$4x_1 + 6x_2 \geq 72$$
$$7x_1 + 5x_2 \geq 105$$
$$x_1, x_2 \geq 0$$

(d)   Minimize: $C = 3x_1 + 4x_2$
               Subject to: $6x_1 + 5x_2 \geq 45$
                                $5x_1 + 8x_2 \geq 60$
                                           $x_2 \geq 2$
                                    $x_1 \geq 0$

2. Solve the following linear programming problems.

(a)   Maximize: $Z = 3x_1 + 5x_2$
               Subject to: $13x_1 + 9x_2 \leq 234$
                                $9x_1 + 15x_2 \leq 270$
                                    $x_1, x_2 \geq 0$

(b)   Maximize: $Z = 2x_1 + 5x_2$
               Subject to: $4x_1 + 3x_2 \leq 24$
                                $3x_1 + 5x_2 \leq 30$
                                $x_1 + x_2 \geq 8$
                                    $x_1, x_2 \geq 0$

(c)   Minimize: $C = x_1 + x_2$
               Subject to: $4x_1 + 3x_2 \geq 12$
                                $-3x_1 + 4x_2 \leq 12$
                                    $x_1, x_2 \geq 0$

(d)   Maximize: $Z = x_1 + x_2$
               Subject to: $8x_1 + 5x_2 \geq 40$
                                $-4x_1 + 6x_2 \geq 24$
                                    $x_1, x_2 \geq 0$

3. Use the graphical solution technique to show that the problem
                Maximize: $Z = x_1 + x_2$
                Subject to: $3x_1 + 2x_2 \leq 24$
                                 $4x_1 + 7x_2 \leq 56$
                                 $-5x_1 + 6x_2 \leq 30$
                                     $x_1, x_2 \geq 0$

is equivalent to the problem
                Minimize: $-Z = -x_1 - x_2$
                Subject to: $3x_1 + 2x_2 \leq 24$
                                 $4x_1 + 7x_2 \leq 56$
                                 $-5x_1 + 6x_2 \leq 30$
                                     $x_1, x_2 \geq 0$

4. Piper Farms plans to introduce two new gift packages of fruit for the Christmas market. Box A will contain 20 apples and 15 pears. Box B will contain 40 apples and 20 pears. Piper has 18,000 apples and 12,000 pears available for packaging. They believe that all fruit packaged can be sold. Profits are estimated as \$0.60 for A and \$1.00 for B. Determine the number of boxes of A and B that should be prepared to maximize profits.

5. Radio Manufacturing Company must determine production quantities for this month for two different models, A and B. Data per unit are given in the following table.

Model	Revenue	Subassembly Time (hr)	Final Assembly Time (hr)	Quality Inspection (hr)
A	$10	1.0	0.8	0.5
B	$20	1.2	2.0	0

The maximum time available for these products is 1200 hours for subassembly, 1600 hours for final assembly, and 500 hours for quality inspection. Orders outstanding require that at least 200 units of A and 100 units of B be produced. Determine the quantities of A and B that maximize total revenue.

6. A baseball manufacturer makes two types of baseballs, Major League and Minor League. He has four manufacturing processes that are used in making each type of baseball. These processes are designated as J, K, L, and M. Available time and time required per baseball for each process are listed below.

Process	Major League (time/baseball)	Minor League (time/baseball)	Total Time Available
J	2	3	6,000
K	8	5	20,000
L	1	$3\frac{1}{3}$	5,000
M	3	1	6,000

The contribution to profit is $2 for Major and $1 for Minor League baseballs. Determine the product mix that leads to maximum profit.

7. A company makes two types of leather wallets. Type A is a high-quality wallet and type B is a medium-quality wallet. The profits on the two wallets are $0.80 for type A and $0.60 for type B. The type A wallet requires twice as much time to manufacture as the type B. If all wallets were type B, the company could make 1000 per day. The supply of leather is sufficient for a maximum of 800 wallets per day (both A and B combined). A special process further limits production to a maximum of 450 wallets of type A and 700 wallets of type B. Assuming that all wallets manufactured can be sold, determine the number of each type to maximize profits.

8. Kawar Corporation produces two types of tract homes, the Tempo model and the Trend model. Each type is built on the same size lot and the tract will allow for 120 homes. The building materials used for both types of home are the same except for the framing materials. The Tempo model uses lumber for framing and the Trend model uses aluminum. Fifty thousand board feet of lumber and 75,000 running feet of standard aluminum framing are available during the construction period. Due to architectural styling, the Tempo model requires 600 board feet of lumber and 3000 man-hours, while the Trend model requires 800 running feet of standard aluminum frame and 6000 man hours. Total labor available during the construction period is 500,000 man-hours. The profit margin is $5000 for each Tempo model and $7000 for each Trend model. Determine the product mix for maximum profit.

9. A poultry farmer must supplement the vitamins in the feed he buys. He is considering two supplements, each of which contains the four vitamins required but in differing amounts. He must meet or exceed the minimum vitamin requirements. The vitamin content per ounce of the supplements is given in the following table.

Vitamin	Supplement 1	Supplement 2
1	5 units	25 units
2	25 units	10 units
3	10 units	10 units
4	35 units	20 units

Supplement 1 costs 3 cents per ounce and supplement 2 costs 4 cents per ounce. The feed must contain at least 50 units of vitamin 1, 100 units of vitamin 2, 60 units of vitamin 3, and 180 units of vitamin 4. Determine the mixture that has the minimum cost.

10. The Whoop-Bang Novelty Co. makes three basic types of noisemakers: Toot, Wheet, and Honk. A Toot can be made in 30 minutes and has a feather attached to it. A Wheet requires 20 minutes, has two feathers, and is sprinkled with 0.5 oz of sequin powder. The Honk requires 30 minutes, has three feathers, and 1 oz of sequin powder. The net profit is $0.45 per Toot, $0.55 per Wheet, and $0.70 per Honk. The following resources are available: 80 hours of labor, 90 ounces of sequin powder, and 360 feathers. Set up the linear programming problem to maximize profits.

11. A company manufactures three products: A, B, and C. The products require four operations: grinding, turning, assembly, and testing. The requirements per unit of product in hours for each operation are as follows:

Product	Grinding	Turning	Assembly	Test
A	0.03	0.11	0.30	0.08
B	0.02	0.14	0.20	0.07
C	0.03	0.20	0.26	0.08
Capacity (Hr)	1000	4000	9000	3000

The minimum monthly sales requirements are

	A	9000 units
	B	9000 units
	C	6000 units

The profit per unit sold of each product is $0.15 for A, $0.12 for B, and $0.09 for C. All units produced above the minimum can be sold. What quantities of each product should be produced next month for maximum profit? Set up only.

12. A certain firm has two plants. Orders from four customers have been received. The number of units ordered by each customer and the shipping costs from each plant are shown in the following table.

		Shipping Cost/Unit	
Customer	Units Ordered	From Plant 1	From Plant 2
A	500	$1.50	$4.00
B	300	2.00	3.00
C	1000	3.00	2.50
D	200	3.50	2.00

Each unit of the product must be machined and assembled. These costs, together with the capacities at each plant, are shown below.

	Hours/Unit	Cost/Hour	Hours Available
Plant No. 1:			
Machining	0.10	$4.00	120
Assembling	0.20	3.00	260
Plant No. 2:			
Machining	0.11	4.00	140
Assembling	0.22	3.00	250

Formulate the linear programming problem to minimize cost. Set up only.

13. Crane Feed Company markets two feed mixes for rabbits. The first mix, Fertilex, requires at least twice as much wheat as barley. The second mix, Multiplex, requires at least twice as much barley as wheat. Wheat costs $0.49 per pound, and only 1000 pounds are available this month. Barley costs $0.36 per pound, and 1200 pounds are available. Fertilex sells for $1.59 per pound up to 99 pounds, and each additional pound over 99 sells for $1.43. Multiplex sells at $1.35 per pound up to 99 pounds, and each additional pound over 99 sells for $1.18. Rancho Farms will buy any and all amounts of both mixes Crane Feed Company will mix. Set up the linear programming problem to determine the product mix that results in maximum profits.

14. A farmer must decide how many pounds of each of several types of grain he should purchase in order that his livestock receive the minimum nutrient requirement at the lowest cost possible. The relevant information is as follows:

	One Pound Of				Minimum Nutrient Requirement
	Grain 1	*Grain 2*	*Grain 3*	*Grain 4*	
Nutrient 1	4	5	6	3	1000
Nutrient 2	2	1	0	3	850
Nutrient 3	1	2	3	1	700
Nutrient 4	2	3	1	2	1320
Nutrient 5	0	2	1	1	550
Cost/pound	$0.35	$0.42	$0.45	$0.37	

Set up the linear programming problem to determine the mix of grains that minimizes cost.

15. The Franklin Oil Company produces three oils of different viscosities. Franklin Oil makes its products by blending two grades of oil, each with a different viscosity. The final viscosity is linearly proportional to the blending viscosities. There is an unlimited demand for the three oils.

Blending Component	Viscosity	Cost/Quart	Supply
A	10	$0.20	4000 qt/wk
B	50	$0.30	2000 qt/wk

Brand	Minimum Viscosity	Sales Price/Quart
S	20	$0.43
SS	30	$0.48
SSS	40	$0.53

Formulate the linear programming problem to maximize profits.

16. Western Oil Company makes three brands of gasoline: Super, Low Lead, and Regular. Western makes its gasolines by blending two grades of gasolines and a high-octane lead additive. Each brand of gasoline must have an octane rating of at least the stated minimum. The brands are made by blending the two grades of gasoline and the high-octane lead additive. The relevant data are given below.

Blending Component	Octane	Cost per Gallon	Supply from Refinery (per week)
A	120	$0.12	18,000 gallons
B	80	$0.09	26,000 gallons
L	1200	$1.40	Unlimited

Gasoline Brand	Minimum Octane	Sales Price per Gallon
Super	105	$0.18
Low Lead*	94	$0.16
Regular	96	$0.14

* Low Lead can contain no more than 0.1 percent by volume of blending component L.

Assuming that Western Oil can sell all gasoline produced, set up the linear programming problem to maximize profit.

17. The Hamilton Data Processing Company performs three types of data processing activities: payrolls, accounts receivable, and inventories. The profit and time requirements for key punch, computation, and off-line printing for a "standard job" are shown in the following table.

Job	Profit per Std. Job	Time Requirements (Minutes)		
		Key Punch	Computation	Print
Payroll	$275	1200	20	100
Accounts receivable	$125	1400	15	60
Inventory	$225	800	35	80

Hamilton guarantees overnight completion of the job. Any job scheduled during the day can be completed during the day or during the night. Any job scheduled during the night, however, must be completed during the night. The capacity for both day and night are shown in the following table.

Capacity (minutes)	Key Punch	Operation Computation	Print
Day	4200	150	400
Night	9200	250	650

Set up the linear programming problem to determine the mixture of "standard jobs" that should be accepted during the day and during the night in order to maximize profit.

18. The Trust Department of the Barclay Bank is preparing an investment portfolio for a wealthy customer. The investment alternatives along with their current yields and risk factors are given in the following table.

Investment	Yield	Risk
AAA corporate bonds	6.5%	1.0
Convertible debentures	7.2	1.5
Selected high-grade common stock	8.3	2.0
Preferred stock	6.3	1.8
Growth stock	9.0	3.2

The risk factor is a subjective estimate on a scale from 0 to 10, the lower numbers representing the lower risk investments. On the assumption that the current yields will continue, the Trust Department wishes to design a portfolio with the maximum yield. However, the average risk must not exceed 2.4. In addition, the customer has specified that the investment in growth stock cannot exceed the combined total investment in corporate bonds and convertible debentures. Set up the linear programming problem to determine the proportion of funds in each investment alternative.

19. G and H Advertising, Inc., is preparing a proposal for an advertising campaign for White Chemical Company, manufacturers of Gro-More Lawn products. The advertising copy has been written and approved. The final step in the proposal, therefore, is to recommend an allocation of advertising funds so as to maximize the total number of effective exposures. The characteristics of three alternative publications are shown in the following table.

	House Beautiful	Publication Home & Garden	Lawn Care
Cost per advertisement	$600	$800	$450
Maximum no. of adv.	12	24	12
Minimum no. of adv.	3	6	2
Characteristics:			
Homeowner	80%	70%	20%
Income: $10,000 or more	70%	80%	60%
Occupation: gardener	15%	20%	40%
Audience size	600,000	800,000	300,000

The relative importance of the three characteristics are: homeowner, 0.4; income, 0.2; gardener, 0.4. The advertising budget is $20,000. Formulate the linear programming problem to determine the most effective number of exposures in each magazine.

20. Great Western Airlines, Inc., must determine how many stewardesses to hire and train during the next six months. Their requirements, expressed in terms of flight hours, are 8400 in January, 9100 in February, 9800 in March, 11,200 in April, 11,200 in May, and 11,800 in June.

   One month of training is required before a stewardess can be put on a regular flight. Consequently, a girl must be hired at least a month before she is needed. A maximum of 20 girls can be enrolled in the training program during any month.

   Each stewardess can work up to 70 flight hours per month. Great Western has 140 stewardesses available for scheduling at the beginning of January. Approximately 10 percent of the girls on flight status either quit or are dismissed each month. Girls on standby normally do not quit. The cost of training a girl is $2500, and the salary of the stewardess is $700. Girls not used on flights are put on standby duty and earn their full salary. Formulate the linear programming problem to determine the hiring schedule during the six-month period.

21. Fast Eddie is a pool shark. To keep sharp he must play at least seven hours a day and not more than twelve hours. Of his playing hours, at least four hours must be straight practicing. Eddie's average income from his three games is: snooker, $30 per hour; pocket billiards, $35 per hour; three-rail billiards, $55 per hour.

   Eddie has the following problem: For every hour of "hustling" a given game, he must practice the other two games to remain in top hustling condition. The following table shows the relationships:

Game-1 hour	Minimum Practice Required	
Snooker	$\frac{1}{4}$ hr Pocket Billiards	$\frac{1}{6}$ hr 3-Rail
Pocket billiards	$\frac{1}{6}$ hr Snooker	$\frac{1}{4}$ hr 3-Rail
3-Rail billiards	$\frac{1}{2}$ hr Pocket Billiards	$\frac{1}{3}$ hr Snooker

Eddie also feels that he must practice each of the three games at least one hour apiece. Set up the linear programming problem to determine the mix of hustling and practicing that will maximize Eddie's profit.

22. Next Monday, Gordon Shipley has a term paper due in Marketing and exams in Quantitative Methods and Accounting. Shipley has at most 20 hours of study time available. Shipley has completed the rough draft of the term paper but he has not started the final draft. He estimates that he could turn in the rough draft and receive a grade of 60. With 10 hours of effort on the final draft he could get an 80 and with 20 hours of study effort he could get 100. Without studying he could get a 60 on the Quantitative Methods exam. With 4 hours of study, however, he could get an 80 and 8 hours of study should result in a perfect score. Shipley is a little behind in Accounting. Without any study he would score 40. With 5 hours of study he would score 70, and with 10 hours of study he would expect to score 100 on the exam.

The term paper is 50 percent of the Marketing course grade, the Quantitative Methods exam is 25 percent, and the Accounting exam is 20 percent. Taking into account that Accounting and Quantitative Methods are four-unit courses and Marketing is a two-unit course, how should Shipley allocate his time to maximize his overall GPA (grade point average) and still not receive a grade of less than 70 in any subject? Set up the problem only.

SELECTED REFERENCES

BAUMOL, WILLIAM J., *Economic Theory and Operations Analysis*, 2nd ed. (Englewood Cliffs, N.J.: Prentice-Hall, Inc., 1965).

BOULDING, K. E. AND W. A. SPIVEY, *Linear Programming and the Theory of the Firm* (New York, N.Y.: The Macmillan Company, Inc., 1960).

CHARNES, A. AND W. W. COOPER, *Management Models and Industrial Applications of Linear Programming* (New York, N.Y.: John Wiley and Sons, Inc., 1961).

GARVIN, WALTER W., *Introduction to Linear Programming* (New York, N.Y.: The McGraw-Hill Book Company, Inc., 1960).

GASS, SAUL I., *Linear Programming: Methods and Applications* (New York, N.Y.: The McGraw-Hill Book Company, Inc., 1964).

HADLEY, GEORGE, *Linear Programming* (Reading, Mass.: Addison-Wesley Publishing Company, Inc., 1962).

KIM, CHAIHO, *Introduction to Linear Programming* (New York, N.Y.: Holt, Rinehart, and Winston, Inc., 1971).

LEVIN, R. I. AND RUDY P. LAMONE, *Linear Programming for Management Decisions* (Homewood, Ill.: Richard D. Irwin, Inc., 1969).

NAYLOR, T. H., E. T. BRYNE, AND J. M. VERNON, *Introduction to Linear Programming: Methods and Cases* (Belmont, Ca.: Wadsworth Publishing Company, Inc., 1971).

SPIVEY, W. A. AND R. M. THRALL, *Linear Optimization* (New York, N.Y.: Holt, Rinehart, and Winston, Inc., 1970).

STRUM, JAY E., *Introduction to Linear Programming* (San Francisco, Ca.: Holden-Day, Inc., 1972).

VANDERMEULEN, DANIEL C., *Linear Economic Theory* (Englewood Cliffs, N.J.: Prentice-Hall Inc., 1971).

Chapter 6

The
Simplex Method

The simplex method is a solution algorithm for solving linear programming problems.† The simplex algorithm was developed by George Dantzig in 1947 and made generally available in 1951. The importance of the algorithm is demonstrated by the fact that linear programming has become one of the most widely used quantitative approaches to business decisions.

6.1 System of Equations

The simplex algorithm utilizes the *extreme point theorem* of linear programming. This theorem states that *an optimal solution to a linear programming problem occurs at one of the extreme points of the feasible region.* The graphical solution technique was used to show that extreme points are defined by the intersection of systems of equations. The first step in the simplex algorithm, therefore, is to convert the system of linear inequalities to linear equations. These equations are then used to determine the extreme points.

---

† The term *algorithm* refers to a systematic procedure or series of rules for solving a problem.

6.1.1 CONVERTING INEQUALITIES TO EQUATIONS

A straightforward technique for converting an inequality to an equation involves adding a *slack variable* to the left side of inequalities of the form $ax_1 + bx_2 \leq c$ and subtracting a *surplus variable* from the left side of inequalities of the form $ax_1 + bx_2 \geq c$. To illustrate this technique, consider the system of inequalities

$$3x_1 + 2x_2 \leq 40$$
$$2x_1 + x_2 \geq 10$$

The first inequality is converted to an equation by adding the slack variable x_3. The resulting equation is

$$3x_1 + 2x_2 + x_3 = 40$$

The slack variable x_3 represents the difference between $3x_1 + 2x_2$ and 40. If $3x_1 + 2x_2$ equals 40, then x_3 has the value 0. If, on the other hand, $3x_1 + 2x_2$ is less than 40, x_3 assumes the positive value equal to the difference between $3x_1 + 2x_2$ and 40.

The second inequality is converted to an equation by subtracting the surplus variable x_4. The resulting equation is

$$2x_1 + x_2 - x_4 = 10$$

The surplus variable x_4 represents the difference between the expression $2x_1 + x_2$ and 10. As in the case of slack variables, if $2x_1 + x_2$ equals 10, then the surplus variable x_4 has the value 0. If, however, $2x_1 + x_2$ is greater than 10, then x_4 has a positive value that is equal to the difference between $2x_1 + x_2$ and 10. Since $2x_1 + x_2 \geq 10$, the value of the surplus variable subtracted from the left side of the inequality must be greater than or equal to 0.

Several important characteristics of the new system of equations should be noted. First, both the slack and surplus variables are nonnegative. Second, a slack variable must be added for each "less than or equal to" inequality and a surplus variable subtracted for each "greater than or equal to" inequality. Finally, slack and surplus variables would only coincidently have equivalent values. Consequently, different variables must be used for each inequality.

Example: A manufacturer makes two kinds of bookcases, A and B. Type A requires 2 hours on machine 1 and 4 hours on machine 2. Type B requires 3 hours on machine 1 and 2 hours on machine 2. The machines work no more than 16 hours per day. The profit is $3 per bookcase A and $4 per bookcase B. Formulate the linear programming problem, convert the constraining inequalities to equations, and interpret the meaning of the slack variables.

Let x_1 and x_2 represent the number of bookcases of type A and B, respectively. The linear programming problem is

$$\text{Maximize:} \quad P = 3x_1 + 4x_2$$
$$\text{Subject to:} \quad 2x_1 + 3x_2 \leq 16$$
$$4x_1 + 2x_2 \leq 16$$
$$x_1, x_2 \geq 0$$

The constraining inequalities are converted to equations by adding slack variables. The problem becomes

$$\text{Maximize:} \quad P = 3x_1 + 4x_2$$
$$\text{Subject to:} \quad 2x_1 + 3x_2 + x_3 \qquad = 16$$
$$4x_1 + 2x_2 \qquad + x_4 = 16$$
$$x_1, x_2, x_3, x_4 \geq 0$$

The slack variable x_3 represents unused machine 1 time and the slack variable x_4 represents unused machine 2 time. If both machines are fully utilized, x_3 and x_4 would have values of zero.

Example: American Foods, Inc., is developing a low-calorie, high-protein diet supplement. The linear programming formulation of this problem along with the data is given in Chapter 5, p. 163. Convert the system of inequalities to equations by adding slack and subtracting surplus variables. Interpret the meaning of the slack and surplus variables.

After adding the slack and subtracting the surplus variables, the problem becomes

$$\text{Minimize:} \quad C = 0.15x_1 + 0.20x_2 + 0.12x_3$$
$$\text{Subject to:} \quad 350x_1 + 250x_2 + 200x_3 + x_4 \qquad\qquad\qquad = 300$$
$$250x_1 + 300x_2 + 150x_3 \qquad - x_5 \qquad\qquad = 200$$
$$100x_1 + 150x_2 + 75x_3 \qquad\qquad - x_6 \qquad = 100$$
$$75x_1 + 125x_2 + 150x_3 \qquad\qquad\qquad - x_7 = 100$$
$$x_j \geq 0 \text{ for } j = 1, 2, \ldots, 7$$

The slack variable x_4 represents the difference between the calories in the diet supplement and the maximum number of calories allowed by the specifications. The surplus variables x_5, x_6, and x_7 represent, respectively, the number of units of protein, vitamin A, and vitamin C in excess of that required by the specifications.

6.1.2 BASIC FEASIBLE SOLUTIONS

The addition of slack variables and subtraction of surplus variables enables linear inequalities to be expressed in the form of linear equations. As stated earlier, the reason that this transformation from inequalities to equations is important in the simplex algorithm is that it allows one to solve the resulting systems of equations for extreme points. The objective function in a linear programming problem, according to the extreme point theorem of linear programming, reaches an optimal value at one of the extreme points. Since the extreme points are defined by the intersection of sets of linear equations, the transformation from inequalities to equations is necessary.

To illustrate, consider the linear programming problem

$$\text{Maximize:} \quad P = x_1 + x_2$$
$$\text{Subject to:} \quad 3x_1 + 2x_2 \leq 40$$
$$2x_1 + x_2 \geq 10$$
$$x_1, x_2 \geq 0$$

The system of linear inequalities is transformed to a system of equations by adding a slack variable x_3 and subtracting a surplus variable x_4. This gives the system of linear equations

$$3x_1 + 2x_2 + x_3 = 40$$
$$2x_1 + x_2 - x_4 = 10$$

This system of two equations and four variables does not have a unique solution. In Chapter 2 it is shown that a system of m equations and n variables, where $n > m$, has an infinite number of solutions, provided the system of equations is consistent. Since this system of equations is consistent, it has an infinite number of solutions rather than a unique solution.†

Although the system of equations has an infinite number of solutions, it does not have an infinite number of extreme points. In fact, in this system of two equations and four variables there are only six extreme points. The extreme points can be determined by applying the *basis theorem* of linear programming. The *basis theorem* states that *for a system of m equations and n variables, where n > m, a solution in which at least n − m of the variables have values of zero is an extreme point.* Any solution found by setting $n - m$ of the variables equal to zero and solving the resulting system of m equations for the m remaining variables gives an extreme point. This solution is termed a *basic solution.*

The basic solutions for the preceding system of equations are determined by selecting two of the four variables and equating these variables with zero. The resulting system of two equations and two basic variables is solved simultaneously. For instance, if $x_3 = 0$ and $x_4 = 0$, the system of equations is

$$3x_1 + 2x_2 = 40$$
$$2x_1 + x_2 = 10$$

---

† Consistent systems of equations are discussed in Chapter 2, p. 56 and Chapter 4, p. 135.

and the basic solution is $x_1 = -20$ and $x_2 = 50$. Similarly, if $x_1 = 0$ and $x_2 = 0$, the resulting basic solution is $x_3 = 40$ and $x_4 = -10$. The six basic solutions to the system of equations are shown in Table 6.1.†

Table 6.1

x_1	x_2	x_3	x_4	Objective Function
0	0	40	−10	nonfeasible
0	20	0	10	20
0	10	20	0	10
13.3	0	0	16.6	13.3
5	0	25	0	5
−20	50	0	0	nonfeasible

The extreme point theorem of linear programming can be extended to state that *the objective function is optimal at at least one of the basic solutions.* In the example given in Table 6.1 there are only six extreme points or, alternatively, only six basic solutions. Two of the solutions have negative values for variables and are, therefore, *nonfeasible.* The optimal value of the objective function, therefore, occurs at one of the four *basic feasible solutions* Table 6.1 shows that the maximum value of the objective function occurs when $x_1 = 0$, $x_2 = 20$, $x_3 = 0$, and $x_4 = 10$. The value of the objective function at this basic solution is $P = 20$.

It can be shown that the extreme point and basis theorems of linear programming are related. This relationship is demonstrated through the use of the simple two-variable problem,

$$\begin{aligned} \text{Maximize:} \quad & P = 2x_1 + 3x_2 \\ \text{Subject to:} \quad & 3x_1 + 2x_2 \leq 12 \\ & 2x_1 + 4x_2 \leq 16 \\ & x_1, x_2 \geq 0 \end{aligned}$$

The feasible region for this problem is shown in Fig. 6.1.

The objective function is a maximum at an extreme point. The four extreme points are $E = \{(0, 0), (0, 4), (2, 3), (4, 0)\}$.

This same set of extreme points is generated from the basis theorem. To

---

† The number of basic solutions is given by

$$\frac{n!}{m!(n-m)!}$$

where $n!$, read *n factorial*, equals $n(n-1)(n-2)\ldots 1$. To illustrate the number of basic solutions when $n = 4$ and $m = 2$ is 6.

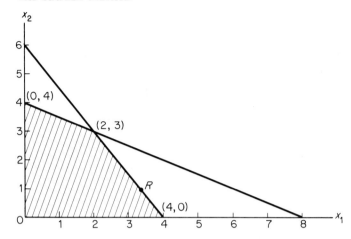

Figure 6.1

apply the basis theorem, the inequalities are first converted to equations by adding the slack variables x_3 and x_4, i.e.,

$$3x_1 + 2x_2 + x_3 = 12$$

$$2x_1 + 4x_2 + x_4 = 16$$

With four variables and two equations, the basic solutions are found by equating two of the four variables to zero. The four basic feasible solutions are given in Table 6.2. These were found by solving the four sets of two simul-

Table 6.2

Solution	x_1	x_2	x_3	x_4
A	0	0	12	16
B	0	4	4	0
C	2	3	0	0
D	4	0	0	8

taneous equations. The basic solutions A, B, C, and D correspond, respectively, to the extreme points $(0, 0)$, $(0, 4)$, $(2, 3)$, and $(4, 0)$.

Nonbasic solutions correspond to points on the boundary of the constraining inequalities. The point R in Fig. 6.1 is an example of a nonbasic solution. The coordinates of R are $x_1 = 3\frac{1}{3}$ and $x_2 = 1$. Substituting these coordinates into the system of equations gives $x_1 = 3\frac{1}{3}$, $x_2 = 1$, $x_3 = 0$, and $x_4 = 5\frac{1}{3}$. This solution is nonbasic, since less than $n - m$ of the variables have values of zero.

This example illustrates the relationship among the basis theorem, the extreme point theorem, and the optimal solution. The extreme point theorem states that the optimal value of the objective function occurs at one of the extreme points of the feasible region. The example shows that basic solutions are extreme points. Consequently, the relationship between the basis theorem, the extreme point theorem, and the optimal solution is that (1) a basic solution occurs for a system of m equations and n variables when at least $n - m$ of the variables have values of zero and (2) the objective function is optimal at one of the basic solutions. We can thus determine the optimal solution by finding all basic solutions and evaluating the objective function at each of these basic solutions.

Example: Determine all basic solutions for the system of equations,

$$2x_1 + x_2 - 2x_3 + x_4 = 300$$
$$3x_1 - 2x_2 + 2x_3 + x_5 = 200$$

Since $m = 2$ equations and $n = 5$ variables, there are

$$\frac{n!}{m!(n-m)!} = \frac{5!}{2!3!} = \frac{5 \cdot 4 \cdot 3 \cdot 2 \cdot 1}{2 \cdot 1 \cdot 3 \cdot 2 \cdot 1} = 10 \text{ basic solutions}$$

The ten basic solutions are found by equating all possible combinations of three of the five variables to zero and solving the resulting systems of two equations and two basic variables. The basic solutions are given in Table 6.3.

Table 6.3

x_1	x_2	x_3	x_4	x_5	Feasible/ Nonfeasible
0	0	0	300	200	feasible
0	0	−150	0	500	nonfeasible
0	0	100	500	0	feasible
0	300	0	0	800	feasible
0	−100	0	400	0	nonfeasible
0	−500	−400	0	0	nonfeasible
150	0	0	0	−250	nonfeasible
66.7	0	0	167.3	0	feasible
100	0	−50	0	0	nonfeasible
114.3	71.4	0	0	0	feasible

The solutions that are feasible are also indicated in the table.

Example: Determine the basic feasible solutions to the following linear programming problem. Show that these solutions correspond to the extreme points in Fig. 5.4.

$$\text{Maximize:} \quad P = 4x_1 + 3x_2$$
$$\text{Subject to:} \quad 21x_1 + 16x_2 \leq 336$$
$$13x_1 + 25x_2 \leq 325$$
$$15x_1 + 18x_2 \leq 270$$
$$x_1, x_2 \geq 0$$

The system of equations is

$$21x_1 + 16x_2 + x_3 \qquad\qquad = 336$$
$$13x_1 + 25x_2 \qquad + x_4 \qquad = 325$$
$$15x_1 + 18x_2 \qquad\qquad + x_5 = 270$$

The basic solutions are given in Table 6.4. The problem was solved graphically in Chapter 5, Fig. 5.4. The extreme points that correspond to the basic feasible solutions are given in the table.

Table 6.4

x_1	x_2	x_3	x_4	x_5	Feasible/ Nonfeasible	Figure 5.4 Extreme Point
0	0	336	325	270	feasible	(0, 0)
0	21	0	−200	−108	nonfeasible	—
0	13	128	0	36	feasible	(0, 13)
0	15	96	−50	0	nonfeasible	—
16	0	0	117	30	feasible	(16, 0)
25	0	−189	0	−105	nonfeasible	—
18	0	−42	91	0	nonfeasible	—
10.08	7.75	0	0	−20.70	nonfeasible	—
12.45	4.57	0	48.90	0	feasible	(12.45, 4.57)
6.38	9.68	47.14	0	0	feasible	(6.38, 9.68)

6.1.3 ALGEBRAIC SOLUTIONS

Before we introduce the simplex algorithm, it will be useful formally to introduce the algebraic solution technique. This technique, although inefficient, shares several characteristics with the simplex algorithm. The most important of these is that it uses the same algebraic procedure for determining basic solutions.

The algebraic solution technique requires the following steps: (1) determine all basic solutions; (2) evaluate each basic feasible solution for optimality. To illustrate, consider the linear programming problem

$$\text{Maximize:} \quad P = 3x_1 + 4x_2$$
$$\text{Subject to:} \quad 2x_1 + 3x_2 \leq 16$$
$$4x_1 + 2x_2 \leq 16$$
$$x_1, x_2 \geq 0$$

The inequalities are converted to equalities by adding slack variables. The linear programming problem becomes

Maximize: $P = 3x_1 + 4x_2 + 0x_3 + 0x_4$
Subject to: $2x_1 + 3x_2 + 1x_3 + 0x_4 = 16$
$4x_1 + 2x_2 + 0x_3 + 1x_4 = 16$
$x_1, x_2, x_3, x_4 \geq 0$

The slack variables are included in the objective function with zero coefficients. The zero coefficients show that unused resource, or slack, does not contribute to profits.

The basic solutions are determined by selecting all possible combinations of two of the four variables and equating these variables with zero. The solutions of the resulting systems of two equations and two variables are basic solutions. For each basic solution, the variables that are equated to zero are termed *not in solution*, or alternatively, *not in the basis*. Conversely, those variables that are not equated to zero are said to be *in solution, in the basis,* or alternatively, *basic variables.*†

As a starting point, assume that the slack variables are in the basis. If the slack variables are in the basis, the variables x_1 and x_2 are equated with zero. Since the coefficients of the slack variables form an identity matrix, the initial basic solution can be read directly from the system of equations. The initial basic solution is $x_1 = 0$, $x_2 = 0$, $x_3 = 16$, and $x_4 = 16$. Since $x_1, x_2, x_3,$ and x_4 are greater than or equal to zero, the solution is feasible. The value of the objective function at this initial basic feasible solution is 0, i.e.,

$$P = 3(0) + 4(0) + 0(16) + 0(16) = 0$$

For a second solution, let x_1 and x_4 be in the basis. According to the basis theorem of linear programming, x_2 and x_3 are not in the basis and are valued at zero. The solution to the system of two equations and two basic variables can be obtained by using the row operations of matrix algebra.‡

The objective is to solve the system of equations for x_1 and x_4. The system could be written as

$$2x_1 + 0x_4 = 16$$

$$4x_1 + 1x_4 = 16$$

Alternatively, the system could be written in matrix form as

$$\begin{pmatrix} 2 & 0 \\ 4 & 1 \end{pmatrix} \begin{pmatrix} x_1 \\ x_4 \end{pmatrix} = \begin{pmatrix} 16 \\ 16 \end{pmatrix}$$

---

† It is possible for the solution value of a basic variable to be zero. When this occurs, the solution is termed *degenerate*. A degenerate solution is treated the same as any other basic solution in the algebraic or simplex algorithms.

‡ Chapter 3, pp. 86–91.

Since x_2 and x_3 have values of zero, both formulations are equivalent to the original system of equations,

$$2x_1 + 3x_2 + 1x_3 + 0x_4 = 16$$

$$4x_1 + 2x_2 + 0x_3 + 1x_4 = 16$$

In solving the system of equations, it is customary to apply row operations to the entire system of equations, rather than to only the coefficients of the basic variables. This has the advantage of reducing the number of arithmetic calculations required in moving from one basic solution to another.

The system of equations can be solved for x_1 and x_4 by using row operations to obtain an identity matrix as the coefficient matrix of x_1 and x_4. Multiplying the first equation by $\frac{1}{2}$ gives

$$1x_1 + \tfrac{3}{2}x_2 + \tfrac{1}{2}x_3 + 0x_4 = 8$$

By multiplying the new equation by -4 and adding this product to the second equation, the second equation becomes

$$0x_1 - 4x_2 - 2x_3 + 1x_4 = -16$$

and the system of equations is

$$1x_1 + \tfrac{3}{2}x_2 + \tfrac{1}{2}x_3 + 0x_4 = 8$$

$$0x_1 - 4x_2 - 2x_3 + 1x_4 = -16$$

The coefficients of the basic variables form an identity matrix. The nonbasic variables are valued at zero, and the basic solution can be read directly as $x_1 = 8$, $x_2 = 0$, $x_3 = 0$, and $x_4 = -16$. This solution is nonfeasible.

To determine a third solution, let x_1 and x_2 be basic variables and x_3 and x_4 be nonbasic variables. The values of x_1 and x_2 can be found by solving the original two equations simultaneously. An alternative approach, used in the simplex method, is to use the equations from the preceding iteration in solving for x_1 and x_2. The equations were

$$1x_1 + \tfrac{3}{2}x_2 + \tfrac{1}{2}x_3 + 0x_4 = 8$$

$$0x_1 - 4x_2 - 2x_3 + 1x_4 = -16$$

Multiplying the second equation by $-\frac{1}{4}$ gives

$$0x_1 + 1x_2 + \tfrac{1}{2}x_3 - \tfrac{1}{4}x_4 = 4$$

Multiplying the new second equation by $\frac{3}{2}$ and subtracting it from the first equation gives

$$1x_1 + 0x_2 - \tfrac{1}{4}x_3 + \tfrac{3}{8}x_4 = 2$$

The system of equations is

$$1x_1 + 0x_2 - \tfrac{1}{4}x_3 + \tfrac{3}{8}x_4 = 2$$

$$0x_1 + 1x_2 + \tfrac{1}{2}x_3 - \tfrac{1}{4}x_4 = 4$$

The coefficients of x_1 and x_2 form an identity matrix and x_3 and x_4 are not in the basis. The solution can be read directly as $x_1 = 2$, $x_2 = 4$, $x_3 = 0$, and $x_4 = 0$. The solution is feasible. The value of the objective function is 22, i.e.,

$$P = 3(2) + 4(4) + 0(0) + 0(0) = 22$$

A fourth solution can be found by assuming that x_1 and x_3 are in the basis and x_2 and x_4 are not. Rather than solving the original system of equations, we use the equations from the preceding iteration. The system of equations

$$1x_1 + 0x_2 - \tfrac{1}{4}x_3 + \tfrac{3}{8}x_4 = 2$$

$$0x_1 + 1x_2 + \tfrac{1}{2}x_3 - \tfrac{1}{4}x_4 = 4$$

with basic variables x_1 and x_3 can be solved by multiplying the second equation by 2 and adding the product of $\tfrac{1}{4}$ the new second equation to the first equation. This results in an identity matrix as the coefficient matrix for the basic variables x_1 and x_3.

$$1x_1 + \tfrac{1}{2}x_2 + 0x_3 + \tfrac{1}{4}x_4 = 4$$

$$0x_1 + 2x_2 + 1x_3 - \tfrac{1}{2}x_4 = 8$$

The solution is $x_1 = 4$, $x_2 = 0$, $x_3 = 8$, $x_4 = 0$. The solution is feasible, and the value of the objective function is $P = 12$.

For a fifth solution, let x_2 and x_3 be the basic variables. The system of equations from the preceding iteration,

$$1x_1 + \tfrac{1}{2}x_2 + 0x_3 + \tfrac{1}{4}x_4 = 4$$

$$0x_1 + 2x_2 + 1x_3 - \tfrac{1}{2}x_4 = 8$$

can be solved for x_2 and x_3 by multiplying the first equation by 2 and subtracting the product of 2 times the new first equation from the second equation. This gives

$$2x_1 + 1x_2 + 0x_3 + \tfrac{1}{2}x_4 = 8$$

$$-4x_1 + 0x_2 + 1x_3 - \tfrac{3}{2}x_4 = -8$$

The solution, $x_1 = 0$, $x_2 = 8$, $x_3 = -8$, $x_4 = 0$, is not feasible.

The final solution occurs when x_2 and x_4 are in the basis. The system of equations

$$2x_1 + 1x_2 + 0x_3 + \tfrac{1}{2}x_4 = 8$$

$$-4x_1 + 0x_2 + 1x_3 - \tfrac{3}{2}x_4 = -8$$

can be solved for x_2 and x_4 by multiplying the second equation by $-\tfrac{2}{3}$ and subtracting the product of $\tfrac{1}{2}$ times this equation from the first equation. This gives

$$\tfrac{2}{3}x_1 + 1x_2 + \tfrac{1}{3}x_3 + 0x_4 = \tfrac{16}{3}$$

$$\tfrac{8}{3}x_1 + 0x_2 - \tfrac{2}{3}x_3 + 1x_4 = \tfrac{16}{3}$$

The solution is $x_1 = 0$, $x_2 = \frac{16}{3}$, $x_3 = 0$, $x_4 = \frac{16}{3}$. The value of the objective function is $P = 16$.

The six solutions to the linear programming problem are summarized in Table 6.5. The optimal value of the objective function, $P = 22$, occurs when $x_1 = 2$, $x_2 = 4$, $x_3 = 0$, and $x_4 = 0$.

Table 6.5

Solution	x_1	x_2	x_3	x_4	Objective Function
1	0	0	16	16	0
2	8	0	0	−16	Nonfeasible
3	2	4	0	0	22
4	4	0	8	0	12
5	0	8	−8	0	Nonfeasible
6	0	$\frac{16}{3}$	0	$\frac{16}{3}$	16

The algebraic and simplex algorithms have several common characteristics. The most important of these is that both algorithms utilize the extreme point theorem of linear programming. This theorem states that an optimal value of the objective functions occurs at one of the basic solutions to the system of equations. Basic solutions to a system of m equations and n variables are found in both algorithms by equating $n - m$ of the variables to zero and solving the resulting system of m equations and m basic variables. Consequently, much of the arithmetic required for the algebraic solution to a linear programming problem is also required for a simplex solution.

6.2 The Simplex Algorithm

The simplex algorithm is an iterative procedure for determining basic feasible solutions to a system of equations and testing each solution for optimality. The algorithm involves moving from one basic feasible solution to another, always maintaining or improving the value of the objective function, until an optimal solution is reached.

Several of the inefficiencies of the algebraic method are eliminated by the simplex algorithm. First, feasible solutions found by using the algebraic technique cannot be distinguished from nonfeasible solutions until after the solution has been determined. In contrast, all solutions found by using the simplex algorithm are feasible. Since only feasible solutions are considered, the computational requirements are reduced by the simplex algorithm. Second, the algebraic technique requires that all basic feasible solutions be evaluated. In comparison, each iteration of the simplex algorithm gives a solution that improves or maintains the value of the objective function. This means that only a subset of the set of extreme points need be evaluated. This results in a further reduction in the computational requirements.

6.2.1 SIMPLEX TABLEAU

The first step in applying the simplex algorithm is to transform the inequalities of the linear programming problem to equations by adding slack and subtracting surplus variables. The coefficients of the objective function and constraining equations along with the right-hand-side values are then transferred to the simplex tableau. As an example, consider the linear programming problem

$$\begin{aligned}
\text{Maximize:} \quad & P = 3x_1 + 4x_2 + 5x_3 + 4x_4 \\
\text{Subject to:} \quad & 2x_1 + 5x_2 + 4x_3 + 3x_4 \leq 224 \\
& 5x_1 + 4x_2 - 5x_3 + 10x_4 \leq 280 \\
& 2x_1 + 4x_2 + 4x_3 - 2x_4 \leq 184 \\
& x_j \geq 0 \quad \text{for } j = 1, 2, 3, 4
\end{aligned}$$

By adding slack variables, the problem becomes

$$\begin{aligned}
\text{Maximize:} \quad & P = 3x_1 + 4x_2 + 5x_3 + 4x_4 + 0x_5 + 0x_6 + 0x_7 \\
\text{Subject to:} \quad & 2x_1 + 5x_2 + 4x_3 + 3x_4 + 1x_5 + 0x_6 + 0x_7 = 224 \\
& 5x_1 + 4x_2 - 5x_3 + 10x_4 + 0x_5 + 1x_6 + 0x_7 = 280 \\
& 2x_1 + 4x_2 + 4x_3 - 2x_4 + 0x_5 + 0x_6 + 1x_7 = 184 \\
& x_j \geq 0 \quad \text{for } j = 1, 2, \ldots, 7
\end{aligned}$$

The coefficients of the objective function and equations along with the right-hand-side values can now be transferred to the simplex tableau. The initial simplex tableau is shown in Table 6.6.

Table 6.6   Initial Simplex Tableau

c_b	c_j Basis	3 x_1	4 x_2	5 x_3	4 x_4	0 x_5	0 x_6	0 x_7	Solution
0	x_5	2	5	4	3	1	0	0	224
0	x_6	5	4	-5	10	0	1	0	280
0	x_7	2	4	4	-2	0	0	1	184
	z_j	0	0	0	0	0	0	0	0
	$c_j - z_j$	3	4	5	4	0	0	0	

The first two rows of the tableau give the coefficients of the objective function and the column headings. The coefficients of the objective function are copied directly from the objective function. *Both slack and surplus variables have zero coefficients in the objective function.*

The coefficients of the constraining equations are shown in the tableau under the appropriate column heading. The coefficients of the variable x_1 in the constraining equations are given in the column labeled x_1, the coefficients of x_2 are given in the column labeled x_2, etc. The right-hand-side values are listed under the column labeled Solution.

The column labeled Basis contains the basic variables, i.e., the variables that are "in the basis." In the initial tableau, the variables x_5, x_6, and x_7 are in the basis. This means that the variables x_1, x_2, x_3, and x_4 are not in the basis and, according to the basis theorem of linear programming, are valued at zero. By referring to the system of linear equations, it can be seen that if x_1, x_2, x_3, and x_4 equal zero, the system reduces to

$$2(0) + 5(0) + 4(0) + 3(0) + 1x_5 + 0x_6 + 0x_7 = 224$$

$$5(0) + 4(0) - 5(0) + 10(0) + 0x_5 + 1x_6 + 0x_7 = 280$$

$$2(0) + 4(0) + 4(0) - 2(0) + 0x_5 + 0x_6 + 1x_7 = 184$$

The solution to this system of three equations and three basic variables can be read from the tableau. The variables x_1, x_2, x_3, and x_4 are not in the basis and have values of zero. Furthermore, the coefficients of the basic variables x_5, x_6, and x_7 form an identity matrix. The values of the variables in the basis can thus be read directly from the Solution column as $x_5 = 224$, $x_6 = 280$, and $x_7 = 184$.

The column labeled c_b contains the objective function coefficients of the basic variables. Since the basic variables in the initial tableau of our example problem are slack variables, the column labeled c_b contains zeros in the initial tableau.

The row labeled z_j contains the sum of the products of the numbers in the c_b column times the coefficients in the x_j column. In the example,

$$z_1 = 0(2) + 0(5) + 0(2) = 0$$

$$z_2 = 0(5) + 0(4) + 0(4) = 0$$

$$z_3 = 0(4) + 0(-5) + 0(4) = 0, \text{etc.}$$

The values of z_j in this example are zero for each of the j columns in the initial tableau. This, of course, is because of the fact that the objective function coefficients of the slack variables in the c_b column are zero. Many of the z_j values will be nonzero after the initial tableau.

The $c_j - z_j$ row is determined by subtracting the z_j value for the jth column from the objective function coefficient for that column. Since the z_j values are zero in the initial tableau, the $c_j - z_j$ values in this example are the same as the values in the c_j row.

The value of the objective function for the initial basic feasible solution is zero. This value is calculated by summing the products of the objective function coefficients of the basic variables and the solution values of the basic variables. This value is shown in the tableau as the last entry in the Solution column.

6.2.2 CHANGE OF BASIS

The value of the objective function can be increased by including one of the nonbasic variables in the basis. In Table 6.6, the nonbasic variables are x_1, x_2, x_3, and x_4. Since the number of variables in the basis remains constant, adding a nonbasic variable means that one of the current basic variables must be removed. Thus, x_5, x_6, or x_7 must leave the basis. The process of adding a variable and, concurrently, removing a variable is termed a *change of basis*.

The $c_j - z_j$ row is used to determine the variable to include in the basis. The numbers in this row show the change in the objective function resulting from including one unit of variable x_j in the basis. In the initial tableau, the $c_j - z_j$ row shows that the objective function will increase by 3 for each unit of x_1, by 4 for each unit of x_2, by 5 for each unit of x_3, etc. Since the objective is to maximize P, the variable that results in the largest per unit increase in the objective function should be placed in the basis. In this example, the objective function increases by 5 for each unit of x_3. Consequently, variable x_3 enters the basis. The rule for deciding the variable that enters the basis is summarized by Simplex Rule 1.

> Simplex Rule 1. *The selection of the variable to enter the basis is based upon the value of $c_j - z_j$. For maximization, the variable selected should have the largest value of $c_j - z_j$. If all values of $c_j - z_j$ are zero or negative, the current basic solution is optimal. For minimization, the variable selected should have the smallest (most negative) value of $c_j - z_j$. If all values of $c_j - z_j$ are zero or positive, the objective function is optimal.*

On the basis of Simplex Rule 1, the variable x_3 is included in the basis.

In certain cases it is possible that there is no single largest $c_j - z_j$ value. For instance, if the $c_j - z_j$ value for x_3 were changed from 5 to 4, the $c_j - z_j$ values for x_2, x_3, and x_4 would be the same. In these cases of ties, it is acceptable arbitrarily to select one of the variables for entry into the basis.

For x_3 to enter the basis, one of the variables currently included in the basis must be removed. Simplex Rule 2 is used to determine this variable.

Simplex Rule 2. *The selection of the variable to leave the basis is made by dividing the numbers in the Solution column by the coefficients in the* x_j *column (i.e., the coefficients of the variable entering the basis). Select the row with the minimum ratio (ignore ratios with zero or negative numbers in the denominator). The variable associated with this row leaves the basis.*†

This rule applies for both maximization and minimization problems. Simplex Rule 2 is illustrated for the example problem in Table 6.7. This table shows the ratios of the Solution column to the coefficients of the variable entering the basis. The row associated with variable x_7 has the minimum ratio, and x_3 therefore replaces x_7 in the basis.

Table 6.7

Basis Variables	Current Solution	÷	Coefficients of Entering Variable		Ratios
x_5	224	÷	4	=	56
x_6	280	÷	-5	=	—
x_7	184	÷	4	=	46

The reason underlying Simplex Rule 2 can be explained by referring to the example problem. This problem was

Maximize: $P = 3x_1 + 4x_2 + 5x_3 + 4x_4$
Subject to: $2x_1 + 5x_2 + 4x_3 + 3x_4 \leq 224$
$5x_1 + 4x_2 - 5x_3 + 10x_4 \leq 280$
$2x_1 + 4x_2 + 4x_3 - 2x_4 \leq 184$
$x_j \geq 0$ for $j = 1, 2, 3, 4$

To illustrate the rule, assume that this is a product mix problem and that the variables represent quantities of product. By applying Simplex Rule 1 to the initial tableau, variable x_3 is selected to enter the basis. The limitation on the maximum amount of x_3 that can be produced comes from the third constraint. That is, $\frac{184}{4}$ or 46 units of product 3 can be produced without exceeding the third constraint, while $\frac{224}{4}$ or 56 units can be produced without exceeding the first constraint. Notice that 46 units of x_3 is a feasible solution, while 56 units is not (i.e., $x_3 = 56$ violates the third constraint). If it is assumed that 46 units of x_3 are produced, the values of the four product variables are $x_1 = 0$, $x_2 = 0$,

---

† In case of ties among the ratios, it is acceptable arbitrarily to select one of the tied variables to leave the basis.

$x_3 = 46$, and $x_4 = 0$. The values of the slack variables are $x_5 = 40$, $x_6 = 510$, and $x_7 = 0$. Since $x_7 = 0$, it is removed from the basis and is replaced by x_3.

The second tableau has the variables x_5, x_6, and x_3 in the basis. The system of equations is

$$2x_1 + 5x_2 + 4x_3 + 3x_4 + 1x_5 + 0x_6 + 0x_7 = 224$$
$$5x_1 + 4x_2 - 5x_3 + 10x_4 + 0x_5 + 1x_6 + 0x_7 = 280$$
$$2x_1 + 4x_2 + 4x_3 - 2x_4 + 0x_5 + 0x_6 + 1x_7 = 184$$

The system of three equations can be solved for the basic variables x_5, x_6, and x_3 by using row operations. Multiply the third row by $\frac{1}{4}$:

$$\tfrac{1}{2}x_1 + 1x_2 + 1x_3 - \tfrac{1}{2}x_4 + 0x_5 + 0x_6 + \tfrac{1}{4}x_7 = 46$$

Next, subtract four times the new third row from the first row and add five times the new third row to the second row. The resulting system of equations is

$$0x_1 + 1x_2 + 0x_3 + 5x_4 + 1x_5 + 0x_6 - 1x_7 = 40$$
$$\tfrac{15}{2}x_1 + 9x_2 + 0x_3 + \tfrac{15}{2}x_4 + 0x_5 + 1x_6 + \tfrac{5}{4}x_7 = 510$$
$$\tfrac{1}{2}x_1 + 1x_2 + 1x_3 - \tfrac{1}{2}x_4 + 0x_5 + 0x_6 + \tfrac{1}{4}x_7 = 46$$

Since the coefficient matrix of the basic variables x_5, x_6, and x_3 is an identity matrix and the nonbasic variables have values of zero, the basic solution can be read directly from the system of equations. The coefficients of this system of equations and the solution values of the basic variables are entered in the second tableau. This tableau is shown in Table 6.8.

Table 6.8 Second Simplex Tableau

c_b	c_j Basis	3 x_1	4 x_2	5 x_3	4 x_4	0 x_5	0 x_6	0 x_7	Solution
0	x_5	0	1	0	⑤	1	0	-1	40
0	x_6	$\frac{15}{2}$	9	0	$\frac{15}{2}$	0	1	$\frac{5}{4}$	510
5	x_3	$\frac{1}{2}$	1	1	$-\frac{1}{2}$	0	0	$\frac{1}{4}$	46
	z_j	$\frac{5}{2}$	5	5	$-\frac{5}{2}$	0	0	$\frac{5}{4}$	230
	$c_j - z_j$	$\frac{1}{2}$	-1	0	$\frac{13}{2}$	0	0	$-\frac{5}{4}$	

The use of row operations to solve the systems of equations for the new basic variables is termed *pivoting*. The *pivot element* is the element at the intersection of the column headed by the variable entering the basis and the row associated with the variable leaving the basis. In the example, the pivot element in the initial tableau (Table 6.6) is the element at the intersection of the column labeled x_3 and the row labeled x_7. The pivot element in the second simplex tableau (Table 6.8) is circled and is at the intersection of the column labeled x_4 and the row labeled x_5.

The pivoting process involves obtaining an identity matrix as the coefficient matrix of the basic variables. Since only one new basic variable is entered in the tableau at any iteration, the identity matrix is completed by using row operations to obtain a one as the pivot element and zeros elsewhere in the *pivot column*. The mechanics of pivoting are straightforward. The first row operation is to multiply the *pivot row* by the reciprocal of the pivot element. In the example, the pivot row in the initial tableau is multiplied by $\frac{1}{4}$. This new row is entered in the second tableau, Table 6.8. The remainder of the row operations are made to obtain zeros in the pivot column. The new pivot row is multiplied by -4 and added to the first equation. The resulting equation is entered in the second tableau. The new pivot row is then multiplied by 5 and added to the second equation. This equation is also entered in the second tableau. These operations result in values of 1 for the pivot element and 0 for the remaining elements in the pivot column.

The pivoting process is equivalent to solving the system of equations simultaneously for the new basic variables. Since $x_1 = 0$, $x_2 = 0$, $x_4 = 0$, and $x_7 = 0$, the system of equations shown in the second tableau can be written as

$$0(x_3) + 1(x_5) + 0(x_6) = 40$$
$$0(x_3) + 0(x_5) + 1(x_6) = 510$$
$$1(x_3) + 0(x_5) + 0(x_6) = 46$$

The values of the variables at the new extreme point can be read directly as $x_1 = 0$, $x_2 = 0$, $x_3 = 46$, $x_4 = 0$, $x_5 = 40$, $x_6 = 510$, and $x_7 = 0$.

The solution in the second tableau is optimal if all entries in the $c_j - z_j$ row are less than or equal to zero. The z_j values in Table 6.8 are calculated by summing the products of the entries in the c_b column times the coefficients in the x_j column. In the second tableau the values are

$$z_1 = 0(0) + 0(\tfrac{15}{2}) + 5(\tfrac{1}{2}) = \tfrac{5}{2}$$
$$z_2 = 0(1) + 0(9) + 5(1) = 5$$
$$z_3 = 0(0) + 0(0) + 5(1) = 5$$
$$z_4 = 0(5) + 0(\tfrac{15}{2}) + 5(-\tfrac{1}{2}) = -\tfrac{5}{2}, \text{ etc.}$$

The $c_j - z_j$ values are found by subtracting the z_j values from c_j. The $c_j - z_j$ values for x_1 and x_4 are positive; consequently, at least one more iteration is required.

The selection of the variable to enter the basis is made by using Simplex

Rule 1. For maximization, the variable selected should be that variable that has the largest value of $c_j - z_j$. Since the $c_j - z_j$ value for x_4 is $\frac{13}{2}$ and the $c_j - z_j$ value for x_1 is $\frac{1}{2}$, variable x_4 is selected to enter the basis.

Simplex Rule 2 is used to determine the variable to be removed from the basis. The ratios of the values in the Solution column to the positive entries in the x_4 column are $\frac{40}{5} = 8$ and $510/(15/2) = 68$. For both maximization and minimization problems, the variable leaving the basis is the variable associated with the row that has the smallest ratio. Consequently, x_5 leaves the basis and is replaced by x_4.

The values of the new basic variables are determined by the pivoting calculations. These calculations again involve completing the identity matrix by obtaining a 1 as the pivot element and zeros elsewhere in the pivot column. The pivot element is circled in Table 6.8. The pivot row is multiplied by the reciprocal of the pivot element and entered in the new tableau. The product of $\frac{1}{5}$ times the pivot row is shown as the x_4 row in Table 6.9.

Table 6.9   Third Simplex Tableau

	c_j	3	4	5	4	0	0	0	
c_b	Basis	x_1	x_2	x_3	x_4	x_5	x_6	x_7	Solution
4	x_4	0	$\frac{1}{5}$	0	1	$\frac{1}{5}$	0	$-\frac{1}{5}$	8
0	x_6	$\left(\frac{15}{2}\right)$	$\frac{15}{2}$	0	0	$-\frac{3}{2}$	1	$\frac{11}{4}$	450
5	x_3	$\frac{1}{2}$	$\frac{1}{10}$	1	0	$\frac{1}{10}$	0	$\frac{3}{20}$	50
	z_j	$\frac{5}{2}$	$\frac{63}{10}$	5	4	$\frac{13}{10}$	0	$-\frac{1}{20}$	282
	$c_j - z_j$	$\frac{1}{2}$	$-\frac{23}{10}$	0	0	$-\frac{13}{10}$	0	$\frac{1}{20}$	

The remaining pivot calculations are made to obtain zeros in the pivot column. The new pivot row is multiplied by $-\frac{15}{2}$ and added to the x_6 row. It is then multiplied by $\frac{1}{2}$ and added to the x_3 row. This results in a 1 as the pivot element and zeros elsewhere in the column. The identity matrix associated with the basic variables is now complete. The values of the basic variables at the new extreme point can be read directly from the Solution column in the table. The solution is $x_1 = 0$, $x_2 = 0$, $x_3 = 50$, $x_4 = 8$, $x_5 = 0$, $x_6 = 450$, and $x_7 = 0$.

The $c_j - z_j$ values in Table 6.9 are positive for variables x_1 and x_7. Since the $c_j - z_j$ value for x_1 is the largest, x_1 is entered in the basis. The variable leaving the basis is determined by calculating the ratios of the numbers in the Solution column to the positive numbers in the x_1 column. The ratio for the

x_6 row is the smallest; consequently, x_6 is removed from the basis and replaced by x_1.

The pivot element is circled in Table 6.9. This element is converted to a 1 by multiplying the pivot row by $\frac{2}{15}$. The new pivot row is then used to obtain zeros elsewhere in the pivot column. The results of the pivoting calculations are shown in Table 6.10.

Table 6.10 Fourth Simplex Tableau

	c_j	3	4	5	4	0	0	0	
c_b	Basis	x_1	x_2	x_3	x_4	x_5	x_6	x_7	Solution
4	x_4	0	$\frac{1}{5}$	0	1	$\frac{1}{5}$	0	$-\frac{1}{5}$	8
3	x_1	1	1	0	0	$-\frac{1}{5}$	$\frac{2}{15}$	$\frac{11}{30}$	60
5	x_3	0	$\frac{3}{5}$	1	0	$\frac{1}{5}$	$-\frac{1}{15}$	$-\frac{1}{30}$	20
	z_j	3	$\frac{34}{5}$	5	4	$\frac{6}{5}$	$\frac{1}{15}$	$\frac{2}{15}$	312
	$c_j - z_j$	0	$-\frac{14}{5}$	0	0	$-\frac{6}{5}$	$-\frac{1}{15}$	$-\frac{2}{15}$	

The $c_j - z_j$ values in Table 6.10 are all zero or negative. This indicates that the solution shown in the tableau is optimal. The solution is $x_1 = 60$, $x_2 = 0$, $x_3 = 20$, $x_4 = 8$, $x_5 = 0$, $x_6 = 0$, and $x_7 = 0$. The value of the objective function is $P = 312$.

6.2.3 OPTIMALITY

According to Simplex Rule 1, the solution to a linear maximization problem is optimal if all values of $c_j - z_j$ are zero or negative. Conversely, the solution to the minimization problem is optimal if all values of $c_j - z_j$ are zero or positive. To understand this rule, consider the example problem just completed.

In the initial tableau of the example problem, Table 6.6, slack variables are in the basis. According to Simplex Rule 1, a change of basis should be made if one or more of the $c_j - z_j$ values are positive. Since the $c_j - z_j$ value for x_3 in the initial tableau is greater than the $c_j - z_j$ values for the other variables, x_3 enters the basis. On the basis of Simplex Rule 2, variable x_7 leaves the basis.

The net contribution from including x_3 in the basis and removing x_7 is given by the $c_j - z_j$ row. The $c_j - z_j$ value for x_3 is 5 in the initial tableau. This means that each unit of x_3 will add 5 to the objective function. Since $x_3 = 0$ before the iteration (Table 6.6), and $x_3 = 46$ after the iteration (Table 6.8), a total of 46 units of x_3 has been included in the current solution. The change in the objective function is equal to the net contribution from including x_3 in the basis multiplied by the number of units in the solution, i.e., 5(46) or 230.

The $c_j - z_j$ values in the second tableau, Table 6.8, are positive for x_1 and x_4. Applying the simplex rules, x_4 is entered in the basis and x_5 removed from the basis. The $c_j - z_j$ value for x_4 shows that the net contribution per unit from x_4 is $\frac{13}{2}$. After the pivoting process, the solution is $x_1 = 0$, $x_2 = 0$, $x_3 = 50$, $x_4 = 8$, $x_5 = 0$, $x_6 = 450$, $x_7 = 0$, and $P = 282$. Notice that the increase in the objective function from $P = 230$ to $P = 282$ is equal to the $c_j - z_j$ value of x_4 in the second tableau multiplied by the solution value of x_4 in the third tableau, that is $\frac{13}{2}(8) = 52$.

The $c_j - z_j$ values for x_1 and x_7 are positive in the third tableau. Since the $c_j - z_j$ value of $\frac{1}{2}$ for x_1 is the largest, x_1 is entered in the basis. The solution for the new basic variables is given in Table 6.10 as $x_1 = 60$, $x_2 = 0$, $x_3 = 20$, $x_4 = 8$, $x_5 = 0$, $x_6 = 0$, $x_7 = 0$, and $P = 312$. The increase in the objective function from $P = 282$ to $P = 312$ is again equal to the $c_j - z_j$ value for x_1 multiplied by the solution value of x_1, that is, $\frac{1}{2}(60) = 30$.

The $c_j - z_j$ values in Table 6.10 are less than or equal to zero. This means that introducing any of the nonbasic variables in this tableau into the basis would lead to a decrease in the value of the objective function. For instance, if x_2 is entered in the basis, the objective function will decrease by $\frac{14}{5}$ times the solution value of x_2. Since the objective is to maximize the objective function, and since any change of the basic variables will decrease the objective function, the current solution is optimal.

Example: A firm uses manufacturing labor and assembly labor to produce three different products. There are 120 hours of manufacturing labor and 260 hours of assembly labor available for scheduling. One unit of product 1 requires 0.10 hr of manufacturing labor and 0.20 hr of assembly labor. Product 2 requires 0.25 hr of manufacturing labor and 0.30 hr of assembly labor for each unit produced. One unit of product 3 requires 0.40 hr of assembly labor but no manufacturing labor. The contribution to profit from products 1, 2, and 3 is $3.00, $4.00, and $5.00, respectively. Formulate the linear programming problem and solve, using the simplex algorithm.

Let x_j, for $j = 1, 2, 3$, represent the number of units of products 1, 2, and 3. The linear programming problem is

$$\begin{aligned}
\text{Maximize:} \quad P = \quad & 3x_1 + \quad 4x_2 + \quad 5x_3 \\
\text{Subject to:} \quad & 0.10x_1 + 0.25x_2 + \quad 0x_3 \le 120 \\
& 0.20x_1 + 0.30x_2 + 0.40x_3 \le 260 \\
& x_1, \, x_2, \, x_3 \ge 0
\end{aligned}$$

The inequalities are converted to equalities by the addition of slack variables. This gives

Maximize: $P =$ $3x_1 +$ $4x_2 +$ $5x_3 + 0x_4 + 0x_5$
Subject to: $0.10x_1 + 0.25x_2 +$ $0x_3 + 1x_4 + 0x_5 = 120$
$0.20x_1 + 0.30x_2 + 0.40x_3 + 0x_4 + 1x_5 = 260$
$x_j \geq 0, \quad j = 1, 2, 3, 4, 5$

The slack variables x_4 and x_5 represent unused manufacturing and assembly labor. Since idle labor hours do not contribute to profit, the coefficients of the slack variables in the objective function are zero.

The initial tableau is shown in Table 6.11. The initial basic feasible solution

Table 6.11

Tableau		c_j	3	4	5	0	0	
	c_b	Basis	x_1	x_2	x_3	x_4	x_5	Solution
Initial	0	x_4	0.10	0.25	0	1	0	120
	0	x_5	0.20	0.30	(0.40)	0	1	260 ←
		z_j	0	0	0	0	0	0
		$c_j - z_j$	3	4	5	0	0	

↑

Second	0	x_4	(0.10)	0.25	0	1	0	120 ←
	5	x_3	0.50	0.75	1	0	2.50	650
		z_j	2.50	3.75	5	0	12.50	3250
		$c_j - z_j$	0.50	0.25	0	0	−12.50	

↑

Third	3	x_1	1	2.50	0	10	0	1200
	5	x_3	0	−0.50	1	−5	2.50	50
		z_j	3	5	5	5	12.50	3850
		$c_j - z_j$	0	−1	0	−5	−12.50	

is $x_1 = 0$, $x_2 = 0$, $x_3 = 0$, $x_4 = 120$, $x_5 = 260$, and $P = 0$. By Simplex Rules 1 and 2, variable x_3 enters the basis and x_5 leaves. The solution for the new basic variables is given in the second tableau as $x_1 = 0$, $x_2 = 0$, $x_3 = 650$, $x_4 = 120$, $x_5 = 0$, and $P = \$3250$. Notice that the change in the objective function from $P = 0$ to $P = 3250$ is equal to the $c_j - z_j$ value for x_3 in the initial tableau times the solution value for x_3 after the iteration, that is, $\$5(650) = \$3250$.

We have shown that the $c_j - z_j$ value of variable x_j gives the net contribution to the objective function for each unit of x_j included in the solution. The reason for this relationship can perhaps be better understood by examining the $c_j - z_j$ values for the variables in the second tableau. The $c_j - z_j$ value for x_1 and x_2 in the second tableau are positive. The $c_j - z_j$ value of $\$0.50$ for x_1 is the larger of the two values; therefore, x_1 enters the basis. Notice in the original problem that x_1 and x_3 use assembly hours in the ratio of 1 to 2. Since all assembly hours are being utilized in the second tableau (i.e., the value of the slack variable in the second tableau for assembly hours is $x_5 = 0$), each unit of x_1 included in the solution in the third tableau reduces the production of x_3 by one-half unit. This, of course, results in a decrease in the objective function of $\$2.50$ for each one-half hour of assembly labor diverted from x_3 to x_1. This effect on the objective function of diverting assembly labor from x_3 to x_1 is shown by z_1. The z_1 value in the second tableau of $\$2.50$ means that the production of each unit of x_1 will reduce the contribution to the objective function from the variables currently in the basis by $\$2.50$. Since, however, the per unit contribution of x_1 is $\$3.00$, the net change in the objective function for each unit of x_1 included in the basis in the third tableau is $\$3.00 - \$2.50$ or $\$0.50$. This value is shown in the $c_j - z_j$ row in the second tableau for x_1.

The final tableau contains basic variables x_1 and x_3. All $c_j - z_j$ values are zero or negative; consequently, the solution is optimal. As expected, the increase in the objective function from $P = \$3250$ to $P = \$3850$ is equal to the $c_j - z_j$ value of $\$0.50$ for x_1 times the solution value of x_1, that is, $\$0.50(1200) = \$600$. The solution to the problem is $x_1 = 1200$, $x_2 = 0$, $x_3 = 50$, $x_4 = 0$, $x_5 = 0$, and $P = \$3850$.

6.2.4 MARGINAL VALUE OF A RESOURCE

The linear programming problem has been described as that of allocating scarce resources among competing products or activities. Since these resources are combined to produce a salable product, the resources have a value to the firm. One measure of this value is termed the *marginal value* of the resource.

The marginal value of a resource is given by the change in the objective function resulting from employing one additional unit of the resource. To illustrate, consider the preceding example. In this example the marginal value of manufacturing labor is equal to the change in the objective function resulting from employing an additional hour of manufacturing labor. Similarly, the marginal value of assembly labor is equal to the change in the objective function resulting from employing one additional hour of assembly labor.

The marginal value of a resource can be determined directly from the $c_j - z_j$ row of the final tableau. In the final tableau (Table 6.11) the $c_j - z_j$ values for x_4 and x_5 are -5 and -12.50, respectively. Including x_4 in the basis would, therefore, decrease the objective function by \$5 for each unit of x_4 in the solution. Similarly, including x_5 in the basis would decrease the objective function by \$12.50 for each unit of x_5. Since x_4 represents unused manufacturing labor hours and x_5 represents unused assembly labor hours, the objective function will decrease by \$5.00 for each hour of manufacturing labor withheld from production and by \$12.50 for each hour of assembly labor withheld from production. Conversely, as long as the basic variables do not change, each additional hour of manufacturing labor included in production results in a \$5.00 increase in the objective function, and each additional hour of assembly labor results in a \$12.50 increase. The marginal value of manufacturing and assembly labor is therefore \$5.00 and \$12.50, respectively.

The example illustrates that the marginal value of an additional unit of resource is determined from the $c_j - z_j$ value in the final tableau. The absolute value of $c_j - z_j$ for the slack variable for the resource gives the marginal value of that resource. If the slack variable for a resource is in the basis (i.e., unused resource is available), the marginal value of an additional unit of the resource as shown by $c_j - z_j$ is zero. If, however, the slack variable is not in the basis (i.e., all resource has been allocated to production), the marginal value of an additional unit of resource is positive.

Example: Determine the marginal value of an additional unit of resources 1, 2, and 3 for the following linear programming problem.

Maximize: $P = \$30x_1 + \$20x_2$
Subject to: $2x_1 + \quad x_2 \leq 280$ (resource 1)
$\qquad\qquad 3x_1 + 2x_2 \leq 500$ (resource 2)
$\qquad\qquad\ \ x_1 + 3x_2 \leq 420$ (resource 3)
$\qquad\qquad x_1, x_2 \geq 0$

This problem is solved by adding slack variables x_3, x_4, and x_5 to convert the inequalities to equations and then applying the simplex algorithm. The final tableau for this problem is shown in Table 6.12.

Since x_3 represents slack for resource 1 and the $c_j - z_j$ value for x_3 is -14, the marginal value of resource 1 is \$14. The marginal value of resource 2 is given by the $c_j - z_j$ entry for x_4. Since 24 units of resource 2 are in the optimal

Table 6.12

c_b	c_j Basis	30 x_1	20 x_2	0 x_3	0 x_4	0 x_5	Solution
30	x_1	1	0	$\frac{3}{5}$	0	$-\frac{1}{5}$	84
0	x_4	0	0	$-\frac{7}{5}$	1	$-\frac{1}{5}$	24
20	x_2	0	1	$-\frac{1}{5}$	0	$\frac{2}{5}$	112
	z_j	30	20	14	0	2	4760
	$c_j - z_j$	0	0	-14	0	-2	

tableau as slack, resource 2 is currently in excess supply and the marginal value of resource 2 is zero. Similarly, the marginal value of resource 3 is $2.

The value of the resources to the firm comes from the fact that they are used to produce a product that is sold for a profit. Since the products are made from the resources, it follows that the value of the resources in generating profit should be equivalent to the profit made from their utilization. This is, in fact, the case. To illustrate, the marginal values of resources 1, 2, and 3 in Table 6.12 are $14, $0, and $2, respectively. The sum of the marginal values of the resources multiplied by the quantity of resource gives the total value of the resources; i.e.,

$$\$14(280) + \$0(500) + \$2(420) = \$4760$$

The value of the resources is equal to the value of the objective function. This result is true at all iterations.

Example: Show that the value of the resources is equal to the value of the objective function at each iteration for the allocation problem given on p. 204.

The tableaus for this problem are given by Table 6.11. The value of the resources in the initial tableau is

$$\$0(120) + \$0(260) = \$0$$

The value of the resources in the second tableau is

$$\$0(120) + \$12.50(260) = \$3250$$

The value of the resources in the final tableau is

$$\$5(120) + \$12.50(260) = \$3850$$

These values equal the objective function at each iteration.

The marginal value of the resource has a number of uses in decision making. One obvious use is in evaluating the employment of additional resources.

In the example, an additional hour of assembly labor is worth $12.50 to the firm in terms of increased profits. If assembly labor can be employed for less than $12.50 per hour, additional laborers should be hired.

Another important use of marginal value occurs in establishing transfer prices for resources between divisions in a firm. Resources in the form of labor, materials, and intermediate products are often transferred from one profit center to another. Knowledge of the marginal value of the resource is, therefore, quite helpful in establishing the transfer price.

6.3 Minimization

The simplex algorithm applies to both maximization and minimization problems. The only difference in the algorithm involves the selection of the variable to enter the basis. In the maximization problem, the variable with the largest $c_j - z_j$ value is included in the basis. Conversely, the variable with the smallest (i.e., most negative) $c_j - z_j$ value is selected to enter the basis in the minimization problem. The selection of the variable to be removed from the basis and the pivoting calculations are the same for both maximization and minimization problems. The solution is optimal in the minimization problem when all $c_j - z_j$ values are zero or positive.

To illustrate minimization, consider the problem

$$\text{Minimize:} \quad C = 20x_1 + 10x_2$$
$$\text{Subject to:} \quad x_1 + 2x_2 \leq 40$$
$$3x_1 + x_2 = 30$$
$$4x_1 + 3x_2 \geq 60$$
$$x_1, x_2 \geq 0$$

To apply the simplex algorithm, the inequalities must be converted to equalities. Adding a slack variable to the first inequality and subtracting a surplus variable from the third inequality, we obtain

$$\text{Minimize:} \quad C = 20x_1 + 10x_2 + 0x_3 + 0x_4$$
$$\text{Subject to:} \quad 1x_1 + 2x_2 + 1x_3 + 0x_4 = 40$$
$$3x_1 + 1x_2 + 0x_3 + 0x_4 = 30$$
$$4x_1 + 3x_2 + 0x_3 - 1x_4 = 60$$
$$x_1, x_2, x_3, x_4 \geq 0$$

The simplex algorithm begins with an initial basic feasible solution.† In the examples in the preceding sections, the initial feasible solution was given by including the slack variables in the basis. This procedure does not give a

---

† The exceptions to this requirement are discussed in Chapter 7.

feasible solution for this particular problem. To illustrate, if x_1 and x_2 are equated to zero, the system of equations reduces to

$$1x_3 + 0x_4 = 40$$
$$0x_3 + 0x_4 = 30$$
$$0x_3 - 1x_4 = 60$$

The second equation obviously is not true, i.e., zero is not equal to thirty. Consequently, the solution is not feasible.

There is another problem in obtaining the initial basic feasible solution that is not as obvious as that of the equality. For the moment, assume that the equality $3x_1 + x_2 = 30$ is omitted from the original problem. On the basis of this assumption, the problem reduces to

Minimize: $C = 20x_1 + 10x_2$
Subject to: $x_1 + 2x_2 \leq 40$
$ 4x_1 + 3x_2 \geq 60$
$ x_1, x_2 \geq 0$

Adding the slack variable and subtracting the surplus variable, we now have the system of equations

$$1x_1 + 2x_2 + 1x_3 + 0x_4 = 40$$
$$4x_1 + 3x_2 + 0x_3 - 1x_4 = 60$$
$$x_1, x_2, x_3, x_4 \geq 0$$

Designating x_3 and x_4 as basic variables in the initial tableau gives the solution $x_1 = 0$, $x_2 = 0$, $x_3 = 40$, and $x_4 = -60$. This solution violates the requirement in the simplex algorithm of $x_j \geq 0$. Thus, even after the original equality is omitted, the solution obtained by including the slack and surplus variables in the basis is not feasible.

6.3.1 ARTIFICIAL VARIABLES

The simplex algorithm requires an initial basic feasible solution. As illustrated by the preceding example, it is not always possible to obtain this initial basic feasible solution by merely adding a slack variable to each "less than or equal to" inequality and subtracting a surplus variable from each "greater than or equal to" inequality. In these cases the problem must be modified by adding *artificial variables*. The initial basic feasible solution to this modified problem is then used as a starting point for applying the simplex algorithm to the original problem.

An artificial variable has no physical interpretation. It is merely a "dummy" variable that is added to constraining equations or inequalities for the purpose of generating an initial basic feasible solution.

The minimization problem discussed above provides an example of the use of artificial variables. The slack and surplus variables in this problem do not provide an initial basic feasible solution. To apply the simplex algorithm, the problem must be modified by adding artificial variables to the second and third constraints. After adding artificial variables, the system of equations is

$$1x_1 + 2x_2 + 1x_3 + 0x_4 + 0A_1 + 0A_2 = 40$$
$$3x_1 + 1x_2 + 0x_3 + 0x_4 + 1A_1 + 0A_2 = 30$$
$$4x_1 + 3x_2 + 0x_3 - 1x_4 + 0A_1 + 1A_2 = 60$$
$$x_1, x_2, x_3, x_4, A_1, A_2 \geq 0$$

This system of equations differs from the original system in that it includes the artificial variables A_1 and A_2. Solutions that include A_1 and A_2 at positive values have no meaning in the linear programming problem. Consequently, the artificial variables cannot have positive values in the final tableau.

The artificial variables merely provide a convenient vehicle for generating an initial basic feasible solution. This solution, obtained by including x_3, A_1, and A_2 in the initial basis, is $x_1 = 0$, $x_2 = 0$, $x_3 = 40$, $x_4 = 0$, $A_1 = 30$, and $A_2 = 60$.

6.3.2 THE "BIG M" METHOD

The basic feasible solution that includes artificial variables is used as a starting point for applying the simplex algorithm. The problem is to minimize the objective function subject to the three constraints. This is accomplished provided the artificial variables have values of zero in the final tableau. If those variables have values of zero, they will have provided a starting point for the simplex algorithm while not affecting the optimal solution.

A simple method is available for assuring that the artificial variables have values of zero in the final tableau. The method is to make the coefficients of the artificial variables in the objective function in a minimization problem extremely large and, conversely, to make the coefficients of the artificial variables in the objective function in a maximization problem extremely small (i.e., large negative numbers). This is analogous in a problem involving minimizing cost to making the artificial variable extremely expensive to produce. Alternatively, in a problem involving maximizing profit, the large negative coefficient has the effect of making the artificial variable extremely costly to produce. Since there are no constraints requiring production of the artificial variable, the simplex algorithm will assure that the artificial variable is not in the basis in the final tableau.

Rather than assigning some arbitrarily large positive or negative numbers as coefficients of the artificial variables in the objective function, it is customary to use the capital letter M. If we adopt this convention, the linear minimization problem becomes

$$
\begin{aligned}
\text{Minimize:} \quad & C = 20x_1 + 10x_2 + 0x_3 + 0x_4 + MA_1 + MA_2 \\
\text{Subject to:} \quad & 1x_1 + 2x_2 + 1x_3 + 0x_4 + 0A_1 + 0A_2 = 40 \\
& 3x_1 + 1x_2 + 0x_3 + 0x_4 + 1A_1 + 0A_2 = 30 \\
& 4x_1 + 3x_2 + 0x_3 - 1x_4 + 0A_1 + 1A_2 = 60 \\
& x_1, x_2, x_3, x_4, A_1, A_2 \geq 0
\end{aligned}
$$

The tableaus for this problem are given in Table 6.13. Since this is a minimization problem, the variable with the smallest $c_j - z_j$ value is included in the basis. Notice in the initial tableau that $20 - 7M$ is smaller than either $10 - 4M$ or M. Thus, variable x_1 enters the basis. From Simplex Rule 2, variable A_1 leaves the basis. Row operations are then performed to determine the solution values of the variables in the second tableau. The iteration process continues until entries in the $c_j - z_j$ row are zero or positive. The optimal solution is shown in the final tableau as $x_1 = 6$, $x_2 = 12$, $x_3 = 10$, $x_4 = 0$, and $C = 240$.

The linear minimization problem illustrates the types of constraints for which artificial variables are used. These constraints are of the forms $ax_1 + bx_2 \geq c$ and $ax_1 + bx_2 = c$. One surplus and one artificial variable are needed for each greater than or equal to constraint, whereas the equalities require only the artificial variable.

Artificial variables are used to obtain an initial basic feasible solution in both maximization and minimization problems. With the exception that large negative rather than positive numbers are used as the coefficients of the artificial variables in the objective function of a maximization problem, the solution procedure for both maximization and minimization problems is the same. This is illustrated by the following example.

Example:

$$
\begin{aligned}
\text{Maximize:} \quad & P = x_1 + 2x_2 + 4x_3 \\
\text{Subject to:} \quad & x_1 + x_2 + x_3 \leq 12 \\
& 2x_1 - x_2 + x_3 \geq 8 \\
& x_1, x_2, x_3 \geq 0
\end{aligned}
$$

A slack variable is required for the first inequality, while both a surplus and an artificial variable are required for the second inequality. Adding the slack, surplus, and artificial variables gives

Maximize: $P = 1x_1 + 2x_2 + 4x_3 + 0x_4 + 0x_5 - MA_1$
Subject to: $1x_1 + 1x_2 + 1x_3 + 1x_4 + 0x_5 + 0A_1 = 12$
$2x_1 - 1x_2 + 1x_3 + 0x_4 - 1x_5 + 1A_1 = 8$
$x_1, x_2, x_3, x_4, x_5, A_1 \geq 0$

The tableaus for the problem are shown in Table 6.14.

An interesting feature of this example is that a variable enters the basis in one of the intermediate tableaus but is not in the basis in the optimal solution. Variable x_1 enters the basis in the second tableau and leaves in the third.

Table 6.13

Tableau	c_b	c_j Basis	20 x_1	10 x_2	0 x_3	0 x_4	M A_1	M A_2	Solution
Initial	0	x_3	1	2	1	0	0	0	40
	M	A_1	③	1	0	0	1	0	30 ←
	M	A_2	4	3	0	-1	0	1	60
		z_j	$7M$	$4M$	0	$-M$	M	M	$90M$
		$c_j - z_j$	$20 - 7M$	$10 - 4M$	0	M	0	0	
			↑						
Second	0	x_3	0	$\frac{5}{3}$	1	0	$-\frac{1}{3}$	0	30
	20	x_1	1	$\frac{1}{3}$	0	0	$\frac{1}{3}$	0	10
	M	A_2	0	$⑤\over 3$	0	-1	$-\frac{4}{3}$	1	20 ←
		z_j	20	$\frac{20}{3} + \frac{5}{3}M$	0	$-M$	$\frac{20}{3} - \frac{4}{3}M$	M	$200 + 20M$
		$c_j - z_j$	0	$\frac{10}{3} - \frac{5}{3}M$	0	M	$\frac{7}{3}M - \frac{20}{3}$	0	
				↑					
Third	0	x_3	0	0	1	1	1	-1	10
	20	x_1	1	0	0	$\frac{1}{5}$	$\frac{3}{5}$	$-\frac{1}{5}$	6
	10	x_2	0	1	0	$-\frac{3}{5}$	$-\frac{4}{5}$	$\frac{3}{5}$	12
		z_j	20	10	0	-2	4	2	240
		$c_j - z_j$	0	0	0	2	$M - 4$	$M - 2$	

Table 6.14

Tableau		c_j	1	2	4	0	0	$-M$	
	c_b	Basis	x_1	x_2	x_3	x_4	x_5	A_1	Solution
Initial	0	x_4	1	1	1	1	0	0	12
	$-M$	A_1	②	-1	1	0	-1	1	8 $\leftarrow$
		z_j	$-2M$	$+M$	$-M$	0	$+M$	$-M$	$-8M$
		$c_j - z_j$	$1 + 2M$	$2 - M$	$4 + M$	0	$-M$	0	

$\uparrow$

Second	0	x_4	0	$\frac{3}{2}$	$\frac{1}{2}$	1	$\frac{1}{2}$	$-\frac{1}{2}$	8
	1	x_1	1	$-\frac{1}{2}$	②	0	$-\frac{1}{2}$	$\frac{1}{2}$	4 $\leftarrow$
		z_j	1	$-\frac{1}{2}$	$\frac{1}{2}$	0	$-\frac{1}{2}$	$\frac{1}{2}$	4
		$c_j - z_j$	0	$\frac{5}{2}$	$\frac{7}{2}$	0	$\frac{1}{2}$	$-M - \frac{1}{2}$	

$\uparrow$

Third	0	x_4	-1	②	0	1	1	-1	4 $\leftarrow$
	4	x_3	2	-1	1	0	-1	1	8
		z_j	8	-4	4	0	-4	4	32
		$c_j - z_j$	-7	6	0	0	4	$-M - 4$	

$\uparrow$

Fourth	2	x_2	$-\frac{1}{2}$	1	0	$\frac{1}{2}$	②	$-\frac{1}{2}$	2 $\leftarrow$
	4	x_3	$\frac{3}{2}$	0	1	$\frac{1}{2}$	$-\frac{1}{2}$	$\frac{1}{2}$	10
		z_j	5	2	4	3	-1	1	44
		$c_j - z_j$	-4	0	0	-3	1	$-M - 1$	

$\uparrow$

0	x_5	−1	2	0	1	1	−1	4
4	x_3	1	1	1	1	0	0	12
	z_j	4	4	4	4	0	0	48
	$c_j - z_j$	−3	−2	0	−4	0	−M	

Fifth

Similarly, x_2 enters the basis in the fourth tableau and is replaced by x_5 in the fifth. Although not illustrated by this example, it is also possible for a variable to enter and leave the basis and then reenter in a later iteration.

6.4 Special Cases

The graphical solution procedure was used in Chapter 5 to define and illustrate the cases of no feasible solution, multiple optimal solutions, and unbounded solutions. The relationship between these cases and the simplex algorithm is shown in this section. The section also introduces a method to allow variables that are not restricted to zero or positive values, i.e., variables unrestricted in sign.

6.4.1 NO FEASIBLE SOLUTION

Realistic linear programming applications often have more than one hundred variables and constraints. In problems of this size, it is impossible to tell by graphing if the problem has feasible solutions. Instead, feasibility must be determined from the simplex tableau.

To illustrate, consider the problem

$$\text{Maximize:} \quad P = x_1 + 2x_2$$
$$\text{Subject to:} \quad x_1 + x_2 \leq 4$$
$$x_1 + x_2 \geq 6$$
$$x_1, x_2 \geq 0$$

It can be seen that the constraints are mutually exclusive and, consequently, there can be no feasible solution. This conclusion can also be made from the simplex tableaus for this problem. These are shown in Table 6.15.

The simplex algorithm is applied to obtain the tableaus in Table 6.15. The elements in the $c_j - z_j$ row of the second tableau are all zero or negative, indicating for the maximization problem that the solution is optimal. The solution, however, contains an artificial variable. This indicates that the

Table 6.15

Tableau		c_j	1	2	0	0	$-M$	
	c_b	Basis	x_1	x_2	x_3	x_4	A_1	Solution
Initial	0	x_3	1	①	1	0	0	4	←
	$-M$	A_1	1	1	0	-1	1	6
		z_j	$-M$	$-M$	0	M	$-M$	$-6M$
		$c_j - z_j$	$1+M$	$2+M$	0	$-M$	0	

↑

Second	2	x_2	1	1		0	0	4
	$-M$	A_1	0	0	-1	-1	1	2
		z_j	2	2	$2+M$	M	$-M$	$8-2M$
		$c_j - z_j$	-1	0	$-2-M$	$-M$	0	

solution shown in the final tableau is not a feasible solution to the original linear programming problem.

The example illustrates the characteristic form of linear programming problems that have no feasible solution. This characteristic is that one or more artificial variables are in the basis at a nonzero level in the final tableau. In the example the variable A_1 was in the basis and had the value $A_1 = 2$. Since the $c_j - z_j$ values are all zero or negative, the solution is optimal but not feasible.

6.4.2 MULTIPLE OPTIMAL SOLUTIONS

The existence of multiple optimal solutions to a linear programming problem is determined from the $c_j - z_j$ row of the final tableau. The values of $c_j - z_j$ give the net change in the objective function from including one unit of variable x_j in the basis. As shown by previous examples, the $c_j - z_j$ values for the basic variables in the final tableau are zero.

Multiple optimal solutions to the linear programming problem *exist if the $c_j - z_j$ value for one of the nonbasic variables is zero.* A $c_j - z_j$ value of zero for a nonbasic variable means that the variable can be included in the basis without changing the value of the objective function. If a nonbasic variable can be entered in the basis without changing the value of the objective function, the solution given by including the new variable in the basis is also optimal.

This is illustrated by the problem

$$\text{Maximize:} \quad P = \quad 3x_1 + \quad 5x_2 + \quad 5x_3$$
$$\text{Subject to:} \quad 0.10x_1 + 0.25x_2 + \quad 0x_3 \le 120$$
$$0.20x_1 + 0.30x_2 + 0.40x_3 \le 260$$
$$x_1, x_2, x_3 \ge 0$$

The solution is given in Table 6.16. The $c_j - z_j$ row of the final tableau shows that variable x_1 can enter the basis without changing the value of the objective

Table 6.16

Tableau	c_b	c_j	3	5	5	0	0	
		Basis	x_1	x_2	x_3	x_4	x_5	Solution
Initial	0	x_4	0.10	0.25	0	1	0	120
	0	x_5	0.20	0.30	0.40	0	1	260 ←
		z_j	0	0	0	0	0	0
		$c_j - z_j$	3	5	5	0	0	
					↑			
Second	0	x_4	0.10	0.25	0	1	0	120 ←
	5	x_3	0.50	0.75	1	0	2.50	650
		z_j	2.50	3.75	5	0	12.50	3250
		$c_j - z_j$	0.50	1.25	0	0	−12.50	
				↑				
Third	5	x_2	0.40	1	0	4	0	480
	5	x_3	0.20	0	1	−3	2.50	290
		z_j	3	5	5	5	12.50	3850
		$c_j - z_j$	0	0	0	−5	−12.50	

function. To demonstrate, the tableau that includes variable x_1 in the basis is given by Table 6.17. The value of the objective function is unchanged from the tableau in Table 6.16.

Table 6.17

c_b	Basis	c_j	3	5	5	0	0	
			x_1	x_2	x_3	x_4	x_5	Solution
3	x_1		1	2.50	0	10	0	1200
5	x_3		0	−0.50	1	−5	2.50	50
	z_j		3	5	5	5	12.50	3850
	$c_j - z_j$		0	0	0	−5	−12.50	

The graphical analysis of multiple optimal solutions showed that if more than one optimal solution exists, then an infinite number of optimal solutions exist. These solutions are given by forming a linear combination of the basic solutions.† In the example the basic solutions were $x_1 = 0$, $x_2 = 480$, $x_3 = 290$, $x_4 = 0$, $x_5 = 0$ and $x_1 = 1200$, $x_2 = 0$, $x_3 = 50$, $x_4 = 0$, $x_5 = 0$. The linear combination of these solutions is

$$x_1 = b(0) + (1 - b)1200$$
$$x_2 = b(480) + (1 - b)0$$
$$x_3 = b(290) + (1 - b)50$$
$$x_4 = b(0) + (1 - b)0$$
$$x_5 = b(0) + (1 - b)0$$

where b is a weighting factor with domain $0 \le b \le 1$. If, for instance, $b = 0.4$, the solution is $x_1 = 720$, $x_2 = 192$, $x_3 = 146$, $x_4 = 0$, and $x_5 = 0$. This solution satisfies the original constraints and has the optimal objective function value of $P = 3850$.

6.4.3 UNBOUNDED SOLUTIONS

The concept of an unbounded solution was introduced in Chapter 5. The problem used to illustrate the concept was

---

† In large-scale applications it is possible to have more than two optimal basic solutions.

$$\text{Maximize:} \quad P = \quad x_1 + 2x_2$$
$$\text{Subject to:} \quad -x_1 + \quad x_2 \leq 2$$
$$x_1 + \quad x_2 \geq 4$$
$$x_1, x_2 \geq 0$$

The effect of an unbounded solution on the simplex algorithm is shown in Table 6.18. The final tableau of the table shows that variable x_4 should enter the basis. The elements in the x_4 column of this tableau are, however, negative. According to Simplex Rule 2, negative elements are ignored in forming the ratios used in selecting the variable to leave the basis. Since neither of the

Table 6.18

c_b	Basis	c_j				$-M$	
		1	2	0	0		
		x_1	x_2	x_3	x_4	A_1	Solution
0	x_3	-1	①	1	0	0	2 ←
$-M$	A_1	1	1	0	-1	1	4
	z_j	$-M$	$-M$	0	M	$-M$	$-4M$
	$c_j - z_j$	$1 + M$	$2 + M$	0	$-M$	0	

↑

c_b	Basis	x_1	x_2	x_3	x_4	A_1	Solution
2	x_2	-1	1	1	0	0	2
$-M$	A_1	②	0	-1	-1	1	2 ←
	z_j	$-2 - 2M$	2	$2 + M$	M	$-M$	$4 - 2M$
	$c_j - z_j$	$3 + 2M$	0	$-2 - M$	$-M$	0	

↑

c_b	Basis	x_1	x_2	x_3	x_4	A_1	Solution
2	x_2	0	1	$\frac{1}{2}$	$-\frac{1}{2}$	$\frac{1}{2}$	3
1	x_1	1	0	$-\frac{1}{2}$	$-\frac{1}{2}$	$\frac{1}{2}$	1
	z_j	1	2	$\frac{1}{2}$	$-\frac{3}{2}$	$\frac{3}{2}$	7
	$c_j - z_j$	0	0	$-\frac{1}{2}$	$\frac{3}{2}$	$-M - \frac{3}{2}$	

↑

basic variables in the final tableau can leave the basis, the change of basis required by Simplex Rule 1 cannot be made.

The condition illustrated by this example occurs for linear programming problems with unbounded solutions. On the basis of Simplex Rule 1, the final solution in Table 6.18 is not optimal and a change of basis is required. The change of basis cannot be made, however, because all the elements in the column headed by the entering variable are zero or negative. If there are no positive elements in this column, the optimal solution is unbounded.

6.4.4 UNRESTRICTED VARIABLES

In certain special applications of linear programming it is possible for variables to have a negative value. A negative-valued variable may, for instance, represent an increase in the inventory of a product. Alternatively, a negative variable could represent returns from a retail outlet to the factory. Variables that are permitted to assume negative values are termed *unrestricted* or *free variables*.

Several techniques are available for incorporating unrestricted variables in the simplex algorithm. One of the widely used methods involves replacing the unrestricted variable by two nonnegative variables. The unrestricted variable x_j is replaced by $x_j = x_j' - x_j''$ and the simplex method is then used without alteration.

To illustrate this technique, assume that the variable x_1 is unrestricted in sign in the linear programming problem

$$\text{Maximize:} \quad P = 3x_1 + 5x_2$$
$$\text{Subject to:} \quad 2x_1 + 5x_2 \leq 132$$
$$3x_1 + 2x_2 \leq 100$$
$$x_2 \geq 0, \ x_1 \text{ unrestricted in sign}$$

The variable x_1 is replaced by $x_1 = x_1' - x_1''$. This gives

$$\text{Maximize:} \quad P = 3x_1' - 3x_1'' + 5x_2$$
$$\text{Subject to:} \quad 2x_1' - 2x_1'' + 5x_2 \leq 132$$
$$3x_1' - 3x_1'' + 2x_2 \leq 100$$
$$x_1', x_1'', x_2 \geq 0$$

The simplex tableaus for this problem are shown in Table 6.19.

The optimal solution is $x_1' = 21.44$, $x_1'' = 0$, and $x_2 = 17.82$. Since $x_1 = x_1' - x_1''$, the solution reduces to $x_1 = 21.44$ and $x_2 = 17.82$.

6.5 Matrix Representation

The linear programming problem can be stated in matrix form. The linear programming problem

Maximize: $P = c_1x_1 + c_2x_2 + c_3x_3$

Subject to:

$$a_{11}x_1 + a_{12}x_2 + a_{13}x_3 \leq r_1$$
$$a_{21}x_1 + a_{22}x_2 + a_{23}x_3 \leq r_2$$
$$a_{31}x_1 + a_{32}x_2 + a_{33}x_3 \leq r_3$$
$$x_1, x_2, x_3 \geq 0$$

can be represented in matrix form as

Maximize: CX

Subject to: $AX \leq R$

$$X \geq 0$$

Table 6.19

c_b	c_j Basis	3 x_1'	-3 x_1''	5 x_2	0 x_3	0 x_4	Solution
0	x_3	2	-2	⑤	1	0	132 ←
0	x_4	3	-3	2	0	1	100
	z_j	0	0	0	0	0	0
	$c_j - z_j$	3	-3	5	0	0	

↑

c_b	Basis	x_1'	x_1''	x_2	x_3	x_4	Solution
5	x_2	0.40	-0.40	1	0.20	0	26.40
0	x_4	②.20	-2.20	0	-0.40	1	47.20 ←
	z_j	2	-2	5	1	0	132
	$c_j - z_j$	1	-1	0	-1	0	

↑

c_b	Basis	x_1'	x_1''	x_2	x_3	x_4	Solution
5	x_2	0	0	1	0.273	-0.182	17.82
3	x_1'	1	-1	0	-0.182	0.454	21.44
	z_j	3	-3	5	0.816	0.404	153.42
	$c_j - z_j$	0	0	0	-0.816	-0.404	

where C is the row vector of objective function coefficients, X is the column vector of variables, A is the matrix of constraint coefficients, and R is the column vector of right-hand-side values.

To illustrate matrix representation of a linear programming problem, consider the following problem.

$$\text{Maximize:} \quad P = 3x_1 + 4x_2 + 5x_3 + 4x_4$$
$$\text{Subject to:} \quad 2x_1 + 5x_2 + 4x_3 + 3x_4 \leq 224$$
$$5x_1 + 4x_2 - 5x_3 + 10x_4 \leq 280$$
$$2x_1 + 4x_2 + 4x_3 - 2x_4 \leq 184$$
$$x_j \geq 0 \quad \text{for } j = 1, 2, 3, 4$$

The initial tableau for this problem was given in Table 6.6 and is repeated in Table 6.20. The matrix format for this tableau is given in Table 6.21.

Table 6.20   Tableau Representation of Linear Programming Problem

c_b	Basis	c_j							
		3	4	5	4	0	0	0	
		x_1	x_2	x_3	x_4	x_5	x_6	x_7	Solution
0	x_5	2	5	4	3	1	0	0	224
0	x_6	5	4	-5	10	0	1	0	280
0	x_7	2	4	4	-2	0	0	1	184
	z_j	0	0	0	0	0	0	0	0
	$c_j - z_j$	3	4	5	4	0	0	0	

Table 6.21   Matrix Representation of Simplex Tableau

c_b	X_b	C		R
		A	I	R
	z_j	0		
	$c_j - z_j$	C		

The matrix I in Table 6.21 is an m by m identity matrix and 0 is a row vector of zeros. A is, of course, the matrix of constraint coefficients, C is the row vector of objective function coefficients, X_b is the column vector of basic variables, and C_b the column vector of objective function coefficients of the basic variables.

The intermediate and final simplex tableaus can also be represented by matrices. Following any iteration of the simplex algorithm, the simplex tableau is given by Table 6.22. This table shows that the solution vector for the basic variables is $B^{-1}R$, the value of the objective function is $C_b{}^tB^{-1}R$, etc.

Table 6.22

	Basis	C		Solution
C_b	X_b	$B^{-1}A$	B^{-1}	$B^{-1}R$
	z_j	$C_b{}^tB^{-1}A$	$C_b{}^tB^{-1}$	$C_b{}^tB^{-1}R$
	$c_j - z_j$	$C - C_b{}^tB^{-1}A$	$- C_b{}^tB^{-1}$	

The matrices A, R, C_b, and C were defined earlier. The matrix B^{-1} has, however, not yet been mentioned. B^{-1} is the inverse of the matrix of constraining coefficients of the current basic variables. The matrix of constraining coefficients of the current basic variables, B, is an m by m matrix. The first column of B contains the coefficients from the initial constraints of the basic variable shown in the first row of the simplex tableau. The second column of B contains the coefficients from the initial constraints of the basic variable shown in the second row of the simplex tableau, etc. The inverse of the matrix B occupies the position of the matrix I in the original tableau. Since the basic variables change at each iteration, the components of B differ at each iteration of the simplex algorithm. This, of course, also means that B^{-1} changes at each iteration.

The relationships shown in Table 6.22 can be illustrated by using the simplex tableaus for the example problem. The problem was

$$\text{Maximize:} \quad P = 3x_1 + 4x_2 + 5x_3 + 4x_4$$
$$\text{Subject to:} \quad 2x_1 + 5x_2 + 4x_3 + 3x_4 \leq 224$$
$$5x_1 + 4x_2 - 5x_3 + 10x_4 \leq 280$$
$$2x_1 + 4x_2 + 4x_3 - 2x_4 \leq 184$$
$$x_j \geq 0 \quad \text{for } j = 1, 2, 3, 4$$

The simplex tableaus for this problem were given earlier. The optimal tableau was given in Table 6.10 and is repeated below in Table 6.23.

The basic variables in the final tableau are x_4, x_1, and x_3. The matrix B contains the column vectors from the original tableau corresponding to these basic variables, i.e.,

$$\begin{matrix} & x_4 & x_1 & x_3 \\ B = & \begin{pmatrix} 3 & 2 & 4 \\ 10 & 5 & -5 \\ -2 & 2 & 4 \end{pmatrix} \end{matrix}$$

Table 6.23

c_j		3	4	5	4	0	0	0	
c_b	Basis	x_1	x_2	x_3	x_4	x_5	x_6	x_7	Solution
4	x_4	0	$\frac{1}{5}$	0	1	$\frac{1}{5}$	0	$-\frac{1}{5}$	8
3	x_1	1	1	0	0	$-\frac{1}{5}$	$\frac{2}{15}$	$\frac{11}{30}$	60
5	x_3	0	$\frac{3}{5}$	1	0	$\frac{1}{5}$	$-\frac{1}{15}$	$-\frac{1}{30}$	20
	z_j	3	$\frac{34}{5}$	5	4	$\frac{6}{5}$	$\frac{1}{15}$	$\frac{2}{15}$	312
	$c_j - z_j$	0	$-\frac{14}{5}$	0	0	$-\frac{6}{5}$	$-\frac{1}{15}$	$-\frac{2}{15}$	

The inverse of B is located in Table 6.23 in the position occupied by the identity matrix in the initial tableau (Table 6.20). Using the methods introduced in Chapter 3, we can verify that

$$B^{-1} = \begin{pmatrix} \frac{1}{5} & 0 & -\frac{1}{5} \\ -\frac{1}{5} & \frac{2}{15} & \frac{11}{30} \\ \frac{1}{5} & -\frac{1}{15} & -\frac{1}{30} \end{pmatrix}$$

The solution values for the basic variables are given by $X_b = B^{-1}R$. Thus,

$$X_b = \begin{pmatrix} \frac{1}{5} & 0 & -\frac{1}{5} \\ -\frac{1}{5} & \frac{2}{15} & \frac{11}{30} \\ \frac{1}{5} & -\frac{1}{15} & -\frac{1}{30} \end{pmatrix} \begin{pmatrix} 224 \\ 280 \\ 184 \end{pmatrix} = \begin{pmatrix} 8 \\ 60 \\ 20 \end{pmatrix}$$

The value of the objective function is given by $P = C_b{}^t B^{-1} R$

$$P = (4, 3, 5) \begin{pmatrix} \frac{1}{5} & 0 & -\frac{1}{5} \\ -\frac{1}{5} & \frac{2}{15} & \frac{11}{30} \\ \frac{1}{5} & -\frac{1}{15} & -\frac{1}{30} \end{pmatrix} \begin{pmatrix} 224 \\ 280 \\ 184 \end{pmatrix} = 312$$

The columns of the tableau are given by $B^{-1}A$,

$$B^{-1}A = \begin{pmatrix} \frac{1}{5} & 0 & -\frac{1}{5} \\ -\frac{1}{5} & \frac{2}{15} & \frac{11}{30} \\ \frac{1}{5} & -\frac{1}{15} & -\frac{1}{30} \end{pmatrix} \begin{pmatrix} 2 & 5 & 4 & 3 \\ 5 & 4 & -5 & 10 \\ 2 & 4 & 4 & -2 \end{pmatrix} = \begin{pmatrix} 0 & \frac{1}{5} & 0 & 1 \\ 1 & 1 & 0 & 0 \\ 0 & \frac{3}{5} & 1 & 0 \end{pmatrix}$$

Finally, the z_j values are given by $C_b{}^t B^{-1} A$ and $C_b{}^t B^{-1}$ and the $c_j - z_j$ values are found by subtracting the z_j row from the c_j row. It is left as an exercise for the student to show that the second and third tableaus can be generated by using the matrix relationships.

PROBLEMS

1. Solve the following linear programming problems by determining the value of the objective function for all basic solutions to the problem.

(a)   Maximize: $Z = 3x_1 + 2x_2$
                 Subject to: $3x_1 + 2x_2 \leq 24$
                                 $7x_1 + 12x_2 \leq 84$
                                      $x_1, x_2 \geq 0$

(b)   Maximize: $Z = 10x_1 + 14x_2$
                 Subject to: $2x_1 + 3x_2 \leq 12$
                                 $x_1 + 3x_2 \leq 10$
                                   $x_1, x_2 \geq 0$

(c)   Maximize: $Z = 7x_1 + 8x_2 + 5x_3$
                 Subject to: $2x_1 + x_2 + x_3 \leq 10$
                                 $x_1 + 3x_2 + 2x_3 \leq 16$
                                   $x_1, x_2, x_3 \geq 0$

(d)   Minimize: $Z = 4x_1 + 3x_2$
                 Subject to: $x_1 \qquad\quad \geq 4$
                                 $3x_1 + 2x_2 \geq 18$
                                     $x_2 \geq 0$

2. Solve the following linear programming problems, using the simplex algorithm.

(a)   Maximize: $Z = 5x_1 + 2x_2$
                 Subject to: $4x_1 + x_2 \leq 8$
                                 $5x_1 + 2x_2 \leq 12$
                                 $x_1, x_2 \geq 0$

(b)   Maximize: $Z = 5x_1 + 6x_2$
                 Subject to: $x_1 \qquad\quad \leq 5$
                                     $x_2 \leq 4$
                            $10x_1 + 25x_2 \leq 100$
                                 $x_1, x_2 \geq 0$

(c)   Maximize: $Z = 3x_1 + 2x_2 + 2.5x_3$
                 Subject to: $4x_1 + 2x_2 + 2x_3 \leq 12$
                                 $x_2 + 3x_3 \leq 4$
                                 $x_1, x_2, x_3 \geq 0$

(d)   Maximize: $Z = 2x_1 + 4x_2 + 3x_3$
                   Subject to: $x_1 + 2x_2 \qquad \leq 80$
                                    $x_1 + 4x_2 + 2x_3 \leq 120$
                                    $x_1, x_2, x_3 \geq 0$

3. Solve the following linear programming problems, using the simplex algorithm.

(a)   Maximize: $Z = 20x_1 + 6x_2 + 8x_3$
                   Subject to: $8x_1 + 2x_2 + 3x_3 \leq 200$
                                    $4x_1 + 3x_2 \qquad \leq 100$
                                    $x_3 \leq 20$
                                    $x_1, x_2, x_3 \geq 0$

(b)   Maximize: $Z = 12x_1 + 20x_2 + 16x_3$
                   Subject to: $4x_1 + 9x_2 + 6x_3 \leq 6000$
                                    $x_1 + x_2 + 3x_3 \leq 4000$
                                    $6x_1 + 5x_2 + 8x_3 \leq 10{,}000$
                                    $x_1, x_2, x_3 \geq 0$

4. Solve the following linear programming problems, using the simplex algorithm.

(a)   Minimize: $Z = x_1 + x_2$
                   Subject to: $5x_1 + 3x_2 \geq 30$
                                    $6x_1 + 17x_2 \geq 103$
                                    $x_1, x_2 \geq 0$

(b)   Minimize: $Z = 2x_1 + 3x_2$
                   Subject to: $12x_1 + 7x_2 \geq 47$
                                    $5x_1 + 12x_2 \geq 65$
                                    $x_1, x_2 \geq 0$

(c)   Minimize: $Z = 2x_1 + 5x_2 + 4x_3$
                   Subject to: $x_1 + 2x_2 + 4x_3 \geq 61$
                                    $3x_1 + x_2 + 5x_3 \geq 78$
                                    $x_1, x_2, x_3 \geq 0$

(d)   Minimize: $Z = 9x_1 + 11x_2 + 7x_3$
                   Subject to: $2x_1 + x_2 + x_3 \geq 50$
                                    $5x_1 + 3x_2 + 6x_3 \geq 153$
                                    $x_1, x_2, x_3 \geq 0$

5. Solve the following linear programming problems, using the simplex algorithm.

(a)   Minimize: $Z = 4x_1 + 12x_2$
                   Subject to: $15x_1 + 44x_2 \leq 660$
                                    $-4x_1 + 8x_2 \geq -32$
                                    $9x_1 + 15x_2 = 270$
                                    $x_1, x_2 \geq 0$

(b) Maximize: $Z = 2x_1 + 3x_2$
Subject to:
$$5x_1 + 4x_2 \leq 40$$
$$11x_1 - 4x_2 = 0$$
$$-4x_1 + 5x_2 \geq 10$$
$$x_1, x_2 \geq 0$$

6. The following linear programming problem has multiple optimal solutions. Use the simplex algorithm to find the basic solutions to the problem. Give the linear combinations of the basic solutions that define the multiple optimal solutions.

Maximize: $Z = 4x_1 + 6x_2 + 5x_3$
Subject to:
$$2x_1 + 4x_2 + 2x_3 \leq 60$$
$$9x_1 + 4x_2 + 16x_3 \leq 240$$
$$5x_1 + 4x_2 + 6x_3 \geq 125$$
$$x_1, x_2, x_3 \geq 0$$

7. The following linear programming problems illustrate the special cases of no feasible solution and unbounded solution. Use the simplex algorithm to specify which problems have no feasible solution and which have unbounded solutions.

(a) Minimize: $Z = x_1 + 2x_2$
Subject to:
$$-10x_1 + 6x_2 \geq 60$$
$$8x_1 + 15x_2 \leq 120$$
$$x_1, x_2 \geq 0$$

(b) Maximize: $Z = 3x_1 + x_2$
Subject to:
$$7x_1 + 5x_2 \geq 140$$
$$-4x_1 + 8x_2 \leq 32$$
$$x_1, x_2 \geq 0$$

(c) Maximize: $Z = 5x_1 + 10x_2$
Subject to:
$$35x_1 + 30x_2 = 1050$$
$$25x_1 + 45x_2 \leq 1125$$
$$40x_1 + 20x_2 \leq 800$$
$$x_1, x_2 \geq 0$$

(d) Minimize: $Z = -2x_1 + 3x_2$
Subject to:
$$9x_1 + 7x_2 \geq 63$$
$$5x_1 + 12x_2 \geq 60$$
$$x_1, x_2 \geq 0$$

8. Solve the following linear programming problems, using the simplex algorithm.

(a) Minimize: $Z = 5x_1 + 10x_2$
Subject to:
$$10x_1 + 7x_2 \geq 36$$
$$4x_1 - 10x_2 \leq 40$$
$$x_1 \geq 0, x_2 \text{ unrestricted in sign}$$

(b)   Minimize: $Z = x_1 + 10x_2$
       Subject to: $6x_1 + 16x_2 \geq 94$
                        $x_1 \qquad \leq 21$
                        x_1, x_2 unrestricted in sign

9. A manufacturer makes three types of decorative tensor lamps; model 1200, model 1201, and model 1202. The cost of raw materials for each lamp is the same; however, the cost of production differs. Each model 1200 lamp requires 0.1 hr of assembly time, 0.2 hr of wiring time, and 0.1 hr of packaging time. The model 1201 requires 0.2 hr of assembly time, 0.1 hr of wiring time, and 0.1 hr of packaging time. The model 1202 requires 0.2 hr of assembly time, 0.3 hr of wiring time, and 0.1 hr of packaging time. The manufacturer makes a profit of $1.20 on each model 1200 lamp, $1.90 on each model 1201 lamp, and $2.10 on each model 1202 lamp. The manufacturer can schedule up to 80 hr of assembly labor, 120 hr of wiring labor, and 100 hr of packaging labor. Assuming that all lamps can be sold, determine the optimal quantities of each model and the marginal values of each resource.

10. A clothing manufacturer is scheduling work for the next week. There are three possible products that can be made: sportcoats, topcoats, and raincoats. The following table gives the profit for each product and the time required in each process.

Product	Profit per Unit	Hours Required per Unit in		
		Cutting	Sewing	Detailing
Sportcoat	$ 5	1.0	1.0	0.5
Topcoat	8	2.0	1.5	1.0
Raincoat	12	2.0	2.0	1.5

The maximum number of hours that can be scheduled for each process are: cutting, 80 hr; sewing, 60 hr; and detailing, 50 hr. Assuming that all garments produced can be sold, determine the optimal quantities of each item and the marginal values of the three resources.

11. Use the simplex algorithm to determine the optimal quantities of Toots, Wheets, and Honks referred to in Problem 10, Chapter 5. What is the value of an additional unit of each resource used in the manufacture of the noisemakers?

12. Use the simplex algorithm to determine the optimal investments for the Barclay Bank in Problem 18, Chapter 5. Use the $c_j - z_j$ row to determine the change in the overall yield of the investment portfolio from increasing the risk factor to 2.5.

13. Determine the solution to Problem 13, Chapter 5.

14. Determine the solution to Problem 14, Chapter 5.
15. Determine the solution to Problem 15, Chapter 5.
16. Determine the solution to Problem 16, Chapter 5.
17. Determine the solution to Problem 17, Chapter 5.
18. Determine the solution to Problem 19, Chapter 5.
19. Determine the solution to Problem 21, Chapter 5.
20. Determine the solution to Problem 22, Chapter 5

SUGGESTED REFERENCES

The references for this chapter are listed in Chapter 5.

Chapter 7

Duality
and
Sensitivity Analysis

Two extensions of linear programming are presented in this chapter. The first, the dual formulation of the linear programming problem, provides a method for solving an alternative form of the linear programming problem. This reformulation has the advantage of reducing the computational burden for certain linear programming problems. More importantly, however, the theorems that permit this alternative formulation provide special algorithms and additional applications of linear programming. These applications are discussed both in this chapter and the chapter that follows.

The second extension discussed in this chapter is sensitivity analysis. Sensitivity analysis provides a method for investigating the effect on the optimal solution of changes in certain parameters of the linear programming problem. These parameters include the coefficients of the objective function and the right-hand-side values of the constraining equations. Since both the coefficients of the objective function and the right-hand-side values of the constraining equations are often only estimates, information concerning the sensitivity of the solution to changes in these estimates is quite valuable. To illustrate, sensitivity analysis is used in a product mix problem to investigate the effect of changes in the contribution margin of one of the products. Similarly, sensitivity analysis is used to determine the change in the optimal product mix caused by a change in the available quantity of a resource. Sensitivity analysis is discussed beginning in Sec. 7.4.

7.1 The Dual Theorem of Linear Programming

For every linear programming *maximization* problem there is a closely related linear programming *minimization* problem. Conversely, for every linear programming *minimization* problem there is a closely related linear programming *maximization* problem. These pairs of closely related problems are called *dual linear programming problems*.

The duality relationships are important for a number of reasons. First, they lead to a number of theorems that add substantially to our knowledge of linear programming. Second, the dual formulation can be usefully employed in the solution of linear programming problems. As stated earlier, the dual formulation of the linear programming problem often results in a significant reduction of the computational burden of solving the linear programming problem. Finally, the dual problem often provides important economic information concerning the value of the scarce resources employed in a firm.

Before giving an economic interpretation to the dual formulation of a linear programming problem, we shall show the mathematical relationship between the pair of linear programming problems. It is important to remember that this relationship provides a method for converting a linear maximization problem into a related dual minimization problem. Conversely, the relationship permits one to convert a minimization problem into a maximization problem. The two related problems are called the *primal* problem and the *dual* problem.

To illustrate the duality relationships, consider the following pair of linear programming problems. For a maximization problem,

$$\text{Maximize:} \quad P = c_1 x_1 + c_2 x_2 + c_3 x_3 \tag{7.1}$$

$$\text{Subject to:} \quad a_{11} x_1 + a_{12} x_2 + a_{13} x_3 \leq r_1$$
$$a_{21} x_1 + a_{22} x_2 + a_{23} x_3 \leq r_2$$
$$a_{31} x_1 + a_{32} x_2 + a_{33} x_3 \leq r_3$$
$$x_1, x_2, x_3 \geq 0$$

the related minimization problem is

$$\text{Minimize:} \quad L = r_1 w_1 + r_2 w_2 + r_3 w_3 \tag{7.2}$$

$$\text{Subject to:} \quad a_{11} w_1 + a_{21} w_2 + a_{31} w_3 \geq c_1$$
$$a_{12} w_1 + a_{22} w_2 + a_{32} w_3 \geq c_2$$
$$a_{13} w_1 + a_{23} w_2 + a_{33} w_3 \geq c_3$$
$$w_1, w_2, w_3 \geq 0$$

where w_i represents the dual variables.

Although the choice is arbitrary, we shall refer to the maximization problem in this example as the primal problem and to the minimization problem as the dual problem. Throughout this chapter we follow the custom of denoting the original statement of the problem as the primal and the alternative formulation as the dual.

The dual formulation of the primal problem was obtained by:

1. Replacing the variables x_j in the primal by w_i in the dual.
2. Transposing the rows in the primal coefficient matrix to columns in the dual coefficient matrix.
3. Writing the coefficients of the objective function of the primal as the right-hand-side (resource) values in the dual.
4. Writing the right-hand-side values of the primal as the objective function coefficients in the dual.
5. Reversing the direction of the inequalities from the primal to the dual, i.e., if the primal inequalities are $\leq$, the dual inequalities are $\geq$, and vice versa.
6. Reversing the sense of optimization from the primal to the dual, i.e., a primal maximization problem becomes a dual minimization problem and vice versa.

The dual problem can be solved by using the simplex algorithm. One of the important dual theorems is that the optimal value of the primal objective function is equal to the optimal value of the dual objective function. It is also true that if either the primal or the dual problem has an optimal solution, then the related problem must also have an optimal solution.

To illustrate the relationships between the primal and dual problems, consider the problem

$$\begin{aligned}
\text{Minimize:} \quad & Z = 30x_1 + 15x_2 \\
\text{Subject to:} \quad & 3x_1 + x_2 \geq 3 \\
& 4x_1 + 3x_2 \geq 6 \\
& x_1 + 2x_2 \geq 2 \\
& x_1, x_2 \geq 0
\end{aligned}$$

The dual formulation of this minimization problem is

$$\begin{aligned}
\text{Maximize:} \quad & P = 3w_1 + 6w_2 + 2w_3 \\
\text{Subject to:} \quad & 3w_1 + 4w_2 + w_3 \leq 30 \\
& w_1 + 3w_2 + 2w_3 \leq 15 \\
& w_1, w_2, w_3 \geq 0
\end{aligned}$$

In the illustration the primal problem has two decision variables and three inequalities. To solve the primal problem using the simplex algorithm requires three surplus variables and three artificial variables. After these variables are added, the problem becomes

Minimize: $Z = 30x_1 + 15x_2 + 0x_3 + 0x_4 + 0x_5 + MA_1 + MA_2 + MA_3$
Subject to: $3x_1 + 1x_2 - 1x_3 + 0x_4 + 0x_5 + 1A_1 + 0A_2 + 0A_3 = 3$
                      $4x_1 + 3x_2 + 0x_3 - 1x_4 + 0x_5 + 0A_1 + 1A_2 + 0A_3 = 6$
                      $1x_1 + 2x_2 + 0x_3 + 0x_4 - 1x_5 + 0A_1 + 0A_2 + 1A_3 = 2$
                      $x_1, x_2, x_3, x_4, x_5, A_1, A_2, A_3 \geq 0$

The dual formulation of the problem has three decision variables but only two inequalities. Furthermore, the inequalities do not require artificial variables. Since only three decision variables and two slack variables are needed for the dual problem, the arithmetic necessary to determine the solution for the dual is considerably less than that required for the primal. The dual formulation is

Maximize: $P = 3w_1 + 6w_2 + 2w_3 + 0w_4 + 0w_5$
Subject to: $3w_1 + 4w_2 + 1w_3 + 1w_4 + 0w_5 = 30$
                      $1w_1 + 3w_2 + 2w_3 + 0w_4 + 1w_5 = 15$
                      $w_i \geq 0$ for $i = 1, 2, \ldots, 5$

The solution to the dual problem is found by using the simplex algorithm. The optimal value of the objective function for the dual problem is $P = 36$. Since the optimal value of the objective function for the dual is equal to the optimal value of the objective function for the primal, the solution to the primal problem is also $Z = 36$. The tableaus for the dual problem are given in Table 7.1.

The tableaus for the primal problem are given in Table 7.2. By comparing Tables 7.1 and 7.2, we see that the optimum values of the objective function for the primal and dual formulations to a linear programming problem are equal. The tables also show that the arithmetic required to obtain the solution in this particular problem for the dual is substantially less than that required for the primal.

7.1.1   OBTAINING THE OPTIMAL PRIMAL SOLUTION FROM THE DUAL

The solution to a linear programming problem includes both the value of the objective function and the value of the variables. We have shown that the optimal value of the objective function for the primal is equal to the optimal value of the objective function for the dual and is, therefore, given in the final tableau of the dual.

Table 7.1

Tableau	c_b	c_j Basis	3 w_1	6 w_2	2 w_3	0 w_4	0 w_5	Solution
Initial	0	w_4	3	4	1	1	0	30
	0	w_5	1	③	2	0	1	15
		z_j	0	0	0	0	0	0
		$c_j - z_j$	3	6	2	0	0	
Second	0	w_4	$\frac{5}{3}$	0	$-\frac{5}{3}$	1	$-\frac{4}{3}$	10
	6	w_2	$\frac{1}{3}$	1	$\frac{2}{3}$	0	$\frac{1}{3}$	5
		z_j	2	6	4	0	2	30
		$c_j - z_j$	1	0	-2	0	-2	
Third	3	w_1	1	0	-1	$\frac{3}{5}$	$-\frac{4}{5}$	6
	6	w_2	0	1	1	$-\frac{1}{5}$	$\frac{3}{5}$	3
		z_j	3	6	3	$\frac{3}{5}$	$\frac{6}{5}$	36
		$c_j - z_j$	0	0	-1	$-\frac{3}{5}$	$-\frac{6}{5}$	

The values of the primal variables are also given in the optimal tableau of the dual solution. To illustrate, consider again the primal problem,

$$\begin{aligned}
\text{Minimize:} \quad & Z = 6x_1 + 3x_2 \\
\text{Subject to:} \quad & 3x_1 + x_2 \geq 3 \\
& 4x_1 + 3x_2 \geq 6 \\
& x_1 + 2x_2 \geq 2 \\
& x_1, x_2 \geq 0
\end{aligned}$$

The solution to the dual formulation of this problem is shown in Table 7.1. The values of the primal variables are given in the $c_j - z_j$ row of the final tableau of the dual. The decision variables of the primal are equal to

absolute values of the $c_j - z_j$ entries for the dual slack or surplus variables.†
In the example the values of x_1 and x_2 are given in the $c_j - z_j$ row of the final
tableau in the columns headed by the slack variables w_4 and w_5. The absolute
value of these entries gives $x_1 = \frac{3}{5}$ and $x_2 = \frac{6}{5}$.

The values of the surplus variables in the primal problem are given by the
absolute value of the $c_j - z_j$ entries for the decision variables in the dual formulation. The values of x_3, x_4, and x_5 are equal to the absolute values of
$c_1 - z_1$, $c_2 - z_2$, and $c_3 - z_3$ in Table 7.1 and are $x_3 = 0$, $x_4 = 0$, and $x_5 = 1$.

The values of the artificial variables in the primal problem are not shown
in the dual. Since artificial variables are used only to obtain an initial basic
feasible solution and have no physical interpretation in the problem, the value
of the artificial variables in the optimal feasible solution will always be zero.

The values of the dual variables are found in the same manner in the final
tableau of the primal (Table 7.2). In the example problem the dual decision
variables are w_1, w_2, and w_3. The optimal values of these variables are equal
to the $c_j - z_j$ values of the surplus variables x_3, x_4, and x_5. The values are
$w_1 = 6$, $w_2 = 3$, and $w_3 = 0$. The values of the dual slack variables, w_4 and
w_5, are given by the $c_j - z_j$ values of the primal decision variables. From the
$c_j - z_j$ row of the final tableau of Table 7.2 the values of w_4 and w_5 are $w_4 = 0$
and $w_5 = 0$.

Example: Solve the following linear programming problem, using the
dual theorem of linear programming.

$$\text{Minimize:} \quad L = 100x_1 + 80x_2$$
$$\text{Subject to:} \quad 6x_1 + 2x_2 \geq 24$$
$$x_1 + x_2 \geq 8$$
$$3x_1 + 9x_2 \geq 12$$
$$x_1, x_2 \geq 0$$

The arithmetic required to determine the solution can be reduced by
solving the dual problem. The dual problem is

$$\text{Maximize:} \quad P = 24w_1 + 8w_2 + 12w_3$$
$$\text{Subject to:} \quad 6w_1 + w_2 + 3w_3 \leq 100$$
$$2w_1 + w_2 + 9w_3 \leq 80$$
$$w_1, w_2, w_3 \geq 0$$

The tableaus for this problem are shown in Table 7.3.

The solution to the primal linear programming problem is $x_1 = 2$, $x_2 = 6$,
and $L = 680$. The surplus variables associated with the primal problem have
values $x_3 = 0$, $x_4 = 0$, and $x_5 = 48$.

---

† Decision variables refer to the original variables of the problem as opposed to slack,
surplus, or artificial variables.

Table 7.2

Tableau	c_j		30	15	0	0	0	M	M	M	
	c_b	Basis	x_1	x_2	x_3	x_4	x_5	A_1	A_2	A_3	Solution
Initial	M	A_1	③	1	-1	0	0	1	0	0	3
	M	A_2	4	3	0	-1	0	0	1	0	6
	M	A_3	1	2	0	0	-1	0	0	1	2
		z_j	8M	6M	-M	-M	-M	M	M	M	11M
		$c_j - z_j$	30 - 8M	15 - 6M	M	M	M	0	0	0	
Second	30	x_1	1	$\frac{1}{3}$	$-\frac{1}{3}$	0	0	$\frac{1}{3}$	0	0	1
	M	A_2	0	$\frac{5}{3}$	$\frac{4}{3}$	-1	0	$-\frac{4}{3}$	1	0	2
	M	A_3	0	$⑤\!/\!₃$	$-\frac{1}{3}$	0	-1	$-\frac{1}{3}$	0	1	1
		z_j	30	$10 + \frac{10M}{3}$	$-10 + \frac{5M}{3}$	-M	-M	$10 - \frac{5M}{3}$	M	M	$30 + 3M$
		$c_j - z_j$	0	$5 - \frac{10M}{3}$	$10 - \frac{5M}{3}$	M	M	$-10 + \frac{8M}{3}$	0	0	

236

Third

c_j	Basis									
30	x_1	1	0	$-\frac{2}{5}$	0	$\frac{1}{5}$	$\frac{2}{5}$	0	$-\frac{1}{5}$	$\frac{4}{5}$
M	A_2	0	0	1	-1	$\textcircled{1}$	-1	1	-1	1
15	x_2	0	1	$\frac{1}{5}$	0	$-\frac{3}{5}$	$-\frac{1}{5}$	0	$\frac{3}{5}$	$\frac{3}{5}$
	z_j	30	15	$-9+M$	$-M$	$-3+M$	$9-M$	M	$3-M$	$33+M$
	$c_j - z_j$	0	0	$9-M$	M	$3-M$	$-9+2M$	0	$-3+2M$	

Fourth

c_j	Basis									
30	x_1	1	0	$-\frac{3}{5}$	$\frac{1}{5}$	0	$\frac{3}{5}$	$-\frac{1}{5}$	0	$\frac{3}{5}$
0	x_5	0	0	1	-1	-1	-1	1	-1	1
15	x_2	0	1	$\frac{4}{5}$	$-\frac{3}{5}$	0	$-\frac{4}{5}$	$\frac{3}{5}$	0	$\frac{6}{5}$
	z_j	30	15	-6	-3	0	6	3	0	36
	$c_j - z_j$	0	0	6	3	0	$M-6$	$M-3$	M	

Table 7.3

Tableau	c_b	c_j Basis	24 w_1	8 w_2	12 w_3	0 w_4	0 w_5	Solution
Initial	0	w_4	⑥	1	3	1	0	100
	0	w_5	2	1	9	0	1	80
		z_j	0	0	0	0	0	0
		$c_j - z_j$	24	8	12	0	0	
Second	24	w_1	1	$\frac{1}{6}$	$\frac{1}{2}$	$\frac{1}{6}$	0	$\frac{100}{6}$
	0	w_5	0	⟨$\frac{2}{3}$⟩	8	$-\frac{1}{3}$	1	$\frac{140}{3}$
		z_j	24	4	12	4	0	400
		$c_j - z_j$	0	4	0	-4	0	
Third	24	w_1	1	0	$-\frac{3}{2}$	$\frac{1}{4}$	$-\frac{1}{4}$	5
	8	w_2	0	1	12	$-\frac{1}{2}$	$\frac{3}{2}$	70
		z_j	24	8	60	2	6	680
		$c_j - z_j$	0	0	-48	-2	-6	

In solving a dual linear programming problem, it is necessary that all inequalities in the primal problem have the same form. This means that the inequalities in a primal maximization problem must all be $\geq$ and, conversely, the inequalities in a primal minimization problem must all be $\leq$. To illustrate, consider the following linear programming problem.

$$\text{Minimize:} \quad C = 30x_1 + 24x_2$$
$$\text{Subject to:} \quad x_1 + 2x_2 \geq 80$$
$$-2x_1 + x_2 \leq 10$$
$$x_1 + x_2 \geq 30$$
$$x_1, x_2 \geq 0$$

In order to obtain the dual, we must reverse the sense of the second inequality. This is done by multiplying the second inequality by -1. The primal problem now becomes

$$\text{Minimize:} \quad C = 30x_1 + 24x_2$$
$$\text{Subject to:} \quad x_1 + 2x_2 \geq 80$$
$$2x_1 - x_2 \geq -10$$
$$x_1 + x_2 \geq 30$$
$$x_1, x_2 \geq 0$$

The dual linear programming problem is

$$\text{Maximize:} \quad P = 80w_1 - 10w_2 + 30w_3$$
$$\text{Subject to:} \quad w_1 + 2w_2 + w_3 \leq 30$$
$$2w_1 - w_2 + w_3 \leq 24$$
$$w_1, w_2, w_3 \geq 0$$

The tableaus for the dual linear programming problem are given in Table 7.4. The solution to the primal linear programming problem is $x_1 = 12$, $x_2 = 34$ and $C = 1176$. The surplus variables in the primal problem are $x_3 = 0$, $x_4 = 0$, and $x_5 = 16$.

7.1.2 EQUALITIES AND UNRESTRICTED VARIABLES

In our discussion of the relationship between the primal and dual formulations of a linear programming problem, we have considered only those problems in which the constraints are inequalities. To complete the discussion, we must include the case of constraining equations. Assume that the primal problem has the form

$$\text{Maximize:} \quad P = c_1x_1 + c_2x_2 + c_3x_3$$
$$\text{Subject to:} \quad a_{11}x_1 + a_{12}x_2 + a_{13}x_3 \leq r_1$$
$$a_{21}x_1 - a_{22}x_2 + a_{23}x_3 \geq r_2$$
$$a_{31}x_1 + a_{32}x_2 + a_{33}x_3 = r_3$$
$$x_1, x_2, x_3 \geq 0$$

The primal maximization problem includes one "less than" inequality, one "greater than" inequality, and one equation. Before the primal problem can be written in the dual format, the sense of the second inequality must be changed to "less than or equal to." This is done by multiplying the inequality by -1. The dual linear programming problem is

$$\text{Minimize:} \quad L = r_1w_1 - r_2w_2 + r_3w_3$$
$$\text{Subject to:} \quad a_{11}w_1 - a_{21}w_2 + a_{31}w_3 \geq c_1$$
$$a_{12}w_1 + a_{22}w_2 + a_{32}w_3 \geq c_2$$
$$a_{13}w_1 - a_{23}w_2 + a_{33}w_3 \geq c_3$$
$$w_1, w_2 \geq 0$$
$$w_3 \text{ unrestricted in sign}$$

Table 7.4

Tableau		c_j	80	-10	30	0	0	
	c_b	Basis	w_1	w_2	w_3	w_4	w_5	Solution
Initial	0	w_4	1	2	1	1	0	30
	0	w_5	②	-1	1	0	1	24
		z_j	0	0	0	0	0	0
		$c_j - z_j$	80	-10	30	0	0	
	0	w_4	0	$\frac{5}{2}$	$\frac{1}{2}$	1	$-\frac{1}{2}$	18
Second	80	w_1	1	$-\frac{1}{2}$	$\frac{1}{2}$	0	$\frac{1}{2}$	12
		z_j	80	-40	40	0	40	960
		$c_j - z_j$	0	30	-10	0	-40	
	-10	w_2	0	1	$\frac{1}{5}$	$\frac{2}{5}$	$-\frac{1}{5}$	$\frac{36}{5}$
Third	80	w_1	1	0	$\frac{3}{5}$	$\frac{1}{5}$	$\frac{2}{5}$	$\frac{78}{5}$
		z_j	80	-10	46	12	34	1176
		$c_j - z_j$	0	0	-16	-12	-34	

The formulation shows that if the ith constraint in the primal is an equation, then the ith dual variable is unrestricted in sign. In the preceding formulation, the third constraint is an equation. Consequently, the third dual variable is unrestricted in sign. The duality relationships are summarized in Table 7.5.

Example: Solve the following linear programming problem, using the duality relationships of linear programming.

$$\text{Minimize:} \quad C = 20x_1 + 10x_2$$
$$\text{Subject to:} \quad x_1 + 2x_2 \leq 40$$
$$3x_1 + x_2 = 30$$
$$4x_1 + 3x_2 \geq 60$$
$$x_1, x_2 \geq 0$$

Table 7.5

Primal (maximize)	*Dual (minimize)*
Objective function	Right-hand side
Right-hand side	Objective function
jth column of coefficients	jth row of coefficients
ith row of coefficients	ith column of coefficients
jth variable nonnegative	jth constraint an inequality ($\geq$)
jth variable unrestricted in sign	jth constraint an equality
ith constraint an inequality ($\leq$)	ith variable nonnegative
ith constraint an equality	ith variable unrestricted in sign

The problem was introduced in Chapter 6, p. 209, and the tableaus for the primal problem were given in Table 6.13. To solve the problem, using the dual theorem, we must first reverse the sense of the inequality in the first constraint. This gives

$$\text{Minimize:} \quad C = 20x_1 + 10x_2$$
$$\text{Subject to:} \quad -x_1 - 2x_2 \geq -40$$
$$3x_1 + x_2 = 30$$
$$4x_1 + 3x_2 \geq 60$$
$$x_1, x_2 \geq 0$$

The dual formulation for the problem is

$$\text{Maximize:} \quad P = -40w_1 + 30w_2 + 60w_3$$
$$\text{Subject to:} \quad -w_1 + 3w_2 + 4w_3 \leq 20$$
$$-2w_1 + w_2 + 3w_3 \leq 10$$
$$w_1, w_3 \geq 0$$
$$w_2 \text{ unrestricted in sign}$$

Using the method introduced in Sec. 6.4.4 for solving linear programming problems with unrestricted variables, we replace w_2 by $w_2' - w_2''$ and the problem becomes

$$\text{Maximize:} \quad P = -40w_1 + 30w_2' - 30w_2'' + 60w_3$$
$$\text{Subject to:} \quad -w_1 + 3w_2' - 3w_2'' + 4w_3 \leq 20$$
$$-2w_1 + w_2' - w_2'' + 3w_3 \leq 10$$
$$w_1, w_2', w_2'', w_3 \geq 0$$

The tableaus for this problem are given in Table 7.6. The solution to the primal problem is $x_1 = 6$, $x_2 = 12$, and $C = 240$. The slack variable associated with the first constraint of the primal is given by the $c_j - z_j$ entry for w_1 and is $x_3 = 10$. The surplus variable associated with the third constraint is given by the $c_j - z_j$ entry for w_3 and $x_4 = 0$. The $c_j - z_j$ values for w_2' and w_2'' sum to zero in each tableau, indicating that the second constraint in the primal problem is an equation rather than an inequality.

Table 7.6

Tableau		c_j	-40	30	-30	60	0	0	
	c_b	Basis	w_1	w_2'	w_2''	w_3	w_4	w_5	Solution
Initial	0	w_4	-1	3	-3	4	1	0	20
	0	w_5	-2	1	-1	③	0	1	10
		z_j	0	0	0	0	0	0	0
		$c_j - z_j$	-40	30	-30	60	0	0	
Second	0	w_4	$\frac{5}{3}$	⑤/3	$-\frac{5}{3}$	0	1	$-\frac{4}{3}$	$\frac{20}{3}$
	60	w_3	$-\frac{2}{3}$	$\frac{1}{3}$	$-\frac{1}{3}$	1	0	$\frac{1}{3}$	$\frac{10}{3}$
		z_j	-40	20	-20	60	0	20	200
		$c_j - z_j$	0	10	-10	0	0	-20	
Third	30	w_2'	1	1	-1	0	$\frac{3}{5}$	$-\frac{4}{5}$	4
	60	w_3	-1	0	0	1	$-\frac{1}{5}$	$\frac{3}{5}$	2
		z_j	-30	30	-30	60	6	12	240
		$c_j - z_j$	-10	0	0	0	-6	-12	

7.2 Economic Interpretation of the Dual Problem

The dual problem has an important economic interpretation in certain applications of linear programming. One of these applications is the product mix problem. To illustrate this application, we shall assume that the primal problem is a standard product mix problem and that the objective is to determine the profit-maximizing output levels for each of the firm's several products. Profit and output are, of course, limited by the availability of resources. For a problem involving two resources and three products, the typical primal formulation is

$$\text{Maximize:} \quad P = c_1x_1 + c_2x_2 + c_3x_3$$
$$\text{Subject to:} \quad a_{11}x_1 + a_{12}x_2 + a_{13}x_3 \leq r_1$$
$$a_{21}x_1 + a_{22}x_2 + a_{23}x_3 \leq r_2$$
$$x_1, x_2, x_3 \geq 0$$

In manufacturing the three products in our typical problem, two resources are required. Since the profit resulting from the sale of the products could not have been earned without using the resources, the resources have a certain value to the firm. Suppose, then, that the businessman attempts to determine this value. One approach in attempting to determine the value of the resources would be to calculate their cost. This approach would involve depreciation charges, labor rates, overhead allocations, etc. Another approach would be to recognize that without the resources the firm could not have earned its profits. On the basis of this premise, it follows that the businessman should be able to *impute* a certain portion of the profit to each resource. The dual formulation of the product mix problem involves determining the imputed values of the resources.

In the dual formulation of the product mix problem, we impute an artificial accounting price or value w_i to each of the i resources.† The values are calculated so that the total value of all resources used is exactly equal to the total profit resulting from the sale of the products. It should be emphasized that the artificial accounting price or value w_i of each resource is not equal to the cost of the resource. Instead, it is an imputed value that comes from the fact that the resources are used to generate the profit of the firm.

The variables in the dual formulation are the artificial accounting prices or values w_i of the i resources. In order to understand the dual formulation, it is important to recognize that the artificial accounting price w_i of the ith resource is equal to the marginal value of the ith resource. The marginal value of a resource, introduced in Sec. 6.2.4, represents the change in the objective function of the primal problem from employing an additional unit of resource. The artificial accounting price represents the value of the resource in generating profits. It is easy to demonstrate that for a particular set of basic variables in the primal problem, the value of each additional unit of resource is constant. This means that the marginal value of the first unit of resource used in the optimal product mix is equal to the marginal value of all other units of that resource used in the same product mix. It also means that the marginal value of the resource and the artificial accounting price (or imputed value) are the same.

The resources have value to the firm because they are used in the manufacture of products. In order to maximize profits, the resources must be allocated to the most profitable combinations of products. This requires that

---

† These variables are also termed *shadow prices*.

we utilize the resources so that the marginal value of additional units of the resources is a minimum. If this were not the case, additional units of a resource would be of greater value than those already employed, a condition that could exist only when we have failed to employ our input resources optimally. The objective in the dual formulation of the product mix problem is, therefore, to allocate the resources to the products so that the total value of the resources is minimized. The objective function is thus

$$\text{Minimize:} \quad L = r_1 w_1 + r_2 w_2$$

where w_1 and w_2 represent the marginal or imputed values of resources 1 and 2. The constraints in the dual formulation of the product-mix problem are

$$a_{11} w_1 + a_{21} w_2 \geq c_1$$

$$a_{12} w_1 + a_{22} w_2 \geq c_2$$

$$a_{13} w_1 + a_{23} w_2 \geq c_3$$

$$w_1, w_2 \geq 0$$

The left side of the jth constraint gives the value to the firm of the resources required for the manufacture of one unit of the jth product. For instance, the left side of the first constraint gives the value of the resources used in the manufacture of one unit of product one. Similarly, the left side of the second constraint gives the value of the resources used in the manufacture of one unit of product two. The constraints show that the value to the firm from allocating the scarce resources to manufacture one unit of a product must at least equal the contribution margin from that product. If we use product one as an example, this means that

$$a_{11} w_1 + a_{21} w_2 \geq c_1$$

To understand this constraint, we must remember that the resources have value to the firm because they are used in the manufacture of products. The direction of the inequality guarantees that the imputed value of the resources consumed in the manufacture of one unit of product one must at least equal the contribution to profits from product one. If the value to the firm from employing the resources to manufacture the product exactly equals the contribution to profits resulting from the sale of that product, the inequality is strictly satisfied, and the product will be included in the optimal tableau of the primal formulation.

A second possibility with regard to the constraint is that the value of the resources required to produce one unit of product one exceeds the contribution margin of that product. This occurs if the firm can more profitably employ the resources in the manufacture of the remaining products. If this is the case, the surplus variable associated with the constraint (i.e., w_3) will be included as a basic variable in the optimal tableau of the dual solution. This means, of course, that the $c_j - z_j$ entry for the surplus variable w_3 is zero.

Since the $c_j - z_j$ value of w_3 is also equal to the value of the primal variable x_1, the primal variable x_1 is zero. To illustrate these concepts, consider the following example.

Example: A firm uses manufacturing labor and assembly labor to produce three different products. There are 120 hours of manufacturing labor and 260 hours of assembly labor available for scheduling. One unit of product 1 requires 0.10 hr of manufacturing labor and 0.20 hr of assembly labor. Product 2 requires 0.25 hr of manufacturing labor and 0.30 hr of assembly labor for each unit produced. One unit of product 3 requires 0.40 hr of assembly labor but no manufacturing labor. The contribution to profit from products 1, 2, and 3 is $3, $4, and $5, respectively. Determine the imputed or marginal values of the resources by solving the dual problem.

The primal formulation for the problem was given in Chapter 6. The dual formulation is

$$\text{Minimize:} \quad V = 120w_1 + 260w_2$$
$$\text{Subject to:} \quad 0.10w_1 + 0.20w_2 \geq 3$$
$$0.25w_1 + 0.30w_2 \geq 4$$
$$0w_1 + 0.40w_2 \geq 5$$
$$w_1, w_2 \geq 0$$

where w_1 and w_2 represent the marginal values of manufacturing and assembly labor, respectively. The tableaus for the primal problem are given in Table 6.11 and those for the dual formulation in Table 7.7.

The optimal tableau of the dual formulation shows that $w_1 = \$5.00$, $w_2 = \$12.50$, $w_3 = \$0$, $w_4 = \$1.00$, $w_5 = \$0$, and $V = \$3850$. In terms of the product mix example, the marginal values of resources 1 and 2 are $5.00 and $12.50, respectively.

These resources have value because they are used to manufacture products. The primal solution (Table 6.11) shows that the profit from the manufacture of products 1 and 3 is $3850. If the resources can be used to generate $3850 in profits, it follows that the value of the resources to the firm is $3850. This is confirmed by $V = \$3850$ for the dual solution; i.e.,

$$120(\$5.00) + 260(\$12.50) = \$3850$$

We explained earlier that the constraints in the dual specify that the value of the resources used to manufacture a product must at least equal the contribution margin from the product. The dual constraint for product 1 in the example is

$$0.10w_1 + 0.20w_2 \geq 3$$

The value of the resources used to manufacture one unit of product 1 is given by $0.10w_1 + 0.20w_2$. Since $w_1 = \$5.00$ and $w_2 = \$12.50$, the value to the firm of the resources required to manufacture one unit of product 1 is

$$0.10(\$5.00) + 0.20(\$12.50) = \$3.00$$

Table 7.7

Tableau	c_b	Basis	120 w_1	260 w_2	0 w_3	0 w_4	0 w_5	M A_1	M A_2	M A_3	Solution
Initial	M	A_1	0.10	0.20	−1	0	0	1	0	0	3
	M	A_2	0.25	0.30	0	−1	0	0	1	0	4
	M	A_3	0	(0.40)	0	0	−1	0	0	1	5
		z_j	0.35M	0.90M	−M	−M	−M	M	M	M	12M
		$c_j - z_j$	$120 - 0.35M$	$260 - 0.90M$	M	M	M	0	0	0	
Second	M	A_1	0.10	0	−1	0	0.50	1	0	−0.50	0.50
	M	A_2	0.25	0	0	−1	(0.75)	0	1	−0.75	0.25
	260	w_2	0	1	0	0	−2.50	0	0	2.50	12.50
		z_j	0.35M	260	−M	−M	$-650 + 1.25M$	M	M	$650 - 1.25M$	$3250 + 0.75M$
		$c_j - z_j$	$120 - 0.35M$	0	M	M	$650 - 1.25M$	0	0	$-650 + 2.25M$	

246

Third

c_B	Basis								Solution
M	A_1	−0.065	0	−1	(0.67)	1	−0.67	0	0.33
0	w_5	0.33	0	0	−1.33	0	1.33	−1	0.33
260	w_2	0.83	1	0	−3.33	0	3.33	0	13.33
	z_j	214+0.065M	120	−M	−865+0.67M	M	865−0.67M	0	3470+0.33M
	$c_j - z_j$	94−0.065M	0	M	865−0.67M	0	−865+1.67M	M	

Fourth

c_B	Basis								Solution
0	w_4	−0.10	0	−1.50	1	0	1.50	0	0.50
0	w_5	(0.20)	0	−2.00	0	1	2.00	−1	1.00
260	w_2	0.50	1	−5.00	0	0	5.00	0	15.00
	z_j	130	120	−1300	0	0	1300	0	3900
	$c_j - z_j$	−10	140	1300	0	0	M−1300	M	

Fifth

c_B	Basis								Solution
0	w_4	0	0	−2.5	0.5	−1	2.50	−0.5	1
120	w_1	1	0	−10	5	0	10	−5	5
260	w_2	0	1	0	−2.5	0	0	2.5	12.50
	z_j	120	260	−1200	−50	0	1200	50	3850
	$c_j - z_j$	0	0	1200	50	M	M−1200	M−50	

247

The imputed value of the resources used to manufacture product 1 is thus equal to the contribution margin from that product, i.e., \$3.00. Similarly, the third constraint shows that the value of the resources used to manufacture one unit of product 3 is equal to the contribution margin from product 3. Consequently, products 1 and 3 will be manufactured by the firm.

The second constraint describes a product that is not included in the optimal product mix. The constraint for product 2 is

$$0.25w_1 + 0.30w_2 \geq 4$$

The value of the resources required for the manufacture of one unit of product 2 is \$5.00, i.e.,

$$0.25(\$5.00) + 0.30(\$12.50) = \$5.00$$

The contribution margin from manufacturing one unit of product 2 is, however, only \$4.00. Each unit of product 2 manufactured, therefore, reduces profits from the optimal product mix by \$1.00. Consequently, the firm should use the resources to manufacture products 1 and 3 instead of product 2.

The surplus variable for the second constraint is $w_4 = \$1.00$. This verifies that the value of the resources required to manufacture one unit of product 2 exceeds the contribution to profits from that product by $w_4 = \$1.00$. Expressed in a somewhat different manner, the fact that w_4 is positive means that the resources can be more profitably employed to manufacture other products. Furthermore, since w_4 is included as a basis variable in the optimal tableau of the dual, the $c_j - z_j$ entry for w_4 is zero. Remembering that the $c_j - z_j$ entry for w_4 gives the value of x_2, we again conclude that product 2 is not included in the optimal production schedule; i.e., $x_2 = 0$.

Example: McGraw Chemical Company uses nitrates, phosphates, potash, and an inert filler material in the manufacture of three types of chemical fertilizers. McGraw has 1200 tons of nitrates, 2000 tons of phosphates, 1500 tons of potash, and an unlimited supply of the inert filler material on hand. They will not receive additional chemicals during the current production period.

The primal formulation of this problem was given in Chapter 5, p. 161. The objective function and resource constraints for the problem were

$$
\begin{aligned}
\text{Maximize:} \quad & P = 8.00x_1 + 11.40x_2 + 6.20x_3 \\
\text{Subject to:} \quad & 0.05x_1 + 0.05x_2 + 0.08x_3 \leq 1200 \\
& 0.10x_1 + 0.08x_2 + 0.12x_3 \leq 2000 \\
& 0.05x_1 + 0.08x_2 + 0.12x_3 \leq 1500 \\
& \qquad\qquad x_2 \qquad\qquad \geq 8000 \\
& x_1, x_3 \geq 0
\end{aligned}
$$

Management of McGraw Chemical has just learned of a critical shortage of nitrates, phosphates, and potash in the Pharmaceutical Division of the

Company. This shortage means that these chemicals must be transferred from the Fertilizer Division to the Pharmaceutical Division. The cost per ton of nitrates, phosphates, and potash is, respectively, $200, $60, and $100. Since each division is a separate profit center, management realizes that it would be unfair merely to credit the Fertilizer Division with the cost of the chemicals transferred. Rather, they want to establish transfer prices that reflect the value of the resources to the Fertilizer Division. Use the economic interpretation of the dual to establish these transfer prices.

The dual variables represent the marginal value of the resources to the Fertilizer Division. The values of these variables are given in the $c_j - z_j$ row of the optimal primal tableau. This tableau is shown in Table 7.8.

Table 7.8

c_b	Basis	c_j 8.00 x_1	11.40 x_2	6.20 x_3	0 x_4	0 x_5	0 x_6	0 x_7	$-M$ A_1	Solution
0	x_4	0	0	0.005	1	-0.375	-0.25	0	0	75
8.00	x_1	1	0	0	0	20	-20	0	0	10,000
0	x_7	0	0	1.5	0	-12.5	25	1	-1	4,500
11.40	x_2	0	1	1.5	0	-12.5	25	0	0	12,500
	z_j	8.00	11.40	17.10	0	17.5	125	0	0	222,500
	$c_j - z_j$	0	0	-10.90	0	-17.5	-125	0	$-M$	

The marginal values of the resources are given by the $c_j - z_j$ entries for x_4, x_5, and x_6. These entries show that $w_1 = 0$, $w_2 = \$17.50$, and $w_3 = \$125.00$. To determine the transfer price of each chemical, we add the cost of the chemical to the imputed value of the chemical. Since there is an excess supply of nitrates, the transfer price of nitrate is merely the $200 cost of the chemical. The transfer price of phosphates is given by the sum of the cost per ton of the chemical and the marginal value of the chemical. The transfer price of phosphate is, therefore, $77.50 (i.e., $60.00 + $17.50). Similarly, the transfer price of potash is $225.00 (i.e., $100 + $125).

7.3 The Dual Simplex Algorithm

The *dual simplex* algorithm is based on the fact that *the solution to the linear programming problem is optimal when both the primal and dual solutions are feasible.* The algorithm applies directly to the primal problem and should not be confused with the dual formulation of the primal problem. To understand the algorithm, we must first discuss the concepts of primal and dual feasibility.

7.3.1 PRIMAL AND DUAL FEASIBILITY

A solution, it will be remembered, is feasible when all constraints in the linear programming problem are satisfied. The term *primal feasibility* thus refers to a solution in which the primal variables have feasible solution values. Similarly, the term *dual feasible* is used to describe a solution in which the dual variables have feasible values. For a linear programming problem, the solution is primal feasible if the solution values of the primal variables are nonnegative. For a linear programming *maximization* problem, the solution is dual feasible if all $c_j - z_j$ entries are nonpositive. Conversely, for a linear programming *minimization* problem, the solution is dual feasible if the $c_j - z_j$ entries are nonnegative.

To illustrate primal and dual feasibility, consider the following problem.

$$\text{Minimize:} \quad C = 2x_1 + x_2$$
$$\text{Subject to:} \quad 3x_1 + x_2 \geq 6$$
$$4x_1 + 3x_2 \geq 12$$
$$x_1 + 2x_2 \geq 4$$
$$x_1, x_2 \geq 0$$

Assume, for the moment, that we ignore the requirement of nonnegativity for the values of the variables. The initial tableau of the linear programming problem under this assumption is given in Table 7.9. The table was obtained

Table 7.9

c_b	c_j Basis	2 x_1	1 x_2	0 x_3	0 x_4	0 x_5	Solution
0	x_3	-3	-1	1	0	0	-6
0	x_4	-4	-3	0	1	0	-12
0	x_5	-1	-2	0	0	1	-4
	z_j	0	0	0	0	0	0
	$c_j - z_j$	2	1	0	0	0	

by subtracting a surplus variable from each inequality and multiplying the resulting equations by -1. The values of the basic variables in the tableau are $x_3 = -6$, $x_4 = -12$, and $x_5 = -4$.

The initial solution to the primal problem in Table 7.9 is nonfeasible. Interestingly, however, the $c_j - z_j$ entries for the primal problem are nonnegative. Since the primal problem is a minimization problem, the nonnegative $c_j - z_j$ entries indicate that the dual solution is feasible. Remembering that the solution to the dual problem is given by the entries in the $c_j - z_j$ row, we note that the initial basic feasible solution for the dual problem is $w_1 = 0$, $w_2 = 0$, $w_3 = 0$, $w_4 = 2$, and $w_5 = 1$. Table 7.9 illustrates a tableau in which the initial solution is dual feasible but not primal feasible.

One of the duality relationships is that a solution that is both primal and dual feasible is optimal. This relationship provides the basis for the dual simplex algorithm. The idea underlying the algorithm is to restore primal feasibility while retaining dual feasibility. This simply means that in a primal minimization problem we must obtain nonnegative solution values for the primal variables while retaining nonnegative entries in the $c_j - z_j$ row. Conversely, for a primal maximization problem we would eliminate the primal infeasibility by obtaining nonnegative solution values for the primal variables while retaining dual feasibility, i.e., nonpositive entries in the $c_j - z_j$ row. In both maximization and minimization problems, the dual simplex iterations are, therefore, made with the objective of removing the primal infeasibilities while retaining a dual feasible solution.

7.3.2 ADVANTAGES OF THE DUAL SIMPLEX ALGORITHM

There are several advantages to the dual simplex algorithm. One of the advantages is that it eliminates the need for artificial variables for inequalities. This reduces the computational effort required to obtain the solution. A second advantage is that the dual simplex algorithm enables one to add a new constraint directly to the final tableau. This will be shown in Sec. 7.4.

7.3.3 DUAL SIMPLEX RULES

The dual simplex rules for both maximization and minimization problems are the same. The rule for determining which variable leaves the basis is given by Dual Simplex Rule 1.

Dual Simplex Rule 1. *The primal variable whose solution value is most negative leaves the basis. If the solution value for all variables is zero or positive, the current solution is optimal.*

In the example given in Table 7.9, the solution value of x_4 is the most negative. Variable x_4, therefore, leaves the basis. The rule for selecting the variable to enter the basis is given by Dual Simplex Rule 2.

> Dual Simplex Rule 2. *Determine the ratio of the numbers in the* $c_j - z_j$ *row (numerator) with the numbers in the row of the variable leaving the basis (denominator). Ignore ratios with zero or positive numbers in the denominator. The column variable whose ratio has the minimum absolute value enters the basis.*

From Table 7.9, the variable selected to enter the basis is determined by forming the ratios of the $c_j - z_j$ row and the x_4 row. The ratios are $2/-4$, $1/-3$, $0/0$, $0/1$, and $0/0$. The ratios with zero or positive components in the denominator are ignored. From the ratios, the column variable whose ratio has the minimum absolute value enters the basis. Since the absolute value of the ratio of the first column is 0.5 and that of the second column is 0.33, variable x_2 enters the basis. Row operations are used to determine the solution values for the new basic variables. The rules are repeated until the optimal solution is obtained.

The dual simplex algorithm is applied to the example problem in Table 7.10. The primal solution is given in the third tableau as $x_1 = \frac{6}{5}$, $x_2 = \frac{12}{5}$, $x_3 = 0$, $x_4 = 0$, and $x_5 = 2$. The dual solution is $w_1 = \frac{2}{5}$, $w_2 = \frac{1}{5}$, $w_3 = 0$, $w_4 = 0$, and $w_5 = 0$. Since both the primal and dual solutions are feasible, the solution is optimal. The value of the objective function is $C = \frac{24}{5}$.

Example: Solve the following problem, using the dual simplex algorithm.

$$\text{Minimize:} \quad C = 20x_1 + 10x_2$$
$$\text{Subject to:} \quad x_1 + 2x_2 \leq 40$$
$$3x_1 + x_2 = 30$$
$$4x_1 + 3x_2 \geq 60$$
$$x_1, x_2 \geq 0$$

The problem is converted to the simplex format by adding a slack variable to the first inequality, adding an artificial variable to the equation, and subtracting a surplus variable from the final inequality. The solution resulting from this conversion is given by the initial tableau of Table 7.11. Notice that the value of x_4 in the initial tableau is $x_4 = -60$. This, of course, is not a feasible solution for the primal variable. Furthermore, the solution is dual feasible in a primal minimization problem only if all $c_j - z_j$ entries are nonnegative. Since the $c_j - z_j$ entries in the x_1 and x_2 columns are both negative, the initial solution is not dual feasible. Consequently, to apply the dual simplex algorithm, we must first restore dual feasibility. This is accom-

Table 7.10

Tableau	c_b	c_j	2	1	0	0	0	
		Basis	x_1	x_2	x_3	x_4	x_5	Solution
Initial	0	x_3	-3	-1	1	0	0	-6
	0	x_4	-4	-3	0	1	0	-12
	0	x_5	-1	-2	0	0	1	-4
		z_j	0	0	0	0	0	0
		$c_j - z_j$	2	1	0	0	0	
Second	0	x_3	$-\frac{5}{3}$	0	1	$-\frac{1}{3}$	0	-2
	1	x_2	$\frac{4}{3}$	1	0	$-\frac{1}{3}$	0	4
	0	x_5	$\frac{5}{3}$	0	0	$-\frac{2}{3}$	1	4
		z_j	$\frac{4}{3}$	1	0	$-\frac{1}{3}$	0	4
		$c_j - z_j$	$\frac{2}{3}$	0	0	$\frac{1}{3}$	0	
Third	2	x_1	1	0	$-\frac{3}{5}$	$\frac{1}{5}$	0	$\frac{6}{5}$
	1	x_2	0	1	$\frac{4}{5}$	$-\frac{3}{5}$	0	$\frac{12}{5}$
	0	x_5	0	0	1	-1	1	2
		z_j	2	1	$-\frac{2}{5}$	$-\frac{1}{5}$	0	$\frac{24}{5}$
		$c_j - z_j$	0	0	$\frac{2}{5}$	$\frac{1}{5}$	0	

plished by applying the simplex algorithm to the initial tableau. Our objective in applying the simplex algorithm is to obtain nonnegative entries in the $c_j - z_j$ row. The dual simplex algorithm is then used to obtain the optimal solution. For purposes of comparison, the problem was solved by using only the simplex algorithm in Chapter 6, Table 6.13. The problem was again solved in Table 7.6 by applying the simplex algorithm to the dual formulation of the problem.

Table 7.11

Tableau		c_j	20	10	0	M	0	
	c_b	Basis	x_1	x_2	x_3	A_1	x_4	Solution
Initial	0	x_3	1	2	1	0	0	40
	M	A_1	③	1	0	1	0	30	←
	0	x_4	−4	−3	0	0	1	−60
		z_j	$3M$	M	0	M	0	$30M$
		$c_j - z_j$	$20 - 3M$	$10 - M$	0	0	0	
			↑					
Second	0	x_3	0	$\frac{5}{3}$	1	$-\frac{1}{3}$	0	30
	20	x_1	1	$\frac{1}{3}$	0	$\frac{1}{3}$	0	10
	0	x_4	0	$\left(-\frac{5}{3}\right)$	0	$\frac{4}{3}$	1	−20	←
		z_j	20	$\frac{20}{3}$	0	$\frac{20}{3}$	0	200
		$c_j - z_j$	0	$\frac{10}{3}$	0	$M - \frac{20}{3}$	0	
				↑				
Third	0	x_3	0	0	1	1	1	10
	20	x_1	1	0	0	$\frac{3}{5}$	$\frac{1}{5}$	6
	10	x_2	0	1	0	$-\frac{4}{5}$	$-\frac{3}{5}$	12
		z_j	20	10	0	4	−2	240
		$c_j - z_j$	0	0	0	$M - 4$	2	

7.4 Sensitivity Analysis

One of the assumptions in the linear programming problems disussed in the preceding sections is that the parameters of the problem are known. We have assumed, for instance, that the contribution margin for a particular product was given for the problem. Similarly, we have taken for granted that the number of units of a resource available to use in the production of a product was known.

The business practitioner would be quick to point out that these parameters are not always known with certainty. For instance, the contribution margin in a product mix problem includes fixed costs that must be allocated among the products. A change in the method of allocating these costs affects the contribution margin. Similarly, changes in the cost of materials, cost of labor, or the price of a product would cause changes in the contribution margin.

The sensitivity of the optimal solution to changes in the contribution margins of the products is important to the decision maker. The sensitivity is measured by the range over which the contribution margin can vary without causing a change in the optimal solution. To illustrate, assume that the contribution margin for a product is $5.00 and that the product is included as a basic variable in the optimal solution. Furthermore, assume that sensitivity analysis shows that the current solution remains optimal as long as the contribution margin for the product is between $4.00 and $5.50. On the basis of this information, the decision maker knows that a decrease in the contribution margin of more than $1.00 or an increase of more than $0.50 would lead to a revised product mix.

Sensitivity analysis is also used to determine the effect of changes in the resource availabilities on the optimal production schedule. Delayed shipments from suppliers, strikes, spoilage, and other factors all lead to changes in the supply of resources. These changes could, in turn, affect the optimal solution to the linear programming problem.

In addition to investigating the sensitivity of the optimal solution to changes in the objective function coefficients and right-hand-side values, it is often necessary to determine the sensitivity of the optimal solution to changes in products or constraints. For instance, the analyst may want to determine the effect of adding a new product, a new constraint, or of changing the coefficients in the constraining equations. The effect of these types of changes on the optimal solution can also be determined by the techniques presented in this section.

7.4.1 THE OBJECTIVE FUNCTION

The sensitivity of the optimal solution to changes in the coefficients of the objective function is determined by adding a variable δ_j to the objective function coefficient c_j. The new objective function coefficient is $c_j' = c_j + \delta_j$. The domain of δ_j is determined from the requirement that $c_j - z_j$ be nonpositive for a maximization problem and nonnegative for a minimization problem. For the maximization problem, the current solution remains optimal as long as all $c_j - z_j$ entries are nonpositive. Conversely, the current solution is optimal in a minimization problem as long as all $c_j - z_j$ entries are nonnegative. The sensitivity is determined by calculating the values of δ_j for which these criteria are met.

To illustrate the technique, the optimal tableau for the linear programming problem

$$\text{Maximize:} \quad P = 3x_1 + 4x_2 + 5x_3 + 4x_4$$
$$\text{Subject to:} \quad 2x_1 + 5x_2 + 4x_3 + 3x_4 \leq 224$$
$$5x_1 + 4x_2 - 5x_3 + 10x_4 \leq 280$$
$$2x_1 + 4x_2 + 4x_3 - 2x_4 \leq 184$$
$$x_j \geq 0 \quad \text{for } j = 1, 2, 3, 4$$

is given in Table 7.12. The optimal tableau shows that x_1, x_3, and x_4 are basic variables and x_2, x_5, x_6, and x_7 are nonbasic variables.

Table 7.12

c_j		3	4	5	4	0	0	0	
c_b	Basis	x_1	x_2	x_3	x_4	x_5	x_6	x_7	Solution
4	x_4	0	$\frac{1}{5}$	0	1	$\frac{1}{5}$	0	$-\frac{1}{5}$	8
3	x_1	1	1	0	0	$-\frac{1}{5}$	$\frac{2}{15}$	$\frac{11}{30}$	60
5	x_3	0	$\frac{3}{5}$	1	0	$\frac{1}{5}$	$-\frac{1}{15}$	$-\frac{1}{30}$	20
	z_j	3	$\frac{34}{5}$	5	4	$\frac{6}{5}$	$\frac{1}{15}$	$\frac{2}{15}$	312
	$c_j - z_j$	0	$-\frac{14}{5}$	0	0	$-\frac{6}{5}$	$-\frac{1}{15}$	$-\frac{2}{15}$	

We shall first determine the sensitivity of the optimal solution to changes in the objective function coefficients of the nonbasic variables. To illustrate, the sensitivity of the optimal solution to changes in the objective function coefficient of the nonbasic variable x_2 is determined by adding δ_2 to the objective function coefficient c_2. The objective function coefficient becomes $c_2' = 4 + \delta_2$. The current solution remains optimal as long as the $c_j' - z_j'$ entries are nonpositive. Table 7.13 shows that this criterion is met for $\delta_2 \leq \frac{14}{5}$. This means that the objective function coefficient of x_2 can be increased by as much as $\frac{14}{5}$ without causing a change in the optimal solution. If the objective function coefficient of x_2 is increased by more than $\frac{14}{5}$, however, x_2 will enter the basis.

This argument is intuitively plausible, since increasing the objective function coefficient is equivalent to increasing the profit from a product. If profit is sufficiently increased, we would expect to improve the current optimal solution by producing the product.

The sensitivity of the optimal solution to changes in the objective function

Table 7.13

c_j		3	$4 + \delta_2$	5	4	0	0	0	
c_b	Basis	x_1	x_2	x_3	x_4	x_5	x_6	x_7	Solution
4	x_4	0	$\frac{1}{5}$	0	1	$\frac{1}{5}$	0	$-\frac{1}{5}$	8
3	x_1	1	1	0	0	$-\frac{1}{5}$	$\frac{2}{15}$	$\frac{11}{30}$	60
5	x_3	0	$\frac{3}{5}$	1	0	$\frac{1}{5}$	$-\frac{1}{15}$	$-\frac{1}{30}$	20
	z_j'	3	$\frac{34}{5}$	5	4	$\frac{6}{5}$	$\frac{1}{15}$	$\frac{2}{15}$	312
	$c_j' - z_j'$	0	$\delta_2 - \frac{14}{5}$	0	0	$-\frac{6}{5}$	$-\frac{1}{15}$	$-\frac{2}{15}$	

coefficients of the remaining nonbasic variables is determined in the same manner. These values are $\delta_5 \leq \frac{6}{5}$, $\delta_6 \leq \frac{1}{15}$, and $\delta_7 \leq \frac{2}{15}$.

The procedure for determining the sensitivity of the optimal solution to changes in the objective function coefficients of basic variables is illustrated in Table 7.14. The objective function coefficient for the basic variable x_1 has been replaced by $c_1' = 3 + \delta_1$ in this table. For the current solution to remain optimal, all $c_j' - z_j'$ entries in Table 7.14 must remain nonpositive, i.e.,

Table 7.14

c_j'		$3 + \delta_1$	4	5	4	0	0	0	
c_b	Basis	x_1	x_2	x_3	x_4	x_5	x_6	x_7	Solution
4	x_4	0	$\frac{1}{5}$	0	1	$\frac{1}{5}$	0	$-\frac{1}{5}$	8
$3 + \delta_1$	x_1	1	1	0	0	$-\frac{1}{5}$	$\frac{2}{15}$	$\frac{11}{30}$	60
5	x_3	0	$\frac{3}{5}$	1	0	$\frac{1}{5}$	$-\frac{1}{15}$	$-\frac{1}{30}$	20
	z_j'	$3 + \delta_1$	$\frac{34}{5} + \delta_1$	5	4	$\frac{6}{5} - \frac{1}{5}\delta_1$	$\frac{1}{15} + \frac{2}{15}\delta_1$	$\frac{2}{15} + \frac{11}{30}\delta_1$	$312 + 60\delta_1$
	$c_j' - z_j'$	0	$-\frac{14}{5} - \delta_1$	0	0	$-\frac{6}{5} + \frac{1}{5}\delta_1$	$-\frac{1}{15} - \frac{2}{15}\delta_1$	$-\frac{2}{15} - \frac{11}{30}\delta_1$	

$c_j' - z_j' \leq 0$. The values of δ_1 that satisfy this criterion are determined by solving the following system of linear inequalities:

$$-\tfrac{14}{5} - \delta_1 \leq 0$$
$$-\tfrac{6}{5} + \tfrac{1}{5}\delta_1 \leq 0$$
$$-\tfrac{1}{15} - \tfrac{2}{15}\delta_1 \leq 0$$
$$-\tfrac{2}{15} - \tfrac{11}{30}\delta_1 \leq 0$$

By solving each inequality for δ_1 we find that the smallest value of δ_1 that satisfies all inequalities is $\delta_1 = -\frac{4}{11}$ and that the largest value of δ_1 that satisfies all inequalities is $\delta_1 = 6$. The solution, therefore, is $-\frac{4}{11} \le \delta_1 \le 6$. This means that the current variables will remain in the basis for $(3 - \frac{4}{11}) \le c_1 \le (3 + 6)$, or equivalently, $\frac{29}{11} \le c_1 \le 9$.

It is interesting to notice that if the objective function coefficient of a basic variable is either increased or decreased significantly the current basic variables change. Although this result might at first seem odd, it is what one should expect. An increase in the objective function coefficient implies that the product has become more profitable. Resources should then be diverted from the manufacture of other products to the manufacture of the more profitable product. Conversely, if the product becomes less profitable, resources should be diverted from that product to the manufacture of other products. Either case requires a change of basis. In the example, as long as the contribution margin of product 1 is between $\frac{29}{11}$ and 9 the current solution is optimal. If $c_1 < \frac{29}{11}$ or $c_1 > 9$, however, the current solution is no longer optimal and a change of basis is required. The change of basis is made by applying Simplex Rules 1 and 2.

The sensitivity of the optimal solution to changes in the objective function coefficients of each variable is given in Table 7.15. The reader may want to

Table 7.15

Variable	δ_j	c_j
x_1	$-\frac{4}{11} \le \delta_1 \le 6$	$\frac{29}{11} \le c_1 \le 9$
x_2	$\delta_2 \le \frac{14}{5}$	$c_2 \le \frac{34}{5}$
x_3	$-\frac{14}{3} \le \delta_3 \le 1$	$\frac{1}{3} \le c_3 \le 6$
x_4	$-6 \le \delta_4 \le \frac{2}{3}$	$-2 \le c_4 \le \frac{14}{3}$
x_5	$\delta_5 \le \frac{6}{5}$	$c_5 \le \frac{6}{5}$
x_6	$\delta_6 \le \frac{1}{15}$	$c_6 \le \frac{1}{15}$
x_7	$\delta_7 \le \frac{2}{15}$	$c_7 \le \frac{2}{15}$

test his understanding of the technique by calculating the values shown in the table.

Example: Management of the McGraw Chemical Company has expressed concern over the accuracy of the contribution margins for the three chemical fertilizers manufactured by the company. The contribution margins are $8.00 per ton of 5–10–5, $11.40 per ton of 5–8–8, and $6.20 per ton of 8–12–12 (see Chapter 5, p. 161). Determine the sensitivity of the optimal solution to changes in each of the contribution margins.

The linear programming problem was

Maximize: $P = 8.00x_1 + 11.40x_2 + 6.20x_3$

Subject to:
$$0.05x_1 + 0.05x_2 + 0.08x_3 \leq 1200$$
$$0.10x_1 + 0.08x_2 + 0.12x_3 \leq 2000$$
$$0.05x_1 + 0.08x_2 + 0.12x_3 \leq 1500$$
$$x_2 \geq 8000$$
$$x_1, x_2, x_3 \geq 0$$

The optimal solution for the problem was given in Table 7.8, p. 249, and is repeated below. The optimal solution is $x_1 = 10,000$ tons of 5–10–5,

Table 7.16

c_b	Basis	c_j 8.00 x_1	11.40 x_2	6.20 x_3	0 x_4	0 x_5	0 x_6	0 x_7	$-M$ A_1	Solution
0	x_4	0	0	0.005	1	−0.375	−0.25	0	0	75
8.00	x_1	1	0	0	0	20	−20	0	0	10,000
0	x_7	0	0	1.5	0	−12.5	25	1	−1	4,500
11.40	x_2	0	1	1.5	0	−12.5	25	0	0	12,500
	z_j	8.00	11.40	17.10	0	17.5	125	0	0	222,500
	$c_j - z_j$	0	0	−10.90	0	−17.5	−125	0	−M	

$x_2 = 12,500$ tons of 5–8–8, and $x_3 = 0$ tons of 8–12–12 fertilizer. The sensitivity of the nonbasic variable, x_3, can be determined directly from the table. The table shows that the contribution margin of x_3 must increase from $6.20 to $17.10 per ton before x_3 could be profitably manufactured.

The sensitivity of the solution to changes in the contribution margin of x_1 is determined by substituting $c_1' = 8.00 + \delta_1$ for c_1. The $c_j - z_j$ entries remain nonpositive for

$$-17.5 - 20\delta_1 \leq 0$$

and

$$-125 + 20\delta_1 \leq 0$$

The solution to the two inequalities gives

$$-0.88 \leq \delta_1 \leq 6.25$$

Provided that the contribution margins of the other products are not changed, the current solution remains optimal for

$$\$7.12 \leq c_1 \leq \$14.25$$

The sensitivity of the optimal solution to changes in the contribution margin of x_2 is determined in the same manner. The current solution remains optimal for

$$\$6.40 \leq c_2 \leq \$12.80$$

7.4.2 RIGHT-HAND SIDE

The sensitivity of the basic variables to changes in the right-hand-side values of the constraining equations is often important to the analyst. The sensitivity is measured by an upper and a lower bound on a right-hand-side value. To illustrate, the sensitivity of the basic variables to changes in the right-hand-side value or resource r_1 is described by a constraint of the form

$$a \leq r_1 \leq b$$

As long as this inequality is satisfied, the current basic variables remain in the optimal tableau. If, however, the values of r_1 change so that $r_1 < a$ or $r_1 > b$, one or more basic variables are replaced.

To introduce the method for determining the bounds on the resource, consider the problem introduced in Chapter 6, p. 196. The problem was

$$\begin{aligned}
\text{Maximize: } & P = 3x_1 + 4x_2 + 5x_3 + 4x_4 \\
\text{Subject to: } & 2x_1 + 5x_2 + 4x_3 + 3x_4 \leq 224 \\
& 5x_1 + 4x_2 - 5x_3 + 10x_4 \leq 280 \\
& 2x_1 + 4x_2 + 4x_3 - 2x_4 \leq 184 \\
& x_j \geq 0 \quad \text{for } j = 1, 2, 3, 4
\end{aligned}$$

The solution to this problem (Chapter 6, p. 203) is $x_1 = 60$, $x_2 = 0$, $x_3 = 20$, and $x_4 = 8$.

The sensitivity of the basic variables to changes in a right-hand-side value is determined by adding a variable δ_i to the ith resource. The ith resource r_i is replaced by $r_i + \delta_i$. Replacing r_1 by $r_1 + \delta_1$ in the example gives the right-hand-side vector

$$R_1^* = \begin{pmatrix} 224 + \delta_1 \\ 280 \\ 184 \end{pmatrix}$$

The bounds on the resource are determined by solving for the values of δ_1 for which the basic variables x_1, x_3, and x_4 remain in the optimal tableau.

The values of the basic variables are given by $X_b = B^{-1}R$. The upper and lower bounds for r_i are determined from the requirement that the solution vector X_b be nonnegative, i.e., $X_b = B^{-1}R_1^* \geq 0$. The matrix B^{-1} was given in Sec. 6.5 as

$$B^{-1} = \begin{pmatrix} \frac{1}{5} & 0 & -\frac{1}{5} \\ -\frac{1}{5} & \frac{2}{15} & \frac{11}{30} \\ \frac{1}{5} & -\frac{1}{15} & -\frac{1}{30} \end{pmatrix}$$

The values of δ_1 are determined from the system of linear inequalities, $B^{-1}R_1^* \geq 0$. This system is

$$
\begin{pmatrix}
\frac{1}{5} & 0 & -\frac{1}{5} \\
-\frac{1}{5} & \frac{2}{15} & \frac{11}{30} \\
\frac{1}{5} & -\frac{1}{15} & -\frac{1}{30}
\end{pmatrix}
\begin{pmatrix}
224 + \delta_1 \\
280 \\
184
\end{pmatrix}
\geq
\begin{pmatrix}
0 \\
0 \\
0
\end{pmatrix}
$$

or alternatively,

$$8 + \tfrac{1}{5}\delta_1 \geq 0$$
$$60 - \tfrac{1}{5}\delta_1 \geq 0$$
$$20 + \tfrac{1}{5}\delta_1 \geq 0$$

The values of δ_1 that satisfy the three inequalities are $-40 \leq \delta_1 \leq 300$. This means that the basic variables x_1, x_3, and x_4 remain in solution for $184 \leq r_1 \leq 524$. If, however, r_1 is less than 184 or more than 524, a change of basic variables will occur.

The sensitivity of the basic variables to changes in the remaining two resources is determined in the same manner. Replacing the second resource by $280 + \delta_2$ gives the right-hand-side vector

$$
R_2^* =
\begin{pmatrix}
224 \\
280 + \delta_2 \\
184
\end{pmatrix}
$$

The values of δ_2 are determined from the requirement that $B^{-1}R_2^* \geq 0$, i.e.,

$$
\begin{pmatrix}
\frac{1}{5} & 0 & -\frac{1}{5} \\
-\frac{1}{5} & \frac{2}{15} & \frac{11}{30} \\
\frac{1}{5} & -\frac{1}{15} & -\frac{1}{30}
\end{pmatrix}
\begin{pmatrix}
224 \\
280 + \delta_2 \\
184
\end{pmatrix}
\geq
\begin{pmatrix}
0 \\
0 \\
0
\end{pmatrix}
$$

The system of inequalities is

$$8 + 0\delta_2 \geq 0$$
$$60 + \tfrac{2}{15}\delta_2 \geq 0$$
$$20 - \tfrac{1}{15}\delta_2 \geq 0$$

The solution of the system of inequalities gives $-450 \leq \delta_2 \leq 300$. This means that the basic variables x_1, x_3, and x_4 remain in the optimal tableau for $r_1 = 224$, $-170 \leq r_2 \leq 580$, and $r_3 = 184$. For the third resource, the basic variables x_1, x_3, and x_4 remain in the optimal solution for $r_1 = 224$, $r_2 = 280$, and $20 \leq r_3 \leq 224$.

Example: Determine the optimal solution in the preceding problem for $r_1 = 224$, $r_2 = 126$, and $r_3 = 184$.

The sensitivity analysis showed that the basic variables x_1, x_3, and x_4 remain in the optimal tableau for $r_1 = 224$, $-170 \leq r_2 \leq 580$, and $r_3 = 184$. The values of the basic variables are given by $B^{-1}R$. Thus,

$$X_b = B^{-1}R = \begin{pmatrix} \frac{1}{5} & 0 & -\frac{1}{5} \\ -\frac{1}{5} & \frac{2}{15} & \frac{11}{30} \\ \frac{1}{5} & -\frac{1}{15} & -\frac{1}{30} \end{pmatrix} \begin{pmatrix} 224 \\ 126 \\ 184 \end{pmatrix}$$

$$X_b = \begin{pmatrix} 8 \\ 39.5 \\ 30.3 \end{pmatrix}$$

The values of the basic variables are $x_1 = 39.5$, $x_3 = 30.3$, and $x_4 = 8$.

Example: Management of McGraw Chemical Company has received notice from their principal supplier of potash of a pending strike by their employees. McGraw is scheduled to receive 1500 tons of potash from this supplier. If the strike occurs, however, a portion of this shipment could be delayed. In order to plan production during the coming month, McGraw must determine the effect of possible changes in the supply of potash on their products.

The effect of a shortage of potash on the production of the three fertilizers can be determined from the sensitivity of the basic variables to changes in the supply of potash. In the original statement of the problem (Chapter 5, p. 191), McGraw's raw material inventory included 1200 tons of nitrates, 2000 tons of phosphates, and 1500 tons of potash. The optimal solution based on this inventory (Chapter 7, p. 249) was 10,000 tons of 5–10–5 fertilizer and 12,500 tons of 5–8–8 fertilizer. Production of 8–12–12 fertilizer was not scheduled.

The sensitivity of the basic variables to changes in a resource is determined from the requirement that $B^{-1}R \geq 0$. The inverse matrix (Table 7.8, p. 249) is

$$B^{-1} = \begin{pmatrix} 1 & -0.375 & -0.25 & 0 \\ 0 & 20 & -20 & 0 \\ 0 & -12.5 & 25 & -1 \\ 0 & -12.5 & 25 & 0 \end{pmatrix}$$

The right-hand-side vector is

$$R^* = \begin{pmatrix} 1200 \\ 2000 \\ 1500 + \delta_3 \\ 8000 \end{pmatrix}$$

The requirement that the solution values of the basic variables remain nonnegative gives

$$\begin{pmatrix} 1 & -0.375 & -0.25 & 0 \\ 0 & 20 & -20 & 0 \\ 0 & -12.5 & 25 & -1 \\ 0 & -12.5 & 25 & 0 \end{pmatrix} \begin{pmatrix} 1200 \\ 2000 \\ 1500 + \delta_3 \\ 8000 \end{pmatrix} \geq \begin{pmatrix} 0 \\ 0 \\ 0 \\ 0 \end{pmatrix}$$

The system of inequalities is

$$75 - 0.25\delta_3 \geq 0$$
$$10,000 - 20\delta_3 \geq 0$$
$$4500 + 25\delta_3 \geq 0$$
$$12,500 + 25\delta_3 \geq 0.$$

and the solution of this system of inequalities is $-180 \leq \delta_3 \leq 300$.

The system of linear inequalities shows that the basic variables remain in solution for $r_1 = 1200$ tons of nitrates, $r_2 = 2000$ tons of phosphates, and $1320 \leq r_3 \leq 1800$ tons of potash. If the strike limits the supply of potash to less than 1320 tons, one or more of the current basic variables will be replaced in the optimal tableau.

Example: McGraw Chemical has received a shipment of 1000 tons of potash. If no additional chemicals are received during the month, determine that optimal product mix for the company.

From the preceding example, management knows that a change of basic variables is required in the simplex tableau. The values of the basic variables assuming 1000 tons of potash are given by $B^{-1}R$.

$$X_b = B^{-1}R = \begin{pmatrix} 1 & -0.375 & -0.25 & 0 \\ 0 & 20 & -20 & 0 \\ 0 & -12.5 & 25 & -1 \\ 0 & -12.5 & 25 & 0 \end{pmatrix} \begin{pmatrix} 1200 \\ 2000 \\ 1000 \\ 8000 \end{pmatrix} = \begin{pmatrix} 200 \\ 20,000 \\ -8000 \\ 0 \end{pmatrix}$$

Rather than resolving the entire problem, the values of the basic variables can be inserted directly into the optimal tableau of Table 7.8. This is shown in Table 7.17.

Table 7.17

c_b		c_j	8.00	11.40	6.20	0	0	0	0	$-M$	
	Basis		x_1	x_2	x_3	x_4	x_5	x_6	x_7	A_1	Solution
0	x_4		0	0	0.005	1	-0.375	-0.25	0	0	200
8.00	x_1		1	0	0	0	20	-20	0	0	20,000
0	x_7		0	0	1.5	0	-12.5	25	1	-1	-8000
11.40	x_2		0	1	1.5	0	-12.5	25	0	0	0
	z_j		8.00	11.40	17.10	0	17.5	125	0	0	
	$c_j - z_j$		0	0	-10.90	0	-17.5	-125	0	$-M$	

The solution given in Table 7.17 is dual feasible but not primal feasible. By applying the dual simplex algorithm, primal feasibility can be restored. The optimal solution for 1200 tons of nitrates, 2000 tons of phosphates, and 1000 tons of potash is given in Table 7.18 as 7200 tons of 5–10–5 and 8000 tons of 5–8–8.

Table 7.18

		8.00	11.40	6.20	0	0	0	0	$-M$	
c_b	Basis	x_1	x_2	x_3	x_4	x_5	x_6	x_7	A_1	Solution
0	x_4	0	0	−0.04	1	0	−1	−0.03	0.03	440
8.00	x_1	1	0	2.40	0	0	20	1.60	−1.60	7,200
0	x_5	0	0	−0.12	0	1	−2	−0.08	0.08	640
11.40	x_2	0	1	0	0	0	0	−1	1	8,000
	z_j	8.00	11.40	19.20	0	0	160	1.40	−1.40	148,800
	$c_j - z_j$	0	0	−13.00	0	0	−160	−1.40	$-M + 1.40$	

The tableau also shows that 440 tons of nitrates and 640 tons of phosphates are not being used. Profit from this schedule is $148,800.

7.4.3 CHANGES TO A COLUMN

Changes can be made to one or more of the components of a column in a simplex tableau. The column entries, it will be remembered, usually represent the technological relationship between a product and the resources. Should these technological relationships change, the column of technological coefficients must reflect this change. Rather than resolving the entire linear programming problem, the change can be made beginning with the optimal tableau of the original problem.

The procedure for calculating the effect of changing a column vector or from adding an entirely new vector is determined from the matrix relationships introduced in Chapter 6. In Sec. 6.5, A_j represents the jth column in the initial tableau. The corresponding column in any succeeding tableau is given by $B^{-1}A_j$.

The effect of changing a column can be determined by replacing the original column vector A_j by the new column vector A_j^*. The jth column in the original optimal tableau now becomes $B^{-1}A_j^*$. The new optimal solution is found by applying the simplex rules to this modified tableau.

To illustrate this procedure, consider the problem

$$\begin{aligned}
\text{Maximize:} \quad & P = 3x_1 + 4x_2 + 5x_3 + 4x_4 \\
\text{Subject to:} \quad & 2x_1 + 5x_2 + 4x_3 + 3x_4 \leq 224 \\
& 5x_1 + 4x_2 - 5x_3 + 10x_4 \leq 280 \\
& 2x_1 + 4x_2 + 4x_3 - 2x_4 \leq 184 \\
& x_j \geq 0 \quad \text{for } j = 1, 2, 3, 4
\end{aligned}$$

The optimal solution to this problem is given in Chapter 6 in Table 6.10. The effect of changing a column is demonstrated by changing the components of the x_4 column to

$$A_4^* = \begin{pmatrix} 5 \\ 8 \\ -1 \end{pmatrix}$$

The x_4 column in Table 6.10 is replaced by $B^{-1}A_4^*$. The new x_4 column is

$$B^{-1}A_4^* = \begin{pmatrix} \frac{1}{5} & 0 & -\frac{1}{5} \\ -\frac{1}{5} & \frac{2}{15} & \frac{11}{30} \\ \frac{1}{5} & -\frac{1}{15} & -\frac{1}{30} \end{pmatrix} \begin{pmatrix} 5 \\ 8 \\ -1 \end{pmatrix} = \begin{pmatrix} \frac{6}{5} \\ -\frac{3}{10} \\ \frac{1}{2} \end{pmatrix}$$

This column, along with the new z_4 and $c_4 - z_4$ values, is shown by Table 7.18. Since the $c_j - z_j$ value for x_4 is nonpositive in Table 7.19, x_4 remains

Table 7.19

c_b	c_j	3	4	5	4	0	0	0	
	Basis	x_1	x_2	x_3	x_4	x_5	x_6	x_7	Solution
4	x_4	0	$\frac{1}{5}$	0	$\frac{6}{5}$	$\frac{1}{5}$	0	$-\frac{1}{5}$	8
3	x_1	1	1	0	$-\frac{3}{10}$	$-\frac{1}{5}$	$\frac{2}{15}$	$\frac{11}{30}$	60
5	x_3	0	$\frac{3}{5}$	1	$\frac{1}{2}$	$\frac{1}{5}$	$-\frac{1}{15}$	$-\frac{1}{30}$	20
	z_j	3	$\frac{34}{5}$	5	$\frac{32}{5}$	$\frac{6}{5}$	$\frac{1}{15}$	$\frac{2}{15}$	312
	$c_j - z_j$	0	$-\frac{14}{5}$	0	$-\frac{12}{5}$	$-\frac{6}{5}$	$-\frac{1}{15}$	$-\frac{2}{15}$	

as a basic variable in the optimal tableau. The solution values of the basic variables are obtained by pivoting on the first row, fourth column. The optimal solution based on the new column is shown in Table 7.20.

Table 7.20

c_b	Basis	3 x_1	4 x_2	5 x_3	4 x_4	0 x_5	0 x_6	0 x_7	Solution
4	x_4	0	$\frac{1}{6}$	0	1	$\frac{1}{6}$	0	$-\frac{1}{6}$	$\frac{20}{3}$
3	x_1	1	$\frac{21}{20}$	0	0	$-\frac{3}{20}$	$\frac{2}{15}$	$\frac{19}{60}$	62
5	x_3	0	$\frac{31}{60}$	1	0	$\frac{7}{60}$	$-\frac{1}{15}$	$\frac{1}{20}$	$\frac{50}{3}$
	z_j	3	$\frac{32}{5}$	5	4	$\frac{4}{5}$	$\frac{1}{15}$	$\frac{8}{15}$	296
	$c_j - z_j$	0	$-\frac{12}{5}$	0	0	$-\frac{4}{5}$	$-\frac{1}{15}$	$-\frac{8}{15}$	

7.4.4 ADDING A NEW CONSTRAINT

It is sometimes necessary to modify a linear programming problem by adding one or more constraints. Rather than resolving the entire problem, the new constraints can be added directly to the final tableau of the original problem. The dual simplex algorithm is then used to find the solution to the revised problem.

To illustrate this procedure, consider the problem

$$\text{Maximize:} \quad P = 3x_1 + 4x_2 + 5x_3 + 4x_4$$
$$\text{Subject to:} \quad 2x_1 + 5x_2 + 4x_3 + 3x_4 \leq 224$$
$$5x_1 + 4x_2 - 5x_3 + 10x_4 \leq 280$$
$$2x_1 + 4x_2 + 4x_3 - 2x_4 \leq 184$$
$$x_j \geq 0 \quad \text{for } j = 1, 2, 3, 4$$

Suppose that the constraint

$$3x_1 + 6x_2 + 2x_3 + 4x_4 \leq 220$$

is added to this problem. The effect of the constraint on the optimal solution is determined by adding the constraint to the optimal tableau of the original problem. Adding this constraint to the optimal tableau (Table 6.10) gives the system of equations shown in Table 7.21. The solution to this system of equations is found by completing the pivoting process. This involves completing the identity matrix for the basic variables. The revised tableau is shown in Table 7.22.

Adding the new constraint to the optimal tableau of the original problem does not, by itself, change the values of the basic variables. The values of the basic variables continue to be $x_1 = 60$, $x_3 = 20$, and $x_4 = 8$. The additional variable included in the basis in Table 7.22 is x_8. Since x_8 is the slack variable

Table 7.21

c_b	c_j Basis	3 x_1	4 x_2	5 x_3	4 x_4	0 x_5	0 x_6	0 x_7	0 x_8	Solution
4	x_4	0	$\frac{1}{5}$	0	1	$\frac{1}{5}$	0	$-\frac{1}{5}$	0	8
3	x_1	1	1	0	0	$-\frac{1}{5}$	$\frac{2}{15}$	$\frac{11}{30}$	0	60
5	x_3	0	$\frac{3}{5}$	1	0	$\frac{1}{5}$	$-\frac{1}{15}$	$-\frac{1}{30}$	0	20
0	x_8	3	6	2	4	0	0	0	1	220
	z_j	3	$\frac{34}{5}$	5	4	$\frac{6}{5}$	$\frac{1}{15}$	$\frac{2}{15}$	0	312
	$c_j - z_j$	0	$-\frac{14}{5}$	0	0	$-\frac{6}{5}$	$-\frac{1}{15}$	$-\frac{2}{15}$	0	

Table 7.22

c_b	c_j Basis	3 x_1	4 x_2	5 x_3	4 x_4	0 x_5	0 x_6	0 x_7	0 x_8	Solution
4	x_4	0	$\frac{1}{5}$	0	1	$\frac{1}{5}$	0	$-\frac{1}{5}$	0	8
3	x_1	1	1	0	0	$-\frac{1}{5}$	$\frac{2}{15}$	$\frac{11}{30}$	0	60
5	x_3	0	$\frac{3}{5}$	1	0	$\frac{1}{5}$	$-\frac{1}{15}$	$-\frac{1}{30}$	0	20
0	x_8	0	1	0	0	-1	$-\frac{4}{15}$	$-\frac{7}{30}$	1	-32
	z_j	3	$\frac{34}{5}$	5	4	$\frac{6}{5}$	$\frac{1}{15}$	$\frac{2}{15}$	0	312
	$c_j - z_j$	0	$-\frac{14}{5}$	0	0	$-\frac{6}{5}$	$-\frac{1}{15}$	$-\frac{2}{15}$	0	

in the new constraint, the value of x_8 should be given by

$$x_8 = 220 - 3x_1 - 6x_2 - 2x_3 - 4x_4$$

Substituting the values of the basic variables x_1, x_3, and x_4 along with the value of the nonbasic variable x_2 in this equation gives

$$x_8 = 220 - 3(60) - 6(0) - 2(20) - 4(8) = -32$$

The fact that $x_8 = -32$ is also shown by the solution column in Table 7.22.

The revised tableau shown in Table 7.22 is dual feasible but not primal feasible. Primal feasibility can be regained, however, by applying the dual simplex algorithm. The solution is found after one iteration and is shown in Table 7.23.

Table 7.23

c_b	c_j Basis	3 x_1	4 x_2	5 x_3	4 x_4	0 x_5	0 x_6	0 x_7	0 x_8	Solution
4	x_4	0	$\frac{1}{5}$	0	1	$\frac{1}{5}$	0	$-\frac{1}{5}$	0	8
3	x_1	1	$\frac{3}{2}$	0	0	$-\frac{7}{10}$	0	$\frac{1}{4}$	$\frac{1}{2}$	44
5	x_3	0	$\frac{7}{20}$	1	0	$\frac{9}{20}$	0	$\frac{1}{40}$	$-\frac{1}{4}$	28
0	x_6	0	$-\frac{15}{4}$	0	0	$\frac{15}{4}$	1	$\frac{7}{8}$	$-\frac{15}{4}$	120
	z_j	3	$\frac{141}{20}$	5	4	$\frac{19}{20}$	0	$\frac{3}{40}$	$\frac{1}{4}$	304
	$c_j - z_j$	0	$-\frac{61}{20}$	0	0	$-\frac{19}{20}$	0	$-\frac{3}{40}$	$-\frac{1}{4}$	

Example: The Hall Manufacturing Company produces four types of electronic subassemblies for use in aircraft avionic equipment. The parts and labor required for each product along with the resource availabilities and profit are given in Chapter 5, p. 159.

Hall recently received a directive from the Department of Transportation that establishes minimum requirements for the operational life of each subassembly. Although management is confident that their products equal or exceed these requirements, it will nevertheless be necessary to conduct an additional test on each subassembly. Unfortunately, test facilities for the test have a capacity of only 90 units. Determine the revised production schedule based on this new constraint. To simplify the calculations, assume that the original requirement for at least forty units of product 1 no longer applies.

The original linear programming problem was

$$\text{Maximize:} \quad P = 4.25x_1 + 6.25x_2 + 5.00x_3 + 4.50x_4$$

$$
\begin{aligned}
\text{Subject to:} \quad 6.0x_1 + 5.0x_2 + 3.5x_3 + 4.0x_4 &\leq 600 \\
1.0x_1 + 1.5x_2 + 1.2x_3 + 1.2x_4 &\leq 120 \\
4x_1 + 3x_2 + 3x_3 + 3x_4 &\leq 400 \\
2x_1 + 2x_2 + 2x_3 + 3x_4 &\leq 300 \\
x_j \geq 0 \quad \text{for } j = 1, 2, 3, 4
\end{aligned}
$$

The optimal tableau for this problem is given in Table 7.24.

The profit from producing the subassemblies is maximum for $x_1 = 76.9$, $x_2 = 20.5$, $x_3 = 10.3$, and $x_4 = 0$ units. The total of 107.7 units manufactured, however, clearly exceeds the capacity of the test facilities. The requirement for testing each subassembly for minimum operational life can be included in the problem by adding the constraint

$$x_1 + x_2 + x_3 + x_4 \leq 90$$

Table 7.24

c_b	c_j	4.25	6.25	5.00	4.50	0	0	0	0	
	Basis	x_1	x_2	x_3	x_4	x_5	x_6	x_7	x_8	Solution
4.25	x_1	1	0	0	0.115	0.231	−1.154	0.192	0	76.9
6.25	x_2	0	1	0	0.231	0.462	1.026	−0.949	0	20.5
5.00	x_3	0	0	1	0.615	−0.769	0.513	1.026	0	10.3
0	x_8	0	0	0	1.077	0.154	−0.769	−0.538	1	84.6
	z_j	4.25	6.25	5.00	5.01	0.024	4.07	0.015	0	506.46
	$c_j - z_j$	0	0	0	−0.51	−0.024	−4.07	−0.015	0	

to the tableau of Table 7.24. This gives the tableau shown in Table 7.25. The value of the slack variable for the new constraint was found by solving the system of constraining equations for the basic variables, i.e., by completing the identity matrix for the basic variables.

Table 7.25

c_b	c_j	4.25	6.25	5.00	4.50	0	0	0	0	0	
	Basis	x_1	x_2	x_3	x_4	x_5	x_6	x_7	x_8	x_9	Solution
4.25	x_1	1	0	0	0.115	0.231	−1.154	0.192	0	0	76.9
6.25	x_2	0	1	0	0.231	0.462	1.026	−0.949	0	0	20.5
5.00	x_3	0	0	1	0.615	−0.769	0.513	1.026	0	0	10.3
0	x_8	0	0	0	1.077	0.154	−0.769	−0.538	1	0	84.6
0	x_9	0	0	0	0.039	0.076	−0.385	−0.269	0	1	−17.7
	z_j	4.25	6.25	5.00	5.008	0.024	4.073	0.015	0	0	506.46
	$c_j - z_j$	0	0	0	−0.508	−0.024	−4.073	−0.015	0	0	

The tableau shown in Table 7.25 is dual feasible but not primal feasible. Primal feasibility can be restored by the dual simplex algorithm. The solution, obtained after two iterations, is $x_1 = 30$, $x_2 = 60$, $x_3 = 0$, and $x_4 = 0$. Interestingly, the profit for this schedule decreases by only \$3.96 from the previous optimum. Profit, after including the new constraint, is \$502.50.

PROBLEMS

1. Express the following linear programming problems in the dual format.

(a) Minimize: $C = 2x_1 + 3x_2$
     Subject to: $13x_1 + 18x_2 \geq 234$
                   $16x_1 + 14x_2 \geq 214$
                   $x_1, x_2 \geq 0$

(b) Minimize: $C = 4x_1 + 3x_2$
     Subject to: $x_1 \qquad \geq 4$
                   $x_2 \geq 2$
                   $3x_1 + 2x_2 \geq 18$
                   $x_1, x_2 \geq 0$

(c) Minimize: $C = 2x_1 + x_2$
     Subject to: $12x_1 + 7x_2 \geq 42$
                   $5x_1 + 12x_2 \geq 60$
                   $3x_1 + 21x_2 \geq 63$
                   $x_1, x_2 \geq 0$

(d) Maximize: $Z = 2x_1 + 4x_2 + 3x_3$
     Subject to: $x_1 + 2x_2 \qquad \leq 80$
                   $x_1 + 4x_2 + 2x_3 \leq 120$
                   $x_1, x_2, x_3 \geq 0$

2. Determine the optimal values of the primal and dual variables in Problem 1.

3. Express the following problems in the dual format. Determine the optimal values of the primal and dual variables.

(a) Minimize: $C = 40x_1 + 120x_2$
     Subject to: $x_1 - 2x_2 \leq 8$
                   $15x_1 + 44x_2 \leq 660$
                   $3x_1 + 5x_2 = 90$
                   $x_1, x_2 \geq 0$

(b) Maximize: $Z = 8x_1 + 12x_2$
     Subject to: $-4x_1 + 5x_2 \geq 20$
                   $11x_1 - 4x_2 = 0$
                   $5x_1 + 4x_2 \leq 40$
                   $x_1, x_2 \geq 0$

4. Use the dual simplex algorithm to determine the solution to the following linear programming problems.

(a)
$$\begin{aligned}
\text{Minimize:} \quad C &= x_1 + x_2 \\
\text{Subject to:} \quad 5x_1 + 3x_2 &\geq 30 \\
6x_1 + 16x_2 &\geq 110 \\
x_1, x_2 &\geq 0
\end{aligned}$$

(b)
$$\begin{aligned}
\text{Minimize:} \quad C &= 2x_1 + 3x_2 \\
\text{Subject to:} \quad 12x_1 + 8x_2 &\geq 48 \\
5x_1 + 12x_2 &\geq 64 \\
3x_1 + 18x_2 &\geq 60 \\
x_1, x_2 &\geq 0
\end{aligned}$$

(c)
$$\begin{aligned}
\text{Minimize:} \quad C &= 3x_1 + 5x_2 + 2x_3 \\
\text{Subject to:} \quad x_1 + 2x_2 + 3x_3 &\geq 64 \\
3x_1 + 2x_2 + 4x_3 &\geq 86 \\
x_1, x_2, x_3 &\geq 0
\end{aligned}$$

(d)
$$\begin{aligned}
\text{Minimize:} \quad C &= 3x_1 + 5x_2 \\
\text{Subject to:} \quad 5x_1 + 10x_2 &\geq 60 \\
8x_1 + 6x_2 &\geq 54 \\
x_1, x_2 &\geq 0
\end{aligned}$$

5. Solve the following linear programming problems by (1) using the primal simplex algorithm to restore dual feasibility and (2) using the dual simplex algorithm to restore primal feasibility. (*Hint:* Refer to Table 7.11 for an illustration of this approach).

(a)
$$\begin{aligned}
\text{Minimize:} \quad C &= 5x_1 + 6x_2 \\
\text{Subject to:} \quad x_1 + 3x_2 &\geq 76 \\
6x_1 + 7x_2 &\leq 198 \\
x_1, x_2 &\geq 0
\end{aligned}$$

(b)
$$\begin{aligned}
\text{Maximize:} \quad Z &= 5x_1 + 4x_2 + 6x_3 \\
\text{Subject to:} \quad 2x_1 + 3x_2 + 5x_3 &\leq 125 \\
x_1 + x_2 + x_3 &\geq 15 \\
6x_1 + 2x_2 + 2x_3 &\leq 150 \\
x_1, x_2, x_3 &\geq 0
\end{aligned}$$

6. A firm uses three machines in the manufacture of three products. Each unit of product 1 requires 3 hours on machine 1, 2 hours on machine 2, and 1 hour on machine 3. Each unit of product 2 requires 4 hours on machine 1, 1 hour on machine 2, and 3 hours on machine 3. Each unit of product 3 requires 2 hours on machine 1, 2 hours on machine 2, and 2 hours on machine 3. The contribution margin of the three products is $25, $42, and $30 per unit, respectively. Sixty hours of machine 1 time, 36 hours of machine 2 time, and 62 hours of machine 3 time are available for scheduling. The linear programming formulation of this problem is

$$\text{Maximize:} \quad Z = 25x_1 + 42x_2 + 30x_3$$
$$\text{Subject to:} \quad 3x_1 + 4x_2 + 2x_3 \le 60$$
$$2x_1 + x_2 + 2x_3 \le 36$$
$$x_1 + 3x_2 + 2x_3 \le 62$$
$$x_1, x_2, x_3 \ge 0$$

and the optimal tableau for the problem is

	c_j	25	42	30	0	0	0	
c_b	Basis	x_1	x_2	x_3	x_4	x_5	x_6	Solution
42	x_2	$\frac{1}{3}$	1	0	$\frac{1}{3}$	$-\frac{1}{3}$	0	8
30	x_3	$\frac{5}{6}$	0	1	$-\frac{1}{6}$	$\frac{2}{3}$	0	14
0	x_6	$-\frac{5}{3}$	0	0	$-\frac{2}{3}$	$-\frac{1}{3}$	1	10
	z_j	39	42	30	9	6	0	756
	$c_j - z_j$	-14	0	0	-9	-6	0	

(a) Determine the sensitivity of the basic variables to changes in the contribution margins of each of the three products.

(b) Determine the sensitivity of the basic variables to changes in the right-hand-side values of the constraining equations.

7. A fourth product can be added to the product mix described in Problem 6. The contribution margin of this product is $36 and the product requires 2 hours on machine 1, 2 hours on machine 2, and 4 hours on machine 3. Add this product to the tableau in Problem 6 and determine the new solution.

8. A new process is required for the manufacture of the three products in Problem 6. Each unit of product 1 requires 3 hours of process time, each unit of product 2 requires 4 hours of process time, and each unit of product 3 requires 5 hours of process time. Ninety-three hours of process time are available. Add this constraint to the tableau in Problem 6 and determine the new optimal solution.

9. The manager of a large cattle ranch must select a mix of feed for beef cattle. The minimum daily requirements of the cattle of three important nutritional elements and the number of units of each of these elements in two feeds is given in the following table.

Nutritional Element	Units of Nutritional Elements Contained in One Pound of		Minimum Requirement
	Feed 1	Feed 2	
A	6	2	54
B	2	2	30
C	2	4	60
Cost per pound	$0.04	$0.02	

Use the dual simplex algorithm to determine the minimum cost feed mix.

10. The manager of the cattle ranch referred to in Problem 9 has posed the following questions:

   (a) By how much could the price of either feed change before the optimal mix of feeds would change?

   (b) Would a 10 percent increase in the minimum daily requirements of each of the nutritional elements result in a 10 percent increase in cost? Why?

   (c) Suppose that a third feed, costing $0.03 per pound and having 4 units of nutritional element A, 3 units of nutritional element B, and 2 units of nutritional element C, become available. Would this change the optimal mix of feeds?

   (d) Assume that a fourth nutritional element became important. Each pound of feed 1 was found to contain 1.5 units of this element and each pound of feed 2 was found to contain 0.75 unit of the element. The feed mix must contain a minimum of 30 units of the nutritional element. Does this new requirement change the optimal solution in Problem 9? If so, determine the new solution.

SUGGESTED REFERENCES

The references for this chapter are listed in Chapter 5.

Chapter 8

The Transportation
and
Assignment Problems

By this point it should be obvious that linear programming is an important tool for the quantitative analysis of business decisions. The allocation of resources among competing products or activities so that profits are maximized or, alternatively, costs are minimized is a necessary function in a firm. Linear programming provides the manager the information necessary to make an optimal decision in these complex allocation problems.

This chapter considers problems that might be termed derivatives of the general linear programming problem. By this we mean that while certain problems can be formulated and solved as a linear programming problem by using the simplex algorithm, their special mathematical structure allows solution by much more efficient algorithms. Two types of problems for which such special algorithms exist are the transportation and the assignment problems. Examples of these important types of problems, along with the solution algorithms for the problems, are introduced in this chapter.

8.1 The Transportation Problem

The *transportation problem* derives its name from the problem of transporting homogeneous products from various sources of supply to several points of demand. The allocation of the products from the sources of supply to the points or demand is made with the objective of minimizing the total transportation costs or, alternatively, maximizing the profit from the sale of the products. To illustrate, consider a firm that has three factories (i.e., sources of supply) and four warehouses (i.e., points of demand). The cost of shipping from each factory to each warehouse depends on the distance the product must be shipped, freight rates, etc. These costs are shown in Table 8.1.

Table 8.1

Factory (F_i)	Warehouse (W_j)				Factory Capacity
	W_1	W_2	W_3	W_4	
F_1	0.30	0.25	0.40	0.20	100
F_2	0.29	0.26	0.35	0.40	250
F_3	0.31	0.33	0.37	0.30	150
Warehouse Requirement	100	150	200	50	500

The cost of shipping from factory i to warehouse j is represented by c_{ij}. The alternative values of c_{ij} are shown in the table. For instance, the cost of transporting one unit of product from factory 1 to warehouse 2 is $c_{12} = \$0.25$. The factory capacities and warehouse requirements are also given in the table. In this simplified example, the factory capacities and warehouse requirements both sum to 500 units. This represents an example of a *balanced* transportation problem. The more realistic case of the *unbalanced* transportation problem in which factory capacities and warehouse requirements are not equal is introduced later in this section. At this point it is necessary only to observe that an unbalanced transportation problem can be converted to a balanced transportation problem through the addition of an appropriate slack variable.

The objective in this problem is to develop a shipping schedule that minimizes the total transportation cost. The problem can be formulated by letting x_{ij} represent the number of units shipped from factory i to warehouse j. The objective is to minimize the total transportation costs subject to the constraints on factory capacities and warehouse requirements. For the example problem, this can be written as

$$\text{Minimize } C = \quad 0.30x_{11} + 0.25x_{12} + 0.40x_{13} + 0.20x_{14} + 0.29x_{21}$$
$$+ 0.26x_{22} + 0.35x_{23} + 0.40x_{24} + 0.31x_{31} + 0.33x_{32}$$
$$+ 0.37x_{33} + 0.30x_{34}$$

The warehouse constraints are

$$x_{11} + x_{21} + x_{31} = 100$$
$$x_{12} + x_{22} + x_{32} = 150$$
$$x_{13} + x_{23} + x_{33} = 200$$
$$x_{14} + x_{24} + x_{34} = 50$$

The factory constraints are

$$x_{11} + x_{12} + x_{13} + x_{14} = 100$$
$$x_{21} + x_{22} + x_{23} + x_{24} = 250$$
$$x_{31} + x_{32} + x_{33} + x_{34} = 150$$

and the nonnegativity constraints are

$$x_{ij} \geq 0, \qquad \text{for } i = 1, 2, 3 \quad \text{and} \quad j = 1, 2, 3, 4$$

This transportation problem can be solved by using the simplex algorithm. Because of the special structure of the constraints, however, alternative algorithms that reduce the computational burden are available. Before introducing these algorithms, it will be worthwhile to formally state the problem mathematically and to provide additional examples of the transportation problem.

The standard transportation problem is to optimize an objective function

$$Z = \sum_{i=1}^{m} \sum_{j=1}^{n} c_{ij}x_{ij}, \qquad \text{for } i = 1, 2, \ldots, m; j = 1, 2, \ldots, n \quad (8.1)$$

subject to the constraints that

$$\sum_{i=1}^{m} x_{ij} = D_j, \qquad \text{for } j = 1, 2, \ldots, n \qquad (8.2)$$

$$\sum_{j=1}^{n} x_{ij} = S_i, \qquad \text{for } i = 1, 2, \ldots, m \qquad (8.3)$$

and

$$x_{ij} \geq 0 \qquad \text{for all } i \text{ and } j$$

D_j and S_i are nonnegative integers that represent, respectively, the demand at the jth facility (warehouse, retail store, etc.) and the supply at the ith source (factory, warehouse, supplier, etc.). An additional requirement is that the sum of the demands equal the sum of the supplies, i.e.,

$$\sum_{i=1}^{m} S_i = \sum_{j=1}^{n} D_j \qquad (8.4)$$

For the unbalanced transportation problem, the requirement that the sum of the demands equal the sum of the supplies can be satisfied by creating a fictitious demand facility or supply facility. For instance, if supply exceeds demand, a fictitious demand column is established with zero transportation cost to absorb the excess supply. Similarly, if demand exceeds supply, a fictitious supply row is established with zero transportation cost to absorb the unsatisfied demand. The addition of a fictitious supply or demand to convert an unbalanced transportation problem to the required balanced transportation format is illustrated by the following three examples.

Example: A national firm has three factories and four warehouses. The cost of manufacturing a given product is the same at each of the factories. Because of the locations of the warehouses and factories, the cost of shipping the product from the different factories to the warehouses varies. The capacities of factories 1, 2, and 3 are 5000 units per month, 4000 units per month, and 7000 units per month. The requirements of warehouses 1, 2, 3, and 4 are, respectively, 3000 units per month, 2500 units per month, 3500 units per month, and 4000 units per month. The transportation costs, factory capacities, and warehouse requirements are given in the following table.

| Factories | Warehouses | | | | Factory Capacity |
	W_1	W_2	W_3	W_4	
F_1	15	24	11	12	5000
F_2	25	20	14	16	4000
F_3	12	16	22	13	7000
Warehouse Requirements	3000	2500	3500	4000	13,000 16,000

Since the total factory capacity exceeds the total warehouse requirement, the transportation problem is unbalanced. It can be converted to a balanced transportation problem by adding a fictitious warehouse W_5 with requirements of 3000 units. The cost of shipping from each factory to this fictitious warehouse is $0. The balanced tableau is shown below.

Factory	W_1	W_2	W_3	W_4	W_5	Factory Capacity
F_1	15	24	11	12	0	5000
F_2	25	20	14	16	0	4000
F_3	12	16	22	13	0	7000
Warehouse Requirements	3000	2500	3500	4000	3000	16,000

The procedure for converting an unbalanced problem in which the demand exceeds the supply to a balanced problem is similar to that shown in the preceding table. The only difference is that a fictitious row rather than column must be added to the table. The costs associated with this row are again $0. The supply available from this fictitious source equals the number of units required to balance the table.

Example: A firm has decided to expand its product line by producing one or more of five possible products. Three of the firm's current plants have the excess capacity to produce these products. The profit margin on each of the proposed products is the same; consequently, the firm wants to minimize the cost of production. The excess capacity of each plant, the potential sales of each product, and the estimated cost of producing the proposed products at the three plants are given in the following table.

Plant	Product 1	2	3	4	5	Excess Capacity (standard units)
1	22	20	16	23	17	70
2	16	19	14	19	18	80
3	19	16	X	20	21	100
Potential Sales (standard units)	60	70	90	65	85	250 / 370

This problem could be treated as a standard linear programming problem and solved by using the simplex algorithm. If, however, the plants are considered as supply facilities and potential sales of the five products are treated as demands, the problem can also be viewed as a transportation problem.

To convert the table into a balanced transportation table, a fictitious plant capable of supplying 120 units of capacity must be added. This involves adding a fourth row with a capacity of 120 units and production costs of $0. In addition to this modification, the table shows that product 3 cannot be manufactured at plant 3. To insure that the optimal solution does not include the use of plant 3 to produce product 3, we can assign an arbitrarily large cost of production for this plant-product combination. Although any large cost would insure that production of product 3 is not scheduled for plant 3, we shall follow the "Big M" convention introduced in Chapter 6 and assign the letter M as the cost of production. The balanced transportation table is given below.

Plant	Product 1	2	3	4	5	Excess Capacity (Standard Units)
1	22	20	16	23	17	70
2	16	19	14	19	18	80
3	19	16	M	20	21	100
4	0	0	0	0	0	120
Potential Sales (standard units)	60	70	90	65	85	370

Example: Bevitz Furniture Company has requested bids from four furniture manufacturers on five different styles of furniture. The quantities of the five styles of furniture required by Bevitz are shown below.

Style	A	B	C	D	E
Quantity	125	75	50	200	175

The four manufacturers have limited production capacities. The total quantities that can be supplied in the time span available are shown below.

Manufacturer	1	2	3	4
Quantity	275	225	175	200

On the basis of the quotes from each manufacturer, Bevitz estimates that the profit per unit for each item sold will vary as shown in the following table.

		Style A	B	C	D	E
Manufacturer	1	28	35	42	23	15
	2	30	33	45	18	10
	3	25	35	48	20	13
	4	33	28	40	26	18

Determine the allocation that maximizes profit.

This problem can be formulated as a transportation problem. Instead of minimizing transportation cost, however, the objective is to maximize profit.

To apply the transportation algorithm, the problem must be converted to a balanced transportation problem. Since the capacity of the manufacturers exceeds the quantity demanded by Bevitz Furniture, the balance is achieved by adding a fictitious style. The demand for the fictitious style is equal to the unused capacity of the manufacturers, i.e., 250 units. The profit from the fictitious demand is $0. The balanced transportation table is shown below.

Manufacturer	Style						Quantity Supplied
	A	B	C	D	E	F	
1	28	35	42	23	15	0	275
2	30	33	45	18	10	0	225
3	25	35	48	20	13	0	175
4	33	28	40	26	18	0	200
Demand	125	75	50	200	175	250	875

8.2 The Initial Basic Feasible Solution

The first step in solving the transportation problem is to develop an initial basic feasible solution. Three methods for obtaining this solution are illustrated in this section. The first method, termed the *northwest corner rule*, provides a straightforward technique for obtaining the initial solution. It suffers, however, when compared to the *penalty method* and the *minimum (maximum) cell method* in that more iterations are normally required to obtain the optimal solution. The penalty method and the minimum cell method are introduced later in this section.

8.2.1 THE NORTHWEST CORNER RULE

To illustrate the northwest corner rule, consider the transportation problem summarized in Table 8.1, p. 275. The factory capacities and warehouse requirements for the problem are shown in Table 8.2. The transportation costs (i.e., the c_{ij}'s) are not included in this table. These costs are not relevant when one is using the northwest corner rule to determine the initial solution.

The initial solution is found by beginning in the upper left-hand (i.e., northwest) corner of the tableau and allocating the resource of the first row to the cells in the first row until the resource is exhausted. If a resource is exhausted and a requirement is satisfied by a single allocation, a zero is placed in a neighboring cell. We then move to the second row and continue the allocation until the resources of that row are fully allocated. Again, if a single allocation exhausts both the supply of resource and satisfies the requirement

Table 8.2

Factory	Warehouse W_1	W_2	W_3	W_4	Factory Capacity
F_1					100
F_2					250
F_3					150
Requirements	100	150	200	50	500

of a column, a zero must be placed in a neighboring cell. This process is continued until all resources are exhausted and all requirements are satisfied.

In this example, the 100 units of capacity of factory 1 is assigned to warehouse 1. This allocation exhausts the capacity of the first row and satisfies the requirements of the first column; consequently, we place a zero in the row 1, column 2 cell and move to the second row. Since the requirements of warehouse 1 have been met, the next allocation is 150 units of factory 2 capacity to warehouse 2. The remaining 100 units capacity of factory 2 is allocated to warehouse 3. One hundred units of factory 3 capacity completes the requirements of warehouse 3. The final 50 units of factory 3 capacity satisfies the requirements of warehouse 4. The initial solution is shown in Table 8.3.

Table 8.3

Factory	Warehouse W_1	W_2	W_3	W_4	Factory Capacity
F_1	100	0			100
F_2		150	100		250
F_3			100	50	150
Requirements	100	150	200	50	500

By comparing the solution in Table 8.3 with the constraints for the problem given on p. 275, it can be seen that the solution is feasible. Both the warehouse and the factory constraints are satisfied. The cost of this solution is found by multiplying the cell allocations in Table 8.3 by the transportation costs from Table 8.1 and summing these products. The cost of this initial solution is

$0.30(100) + $0.26(150) + $0.35(100) + $0.37(100) + $0.30(50) = $156.00

Example: Use the northwest corner rule to determine an initial solution for the transportation table shown on p. 277.

The factory capacities and the warehouse requirements, including the fictitious warehouse W_5, are given in the transportation table. The allocation is again made by beginning in the upper left-hand corner of the table and assigning capacities to requirements so as to exhaust all capacities and satisfy all requirements. The resulting feasible solution is shown below.

Factories	Warehouses W_1	W_2	W_3	W_4	W_5	Factory Capacity
F_1	3000	2000				5000
F_2		500	3500	0		4000
F_3				4000	3000	7000
Requirements	3000	2500	3500	4000	3000	16,000

The cost of this solution, again found by summing the products of the cell entries and the transportation costs, is $204,000.

The northwest corner rule provides a systematic, easily understandable method of obtaining an initial basic feasible solution. As mentioned earlier, however, it is quite inefficient in comparison with alternative methods for obtaining the initial solution. This inefficiency occurs because the costs (profits) are not considered in determining the initial solution. Although the algorithm used to determine the optimal solution is not introduced until the following section, it is reasonable to conclude that the number of iterations required to obtain the optimal solution is dependent upon how near the initial solution is to being optimal. This implies that the number of iterations can be reduced by beginning the iterative process from a "near optimal" initial solution. Costs (profits) are used to determine the initial solution in both the minimum (maximum) cell method and the penalty method. Consequently, the number of iterations required to obtain the optimal solution is normally reduced by beginning with an initial solution found by using one of these methods.

8.2.2 THE MINIMUM (MAXIMUM) CELL METHOD

To illustrate the minimum (maximum) cell method of obtaining an initial solution, we again consider the transportation problem summarized in Table 8.1. This table, complete with transportation costs, capacities, and requirements, is reproduced as Table 8.4. Notice that the transportation costs from Table 8.1 are placed in the upper left-hand corner of each cell in Table 8.4.

Table 8.4

Factory	Warehouse W_1	W_2	W_3	W_4	Factory Capacity
F_1	0.30	0.25	0.40	0.20	100
F_2	0.29	0.26	0.35	0.40	250
F_3	0.31	0.33	0.37	0.30	150
Requirements	100	150	200	50	500

The initial solution is obtained by sequentially allocating the resources to the cells with the minimum cost (or alternatively, with the maximum profit). As in the northwest corner rule, if a single allocation exhausts both the capacity of a row and satisfies the requirements of a column, a zero is placed in one of the cells that borders the allocation. Referring to Table 8.4, we see that the minimum transportation cost of $0.20 per unit is between factory 1 and warehouse 4. Since the requirements of warehouse 4 are 50 units, this allocation is entered in cell (1, 4). The allocation is shown in Table 8.5 in the lower right-hand corner of cell (1, 4).

Referring again to Table 8.4, we see that the next least costly allocation is the $0.25 per unit transportation cost from factory 1 to warehouse 2. The 50 units of unallocated capacity from factory 1 are, therefore, allocated to warehouse 2. This allocation is shown in cell (1, 2) in Table 8.5.

Again with reference to the table, the next least costly allocation is the $0.26 per unit transportation cost from factory 2 to warehouse 2. The remaining 100 units required by warehouse 2 thus come from factory 2. The allocation is entered in cell (2, 2) in Table 8.5.

The procedure continues by allocating 100 units from factory 2 to warehouse 1, thereby satisfying the demand of warehouse 1. The remaining 50 units capacity of factory 2 are next allocated to warehouse 3. The initial solution is completed by allocating the 150 units of factory 3 capacity to warehouse 3. The total cost of this initial solution is $150.50. This cost is $5.50 less than the cost of the initial solution obtained by using the northwest corner method.

Table 8.5

Factory	Warehouse W_1	W_2	W_3	W_4	Factory Capacity
F_1	0.30	0.25 50	0.40	0.20 50	100
F_2	0.29 100	0.26 100	0.35 50	0.40	250
F_3	0.31	0.33	0.37 150	0.30	150
Requirements	100	150	200	50	500

Example: Use the minimum (maximum) cell method to obtain an initial solution for the transportation problem described on p. 275. Compare this solution with that found by using the northwest corner method.

The transportation table for this problem is reproduced below. The transportation costs are entered in the upper left-hand corner of each cell. The initial allocation of factory capacity to warehouse requirements will be entered in the lower right-hand corner of the cell.

Factory	Warehouse W_1	W_2	W_3	W_4	W_5	Factory Capacity
F_1	15	24	11	12	0	5000
F_2	25	20	14	16	0	4000
F_3	12	16	22	13	0	7000
Requirements	3000	2500	3500	4000	3000	16,000

The transportation costs from each source to the fictitious warehouse W_5 are \$0. Using the minimum cell method, we can allocate units from any of the three factories to W_5. We arbitrarily allocate 3000 units from F_3 to W_5.

The next least costly transportation charge is from factory 1 to warehouse 3. Allocating 3500 units of F_1 capacity to W_3 satisfies the requirements of W_3. The process of allocating the factory capacities to the warehouses with the smallest transportation cost is continued until all factory capacities are exhausted and warehouse requirements are satisfied. The initial solution is shown below.

Factory	W_1	W_2	Warehouse W_3	W_4	W_5	Factory Capacity
F_1	15	24	11 3500	12 1500	0	5000
F_2	25	20 2500	14	16 1500	0	4000
F_3	12 3000	16	22	13 1000	0 3000	7000
Requirements	3000	2500	3500	4000	3000	16,000

The total transportation cost of this initial solution is $179,500. This represents a reduction of $24,500 from the initial solution of $204,000 obtained by using the northwest corner rule.

Example: Use the minimum (maximum) cell method to determine the initial solution for the Bevitz Furniture Company problem.

The balanced transportation table for the Bevitz Furniture Company was given on p. 280. Since the Bevitz problem involves maximizing profit rather than minimizing transportation cost, the allocations are made to the cells with the largest c_{ij} values rather than to those with the smallest. The initial solution using the maximum cell method is given below. The total profit of this initial solution is $16,100.

Manufacturer	A	B	Style C	D	E	F	Quantity Supplied
1	28	35 75	42	23 125	15 75	0	275
2	30	33	45	18	10 100	0 125	225
3	25	35	48 50	20	13	0 125	175
4	33 125	28	40	26 75	18	0	200
Demand	125	75	50	200	175	250	875

8.2.3 THE PENALTY METHOD

The penalty method, also known as *Vogel's approximation method*, involves determining the penalty for not assigning a resource to a requirement. For a minimization problem, the penalties for each resource (row) are determined by subtracting the smallest c_{ij} value in the row from the next smallest. Similarly, the penalty values for each requirement (column) are determined by subtracting the smallest c_{ij} value in each column from the next smallest. The objective is to make the initial allocation such that the penalties from not using the cells are minimized. This can be done by allocating the resources to the requirements so as to avoid large penalties. The rules for determining the initial allocation in a minimization problem are

1. Subtract the smallest c_{ij} value in each row from the next smallest. Place this number at the end of the row.
2. Subtract the smallest c_{ij} value in each column from the next smallest. Place this number at the foot of the column.
3. Determine the largest penalty for either row or column. Allocate as many units as possible to the cell with the largest penalty cost so as to avoid this penalty. If a row resource is exhausted and a column requirement satisfied by a single allocation, place a zero in a neighboring cell.
4. Recalculate the penalty cost, ignoring any row whose resources are exhausted or column whose requirements are satisfied.
5. Repeat Steps 3 and 4 until all allocations have been made.

The procedure for determining the initial allocation in a maximization problem is quite similar to that described above for the minimization problem. The only differences are (1) the penalties are calculated from the largest and next largest c_{ij} values in each row and each column, and (2) the allocations are made so as to maximize, rather than minimize, these penalties.

To illustrate the penalty method, consider again the transportation problem given by Table 8.1. The transportation table together with the initial penalties are shown in Table 8.6.

The largest penalty is that associated with the W_4 column. In order to avoid this $0.10 penalty, units from factory 1 must be allocated to warehouse 4. If this allocation is not made, an additional cost of at least $0.10 per unit will be incurred in transporting units to warehouse 4. This allocation, together with penalties for rows with unexhausted capacity and the columns with unsatisfied demand, is shown in Table 8.7.

The largest penalty in Table 8.7 is the $0.05 associated with factory 1. This penalty can be avoided by allocating factory 1 capacity to warehouse 2. No-

Table 8.6

Factory	Warehouse W_1	W_2	W_3	W_4	Factory Capacity	Penalty
F_1	0.30	0.25	0.40	0.20	100	0.05
F_2	0.29	0.26	0.35	0.40	250	0.03
F_3	0.31	0.33	0.37	0.30	150	0.01
Requirements	100	150	200	50	500	
Penalty	0.01	0.01	0.02	0.10		

Table 8.7

Factory	Warehouse W_1	W_2	W_3	W_4	Factory Capacity	Penalty
F_1	0.30	0.25	0.40	0.20　50	100	0.05
F_2	0.29	0.26	0.35	0.40	250	0.03
F_3	0.31	0.33	0.37	0.30	150	0.02
Requirements	100	150	200	50	500	
Penalty	0.01	0.01	0.02	—		

tice that since warehouse 4 requirements are satisfied, no penalty is given for warehouse 4. Furthermore, the transportation cost from factory 1 to warehouse 4 cannot be used in determining the penalty for factory 1. The allocation of factory 1 capacity to warehouse 2 is shown in Table 8.8.

The largest penalty in Table 8.8 is the $0.07 associated with warehouse 2. This penalty is avoided by allocating factory 2 capacity to warehouse 2. The results are shown in Table 8.9.

The largest remaining penalties are the $0.06 associated with factories 2 and 3. The penalty associated with factory 2 can be avoided by allocating

Table 8.8

Factory	W₁	Warehouse W₂	W₃	W₄	Factory Capacity	Penalty
F_1	0.30	0.25 50	0.40	0.20 50	100	—
F_2	0.29	0.26	0.35	0.40	250	0.03
F_3	0.31	0.33	0.37	0.30	150	0.02
Requirements	100	150	200	50	500	
Penalty	0.01	0.07	0.02	—		

Table 8.9

Factory	W₁	Warehouse W₂	W₃	W₄	Factory Capacity	Penalty
F_1	0.30	0.25 50	0.40	0.20 50	100	—
F_2	0.29	0.26 100	0.35	0.40	250	0.06
F_3	0.31	0.33	0.37	0.30	150	0.06
Requirements	100	150	200	50	500	
Penalty	0.02	—	0.02	—		

factory 2 capacity to warehouse 1. The initial assignment is completed by allocating the unused capacity of factories 2 and 3 to warehouse 3. The initial solution is shown in Table 8.10.

The solutions found by using the penalty method and the minimum cell method are identical in this example. Although this often happens in relatively small problems, this is not a general rule.

The penalty method and the minimum cell method are both used in practice to determine the initial solution. Seeing the number of tableaus required

Table 8.10

Factory	Warehouse				Factory Capacity	Penalty
	W_1	W_2	W_3	W_4		
F_1	0.30	0.25	0.40	0.20	100	—
		50		50		
F_2	0.29	0.26	0.35	0.40	250	—
	100	100	50			
F_3	0.31	0.33	0.37	0.30	150	—
			150			
Requirements	100	150	200	50	500	
Penalty	—	—	—	—		

to obtain the initial solution, the reader may question the need for the penalty method. It should be pointed out that much of the repetition required to explain the technique becomes unnecessary after the technique is understood. To verify this statement, the reader should return to the examples in Sec. 8.1 and determine an initial solution, using this method.

8.3 The Stepping Stone Algorithm

After an initial solution to the transportation problem is obtained, alternative solutions must be evaluated. A straightforward method of calculating the effect of alternative allocations is provided by the *stepping stone* algorithm.

To apply the stepping stone algorithm, we must first verify that the initial solution is a basic solution. The reader will remember that in the linear programming problem, a basic solution is found for a problem of n variables and m constraints by equating $n - m$ of the variables to zero and solving the resulting system of m equations and m variables simultaneously. The procedure differs slightly in a transportation problem.

The transportation problem is constructed so that there are $m + n$ equations, where n represents the number of requirements and m represents the number of resources. It can be shown that one of these $m + n$ equations is redundant and that there are only $m + n - 1$ independent equations. For a solution to be basic, therefore, $m + n - 1$ cells must be occupied. This means that $m + n - 1$ cells in the initial solution must contain either an allocation or a zero.

Example: Show that one of the equations is redundant in the following transportation problem.

Factory	Warehouse W_1	W_2	Factory Capacity
F_1	0.30	0.25	100
F_2	0.29	0.26	150
Requirements	150	100	250

The problem is

Minimize: $C = 0.30x_{11} + 0.25x_{12} + 0.29x_{21} + 0.26x_{22}$
Subject to:

$$x_{11} + x_{12} \qquad\qquad\qquad = 100$$
$$x_{21} + \quad x_{22} = 150$$
$$x_{11} \qquad\qquad + \quad x_{21} \qquad\qquad = 150$$
$$x_{12} \qquad\qquad + \quad x_{22} = 100$$
$$x_{ij} \geq 0$$

Adding the first two equations gives

$$x_{11} + x_{12} + x_{21} + x_{22} = 250$$

Subtracting the third equation from this sum gives

$$x_{12} + x_{22} = 100$$

Since the first three equations can be combined to produce the fourth, one of the equations is redundant and there are only $m + n - 1$ independent equations, i.e., three independent equations. The initial solution to this problem should, therefore, contain $n + m - 1$, or three, occupied cells.

If the initial solution contains less than $n + m - 1$ occupied cells, the solution is termed *degenerate*. Degeneracy occurs only when the resources of a row are exhausted and the requirements of a column are satisfied by a single allocation. It is eliminated by placing a zero in a cell that borders the allocation. The zero-valued cell is then considered occupied when applying the stepping stone algorithm.

The stepping stone algorithm involves transferring one unit from an occupied to an unoccupied cell and calculating the change in the objective function. The transfer must be made so as to retain the column and row equalities of the problem. After all unoccupied cells have been evaluated, a reallocation is made to the cell that provides the greatest per unit change in the objective function. Any degeneracies caused by the transfer of units must be removed by placement of zeros in the appropriate cells. The process of evaluating the empty cells and reallocating the units is continued until no further improvement in the objective function is possible. This final allocation is the optimal solution.

To illustrate the stepping stone algorithm, consider the transportation problem summarized by Table 8.1. The initial solution to this problem, found

Table 8.11

Factory	Warehouse				Factory Capacity
	W_1	W_2	W_3	W_4	
F_1	0.30 100	0.25 0	0.40	0.20	100
F_2	0.29	0.26 150	0.35 100	0.40	250
F_3	0.31	0.33	0.37 100	0.30 50	150
Requirements	100	150	200	50	500

by using the northwest corner rule, was given in Table 8.3. This solution is repeated in Table 8.11. The transportation costs, i.e., the c_{ij}'s, are included in the table.

To determine the effect on the objective function of transferring one unit to an unoccupied cell, we must find a *closed path* between the unoccupied cell and occupied cells. The path consists of a series of steps leading from the unoccupied cell to occupied cells and back to the unoccupied cell. In the case of cell (2, 1), for instance, a closed path consists of the series of steps from this cell to cell (1, 1), from cell (1, 1) to cell (1, 2), from cell (1, 2) to cell (2, 2), and from cell (2, 2) back to cell (2, 1). By following this path, we can determine the effect on the objective function of allocating a unit to cell (2, 1).

To illustrate, assume that one unit is allocated to cell (2, 1). In order to maintain the column and row equalities in the problem, a unit must be subtracted from cell (1, 1), added to cell (2, 1), and subtracted from cell (2, 2). Notice that this reallocation of units follows the closed path for cell (2, 1).

The net change in the objective function from the reallocation of one unit to cell (2, 1) can be found by adding and subtracting the appropriate transportation costs. Again, the closed path is followed; adding one unit to cell (2, 1) increases the objective function by $0.29, subtracting the unit from cell (1, 1) reduces the objective function by $0.30, adding the unit to cell (1, 2) increases the objective function by $0.25, and subtracting the unit from cell

(2, 2) reduces the objective function by \$0.26. The net decrease in the objective function is, therefore, \$0.02. This decrease can be represented in equation form by

$$F_2W_1 = +F_2W_1 - F_1W_1 + F_1W_2 - F_2W_2$$

or

$$F_2W_1 = +0.29 - 0.30 + 0.25 - 0.26 = -\$0.02$$

The net decrease of $-\$0.02$ is entered in the lower right-corner of cell (2, 1) in Table 8.12.

The effect of reallocating one unit to each of the other unoccupied cells is determined in the same manner. The computations are shown below. It is important to remember that the closed paths are established so as to maintain both column and row equalities.

$$F_3W_2 = +F_3W_2 - F_3W_3 + F_2W_3 - F_2W_2$$
$$F_3W_2 = +0.33 - 0.37 + 0.35 - 0.26 = +\$0.05$$
$$F_1W_3 = +F_1W_3 - F_2W_3 + F_2W_2 - F_1W_2$$
$$F_1W_3 = +0.40 - 0.35 + 0.26 - 0.25 = +\$0.06$$
$$F_2W_4 = +F_2W_4 - F_3W_4 + F_3W_3 - F_2W_3$$
$$F_2W_4 = +0.40 - 0.30 + 0.37 - 0.35 = +\$0.12$$
$$F_3W_1 = +F_3W_1 - F_3W_3 + F_2W_3 - F_2W_2 + F_1W_2 - F_1W_1$$
$$F_3W_1 = +0.31 - 0.37 + 0.35 - 0.26 + 0.25 - 0.30 = -\$0.02$$
$$F_1W_4 = +F_1W_4 - F_3W_4 + F_3W_3 - F_2W_3 + F_2W_2 - F_1W_2$$
$$F_1W_4 = +0.20 - 0.30 + 0.37 - 0.35 + 0.26 - 0.25 = -\$0.07$$

The net change in the objective function caused by reallocating one unit to each unoccupied cell is entered in Table 8.12.

Table 8.12

Factory	Warehouse W_1	W_2	W_3	W_4	Factory Capacity
F_1	0.30 100	0.25 0	0.40 +\$0.06	0.20 -\$0.07	100
F_2	0.29 -\$0.02	0.26 150	0.35 100	0.40 +\$0.12	250
F_3	0.31 -\$0.02	0.33 +\$0.05	0.37 100	0.30 50	150
Requirements	100	150	200	50	500

The dollar entries in the lower right-hand corner of each unoccupied cell in Table 8.12 represent the net change in the objective function from reallocating one unit to the cell. These are, in effect, equivalent to the $c_j - z_j$ values in the simplex tableau. The reader will remember that in the simplex algorithm for a minimization problem, the variable with the most negative $c_j - z_j$ entry is introduced into the basis. The same procedure is followed in the stepping stone algorithm. Namely, a reallocation is made to the most favorable evaluation, i.e., the cell that provides the largest per unit decrease in the objective function for a minimization problem and the largest per unit increase in the objective function for a maximization problem. The reallocation follows the closed path used to calculate the change in the objective function. As in the simplex algorithm, as many units as possible are reallocated to the cell.

Referring to Table 8.12, note that the largest per unit decrease in the objective function comes from reallocating units to cell $(1, 4)$. The closed path used to evaluate cell $(1, 4)$ was

$$F_1W_4 = +F_1W_4 - F_3W_4 + F_3W_3 - F_2W_3 + F_2W_2 - F_1W_2$$

The limit on the number of units that can be reallocated to cell $(1, 4)$ is equal to the minimum of the current allocations to cells $(3, 4)$, $(2, 3)$, and $(1, 2)$. The table shows that 50 units can be subtracted from cell $(3, 4)$, 100 units can can be subtracted from cell $(2, 3)$, and 0 units can be subtracted from cell $(1, 2)$. Unfortunately, the closed path used to evaluate cell $(1, 4)$ involved subtracting units from cell $(1, 2)$. Since cell $(1, 2)$ has a zero allocation in the initial solution, allocating units to cell $(1, 4)$ would be equivalent to transferring the 0 entry in cell $(1, 2)$ to cell $(1, 4)$. This, of course, would not decrease the value of the objective function.

Rather than merely transferring the 0 entry from cell $(1, 2)$ to cell $(1, 4)$, units can be reallocated to a cell that decreases the value of the objective function. An allocation to cell $(2, 1)$ or cell $(3, 1)$ would decrease the objective function by \$0.02 per unit. We arbitrarily select cell $(2, 1)$ for reallocation.

The closed path used to evaluate cell $(2, 1)$ was

$$F_2W_1 = +F_2W_1 - F_1W_1 + F_1W_2 - F_2W_2$$

The limit on the number of units that can be added to cell $(2, 1)$ is the 100 units initially assigned to cell $(1, 1)$. This is due to the fact that units must be subtracted from cell $(1, 1)$, and cell $(1, 1)$ contains only 100 units. These units are reallocated to cell $(2, 1)$. In order to maintain the column and row equalities, 100 of the 150 units in cell $(2, 2)$ are reallocated to cell $(1, 2)$. This leaves 50 units in cell $(2, 2)$. The new transportation table is shown by Table 8.13.

The solution shown in Table 8.13 contains six occupied cells. Since there are three rows and four columns in the problem and the number of occupied

Table 8.13

Factory	Warehouse W_1	W_2	W_3	W_4	Factory Capacity
F_1	0.30	0.25 100	0.40	0.20	100
F_2	0.29 100	0.26 50	0.35 100	0.40	250
F_3	0.31	0.33	0.37 100	0.30 50	150
Requirements	100	150	200	50	500

cells is equal to $m + n - 1$, the solution is not degenerate. Therefore, we need not consider any of the blank cells as occupied at a zero level.

The stepping stone algorithm is used to evaluate the unoccupied cells in Table 8.14. The calculations are shown below.

$$F_1W_1 = +F_1W_1 - F_1W_2 + F_2W_2 - F_2W_1 = +\$0.02$$
$$F_3W_1 = +F_3W_1 - F_3W_3 + F_2W_3 - F_2W_1 = \quad \$0.00$$
$$F_3W_2 = +F_3W_2 - F_3W_3 + F_2W_3 - F_2W_2 = +\$0.05$$
$$F_1W_3 = +F_1W_3 - F_2W_3 + F_2W_2 - F_1W_2 = +\$0.06$$
$$F_1W_4 = +F_1W_4 - F_3W_4 + F_3W_3 - F_2W_3 + F_2W_2 - F_1W_2 = -\$0.07$$
$$F_2W_4 = +F_2W_4 - F_3W_4 + F_3W_3 - F_2W_3 = +\$0.12$$

The evaluations are entered in Table 8.14.

Table 8.14

Factory	Warehouse W_1	W_2	W_3	W_4	Factory Capacity
F_1	0.30 +$0.02	0.25 100	0.40 +$0.06	0.20 -$0.07	100
F_2	0.29 100	0.26 50	0.35 100	0.40 +$0.12	250
F_3	0.31 $0.00	0.33 +$0.05	0.37 100	0.30 50	150
Requirements	100	150	200	50	500

The table shows that units should be reallocated to cell (1, 4). The reallocation is made by following the closed path that gave the $0.07 per unit decrease in the objective function for the cell. The reallocation is shown in Table 8.15.

Table 8.15

Factory	Warehouse W_1	W_2	W_3	W_4	Factory Capacity
F_1	0.30	0.25 — 50	0.40	0.20 — 50	100
F_2	0.29 — 100	0.26 — 100	0.35 — 50	0.40	250
F_3	0.31	0.33	0.37 — 150	0.30	150
Requirements	100	150	200	50	500

The stepping stone is again applied to evaluate the unoccupied cells in Table 8.15. The calculations are shown below.

$$F_1W_1 = +F_1W_1 - F_2W_1 + F_2W_2 - F_1W_2 = +\$0.02$$
$$F_3W_1 = +F_3W_1 - F_3W_3 + F_2W_3 - F_2W_1 = \$0.00$$
$$F_3W_2 = +F_3W_2 - F_3W_3 + F_2W_3 - F_2W_2 = +\$0.05$$
$$F_1W_3 = +F_1W_3 - F_2W_3 + F_2W_2 - F_1W_2 = +\$0.06$$
$$F_2W_4 = +F_2W_4 - F_2W_2 + F_1W_2 - F_1W_4 = +\$0.19$$
$$F_3W_4 = +F_3W_4 - F_3W_3 + F_2W_3 - F_2W_2 + F_1W_2 - F_1W_4 = +\$0.07$$

Since all the evaluations are nonnegative, the solution is optimal. This is not, however, a unique solution. The calculations show that the objective function does not change as units are allocated to cell (3, 1). Thus, multiple optimal solutions exist for this problem. The value of the objective function, found by calculating the total transportation cost, is $150.50.

Example: Use the stepping stone algorithm to determine the optimal allocation for the transportation problem introduced on p. 277.

The initial solution for this problem, found by using the northwest corner rule, was determined on p. 287 and is reproduced on p. 296.

| Factory | Warehouse | | | | | Factory Capacity |
	W_1	W_2	W_3	W_4	W_5	
F_1	15 3000	24 2000	11	12	0	5000
F_2	25	20 500	14 3500	16 0	0	4000
F_3	12	16	22	13 4000	0 3000	7000
Requirements	3000	2500	3500	4000	3000	16,000

The unoccupied cells are evaluated by establishing closed paths between each unoccupied cell and the occupied cells. The evaluations are

$$F_1W_3 = +F_1W_3 - F_2W_3 + F_2W_2 - F_1W_2 = -\$7$$

$$F_1W_4 = +F_1W_4 - F_2W_4 + F_2W_2 - F_1W_2 = -\$8$$

$$F_1W_5 = +F_1W_5 - F_3W_5 + F_3W_4 - F_2W_4 + F_2W_2 - F_1W_2 = -\$7$$

$$F_2W_1 = +F_2W_1 - F_1W_1 + F_1W_2 - F_2W_2 = +\$14$$

$$F_2W_5 = +F_2W_5 - F_3W_5 + F_3W_4 - F_2W_4 = -\$3$$

$$F_3W_1 = +F_3W_1 - F_3W_4 + F_2W_4 - F_2W_2 + F_1W_2 - F_1W_1 = +\$4$$

$$F_3W_2 = +F_3W_2 - F_3W_4 + F_2W_4 - F_2W_2 = -\$1$$

$$F_3W_3 = +F_3W_3 - F_3W_4 + F_2W_4 - F_2W_3 = +\$11$$

The most favorable evaluation is the $-\$8$ for cell $(1, 4)$. It is impossible to reallocate units to this cell, however, because of the zero allocation in cell $(2, 4)$.

The next most favorable evaluations are for cells $(1, 3)$ and $(1, 5)$. Reallocating units to cell $(1, 5)$ also requires subtracting units from cell $(2, 4)$. Since this would only mean transferring the zero from cell $(2, 4)$ to cell $(1, 5)$, cell $(1, 3)$ rather than cell $(1, 5)$ is selected to receive the allocation. The result is shown below. The evaluations of the unoccupied cells in the new solution are also given. The reader should verify these evaluations by determining the closed paths for each unoccupied cell and calculating the effect on the objective function of reallocating one unit to the unoccupied cell.

Total transportation cost can be reduced by reallocating units to cell $(3, 1)$. The reallocation, together with the evaluation of the empty cells, is given in the following table.

Factory	Warehouse W_1	W_2	W_3	W_4	W_5	Factory Capacity
F_1	15 3000	24 +$7	11 2000	12 -$1	0 $0	5000
F_2	25 +$7	20 2500	14 1500	16 0	0 -$3	4000
F_3	12 -$3	16 -$1	22 +$11	13 4000	0 3000	7000
Requirements	3000	2500	3500	4000	3000	16,000

Factory	Warehouse W_1	W_2	W_3	W_4	W_5	Factory Capacity
F_1	15 1500	24 +$4	11 3500	12 -$4	0 -$3	5000
F_2	25 +$10	20 2500	14 +$3	16 1500	0 -$3	4000
F_3	12 1500	16 -$1	22 +$14	13 2500	0 3000	7000
Requirement	3000	2500	3500	4000	3000	16,000

Transportation cost can be further reduced by reallocating units to cell (1, 4). This allocation is shown below.

Factory	Warehouse W_1	W_2	W_3	W_4	W_5	Factory Capacity
F_1	15 +$4	24 +$8	11 3500	12 1500	0 +$1	5000
F_2	25 +$10	20 2500	14 -$1	16 1500	0 -$3	4000
F_3	12 3000	16 -$1	22 +$10	13 1000	0 3000	7000
Requirement	3000	2500	3500	4000	3000	16,000

An additional reduction in transportation cost is possible by allocating units to cell (2, 5).

Factory	Warehouse W₁	W₂	W₃	W₄	W₅	Factory Capacity
F_1	15 +$4	24 +$5	11 3500	12 1500	0 +$1	5000
F_2	25 +$13	20 2500	14 +$2	16 +$3	0 1500	4000
F_3	12 3000	16 −$4	22 +$10	13 2500	0 1500	7000
Requirement	3000	2500	3500	4000	3000	16,000

Allocating units to cell (3, 2) gives the following table.

Factory	Warehouse W₁	W₂	W₃	W₄	W₅	Factory Capacity
F_1	15 +$4	24 +$9	11 3500	12 1500	0 +$5	5000
F_2	25 +$9	20 1000	14 −$2	16 −$1	0 3000	4000
F_3	12 3000	16 1500	22 +$10	13 2500	0 +$4	7000
Requirement	3000	2500	3500	4000	3000	16,000

The transportation costs is again reduced by reallocating units to cell (2, 3). The final reallocation, shown on the following page, gives the optimal solution. The total transportation cost is $167,000.

We stated earlier that the number of iterations required to obtain the optimal solution is dependent on how near the initial solution is to being optimal. This is demonstrated by the preceding example. Notice in this example that the fourth tableau (p. 297) contains the same allocations as the initial tableau found using the minimum cell method (p. 285). The iterations in the example leading to the fourth tableau would not have been necessary had we begun with the initial solution obtained by using the minimum cell method instead of the initial solution from the Northwest corner rule.

| Factory | Warehouse | | | | | Factory Capacity |
	W_1	W_2	W_3	W_4	W_5	
F_1	15 +$4	24 +$9	11 2500	12 2500	0 +$3	5000
F_2	25 +$11	20 +$2	14 1000	16 +$1	0 3000	4000
F_3	12 3000	16 2500	22 +$10	13 1500	0 +$2	7000
Requirement	3000	2500	3500	4000	3000	16,000

8.4 The Modified Distribution Algorithm†

The *modified distribution* algorithm, also referred to as the MODI method, offers an alternative approach for evaluating the unoccupied cells in a transportation tableau. The algorithm is introduced for two reasons. First, it provides a shortcut for evaluating the unoccupied cells. Second, and of equal importance, it can be used to demonstrate the relationship between the dual theorem of linear programming and the transportation problem.

To illustrate the modified distribution algorithm, consider the transportation problem first introduced in Table 8.1. The initial solution to this problem, found by using the northwest corner rule, was given in Table 8.3. This solution is repeated in Table 8.16.

In addition to the initial solution, Table 8.16 includes a column labeled u_i and a row labeled v_j. The numerical values of u_i and v_j are calculated from the equation

$$\Delta_{ij} = c_{ij} - u_i - v_j \qquad (8.5)$$

where Δ_{ij} represents the net change in the objective function of the transportation problem from reallocating one unit to cell (i, j) and c_{ij} represents the transportation cost from factory i to warehouse j.

The modified distribution algorithm consists of determining the values of u_i for the $i = 1, 2, \ldots, m$ rows and the values of v_j for the $j = 1, 2, \ldots, n$ columns of the transportation table. After the values of u_i and v_j have been determined, the change in the objective function caused by reallocating units can be determined by calculating the value of Δ_{ij} for each unoccupied cell. As in the stepping stone algorithm, the values of Δ_{ij} are entered in the unoccupied

---

† This section is optional at the discretion of the instructor.

Table 8.16

Factory	Warehouse W_1	W_2	W_3	W_4	Factory Capacity	u_i
F_1	0.30 100	0.25 0	0.40	0.20	100	
F_2	0.29	0.26 150	0.35 100	0.40	250	
F_3	0.31	0.33	0.37 100	0.30 50	150	
Requirement	100	150	200	50	500	
v_j						

cells. Units are then reallocated to the cell with the most favorable Δ_{ij}, i.e., the most favorable evaluation. After the reallocation, the algorithm is re-applied to the subsequent tableau. The procedure continues until no improvement in the objective function through reallocation of units is possible.

The method of determining the values of u_i and v_j is derived from the linear programming formulation of the transportation problem. The reader will remember that the change in the objective function from introducing a variable into the basis is given by the $c_j - z_j$ row of the simplex tableau. The change in the objective function (i.e., the $c_j - z_j$ value) associated with a variable currently in the basis is always zero. Since an occupied cell represents a variable in the current basis, the change in the objective function (i.e., the Δ_{ij} value) associated with that cell is zero. Consequently, the Δ_{ij} values for all occupied cells are zero.

In order to solve for u_i and v_j, we merely need to make Δ_{ij} equal zero for each occupied cell and solve the resulting system of equations simultaneously. The system of equations for the solution shown in Table 8.16 is

$$u_1 + v_1 = 0.30$$
$$u_1 + v_2 = 0.25$$
$$u_2 + v_2 = 0.26$$
$$u_2 + v_3 = 0.35$$
$$u_3 + v_3 = 0.37$$
$$u_3 + v_4 = 0.30$$

The system has six equations and seven variables. In order to obtain a solution, one of the variables must be specified. In the modified distribution algorithm, u_1 is customarily specified as equaling zero. In this example, equating u_1 with zero gives the following solution: $u_1 = \$0.00$; $v_1 = \$0.30$; $v_2 = \$0.25$; $u_2 = \$0.01$; $v_3 = \$0.34$; $u_3 = \$0.03$; and $v_4 = \$0.27$. These values are shown in Table 8.17.

Table 8.17

Factory	W_1	W_2	Warehouse W_3	W_4	Factory Capacity	u_i
F_1	0.30 100	0.25 0	0.40 +\$0.06	0.20 −\$0.07	100	\$0.00
F_2	0.29 −\$0.02	0.26 150	0.35 100	0.40 +\$0.12	250	0.01
F_3	0.31 −\$0.02	0.33 +\$0.05	0.37 100	0.30 50	150	0.03
Requirement	100	150	200	50	500	
v_j	\$0.30	0.25	0.34	0.27		

The change in the objective function from reallocating one unit to the unoccupied cells is also shown in Table 8.17. These values were calculated by formula (8.5), i.e., $\Delta_{ij} = c_{ij} - u_i - v_j$. The reader should note that the evaluations of unoccupied cells are the same for both the stepping stone algorithm, Table 8.12, and the modified distribution algorithm, Table 8.17.

The most favorable evaluation in Table 8.17 is $\Delta_{14} = -\$0.07$. As explained on p. 293, a reallocation to this cell would merely involve reassigning the zero allocation from cell (1, 2) to cell (1, 4). Since cells (2, 1) and (3, 1) were also favorably evaluated, a reallocation can instead be made to one of these cells. We again arbitrarily select cell (2, 1). The resulting tableau is given in Table 8.18.

The values of u_i and v_j in Table 8.19 are again calculated by equating Δ_{ij} to zero for the occupied cells and solving the resulting system of equations. Although the reader may wish to write out the system of equations, the structure of the transportation problem makes this step unnecessary. In Table 8.18, u_1 is specified as \$0.00. Since $c_{12} = u_1 + v_2$, it follows that $v_2 = \$0.25$. Similarly, $c_{22} = u_2 + v_2$, and u_2 is, therefore, \$0.01. The remaining values of u_i and v_j are determined in the same manner. The values of Δ_{ij} for the unoccupied cells are determined to complete the table. The evaluations found by using the modified distribution algorithm, Table 8.18, are again identical to those given by the stepping stone algorithm, Table 8.14.

Table 8.18

Factory	Warehouse				Factory Capacity	u_i
	W_1	W_2	W_3	W_4		
F_1	0.30 +$0.02	0.25 100	0.40 +$0.06	0.20 −$0.07	100	$0.00
F_2	0.29 100	0.26 50	0.35 100	0.40 +$0.12	250	0.01
F_3	0.31 $0.00	0.33 +$0.05	0.37 100	0.30 50	150	0.03
Requirement	100	150	200	50	500	
v_j	$0.28	0.25	0.34	0.27		

The final solution is obtained by reapplying the modified distribution algorithm. One additional tableau is required. This tableau is shown in Table 8.19. The solution is, of course, identical to that given by the stepping stone algorithm in Table 8.15.

Table 8.19

Factory	Warehouse				Factory Capacity	u_i
	W_1	W_2	W_3	W_4		
F_1	0.30 +$0.02	0.25 50	0.40 +$0.06	0.20 50	100	$0.00
F_2	0.29 100	0.26 100	0.35 50	0.40 +$0.19	250	0.01
F_3	0.31 $0.00	0.33 +$0.05	0.37 150	0.30 +$0.07	150	0.03
Requirements	100	150	200	50	500	
v_j	$0.28	0.25	0.34	0.20		

8.5 The Dual Theorem and the Transportation Problem†

In the discussion of the dual theorem in Chapter 7, we pointed out that this theorem provides the basis for many important extensions of linear pro-

---

† This section is optional at the discretion of the instructor.

gramming. Although many of these extensions must be reserved for advanced studies in operations research, we can offer the reader an indication of the importance of this theorem by illustrating the relationship between the dual theorem and the transportation problem.

To introduce the importance of the dual theorem to the transportation problem, consider the transportation problem summarized in Table 8.20.

Table 8.20

Factory	Warehouse W_1	W_2	W_3	Resource
F_1	5	4	4	200
F_2	5	3	6	300
Requirements	100	200	200	500

The linear programming formulation of this problem is

$$\text{Minimize:} \quad Z = 5x_{11} + 4x_{12} + 4x_{13} + 5x_{21} + 3x_{22} + 6x_{23}$$

$$
\begin{aligned}
\text{Subject to:} \quad & x_{11} + x_{12} + x_{13} && = 200 \\
& \qquad\quad x_{21} + x_{22} + x_{23} && = 300 \\
& x_{11} \qquad\quad + x_{21} && = 100 \\
& \quad x_{12} \qquad\quad + x_{22} && = 200 \\
& \qquad\quad x_{13} \qquad\quad + x_{23} && = 200 \\
& x_{ij} \geq 0
\end{aligned}
$$

Five variables are required for the dual formulation of this problem, two for the resource constraints and three for the requirement constraints. In Chapter 7 the dual variables were symbolized by w_i. We depart from this convention in this example by letting u_i represent the dual variables for the resource equalities (i.e., the first two equations) and v_j represent the dual variables for the requirement equalities (i.e., the final three equations). Based on this convention, the dual formulation is

$$\text{Maximize:} \quad P = 200u_1 + 300u_2 + 100v_1 + 200v_2 + 200v_3$$

$$
\begin{aligned}
\text{Subject to:} \quad & u_1 \qquad\quad + v_1 && \leq 5 \\
& u_1 \qquad\qquad\quad + v_2 && \leq 4 \\
& u_1 \qquad\qquad\qquad\quad + v_3 && \leq 4 \\
& \quad u_2 + v_1 && \leq 5 \\
& \quad u_2 \qquad + v_2 && \leq 3 \\
& \quad u_2 \qquad\qquad + v_3 && \leq 6
\end{aligned}
$$

$$u_i, v_j \text{ unrestricted in sign.}$$

The optimal solution could be determined by applying the simplex algorithm to either the dual or the primal formulation of the problem. Both formulations are shown above. Instead, an important relationship between the primal and dual problems is used to develop the modified distribution algorithm.

This relationship can be explained as follows. A basic solution to the primal problem occurs when $n + m - 1$ of the cells are occupied, i.e., when $n + m - 1$ of the primal variables are in the basis. By referring to the discussion of the dual formulation of the primal problem in Chapter 7, the reader will remember that the number of dual inequalities that are strictly satisfied is equal to the number of variables in the primal basis. Therefore, $n + m - 1$ of the dual inequalities are strictly satisfied. This means that the values of the dual variables corresponding to any primal basic solution can be determined by selecting the appropriate subset of $m + n - 1$ dual equations and solving for the $m + n$ dual variables. Since the subset of $m + n - 1$ dual equations has $m + n$ dual variables, a solution can be obtained only if the value of one of the dual variables is specified. As stated earlier, it is customary to specify $u_1 = \$0.00$.

The dual formulation of the transportation problem given in Table 8.20 has six variables and five inequalities. Since there are four primal variables in the basis, four of the dual inequalities are strictly satisfied. The four equations corresponding to the northwest corner solution to the problem are

$$u_1 + v_1 = 5$$
$$u_1 + v_2 = 4$$
$$u_2 + v_2 = 3$$
$$u_2 + v_3 = 6$$

The solution to the system of equations, obtained by specifying $u_1 = 0$, is $u_1 = 0$, $v_1 = 5$, $v_2 = 4$, $u_2 = -1$, and $v_3 = 7$.

The right-hand-side constants in the system of dual inequalities are the c_{ij} values in the transportation table. Suppose that we again define Δ_{ij} by the equation

$$\Delta_{ij} = c_{ij} - u_i - v_j \tag{8.5}$$

Δ_{ij} represents the difference between the right-hand-side constant c_{ij} and the dual variables u_i and v_j. Δ_{ij} is, of course, equal to zero if cell (i, j) is occupied or, alternatively, if variable x_{ij} is in the basis of the primal problem. If cell (i, j) is not occupied, however, Δ_{ij} gives the change in the objective function in the primal problem from reallocating one unit to cell (i, j).

The values of Δ_{ij} in the example are $\Delta_{11} = 0$, $\Delta_{12} = 0$, $\Delta_{13} = -3$, $\Delta_{21} = 1$, $\Delta_{22} = 0$, and $\Delta_{23} = 0$. Since Δ_{13} is negative, the initial northwest corner solution can be improved by reallocating units to cell $(1, 3)$. Following the reallocation, a different subset of four equations is solved for the dual variables. The iterations continue until all Δ_{ij} values are positive.

The reader should by now be aware that the modified distribution algorithm consists of nothing more than solving selected subsets of dual equations to determine the values of the dual variables. At each iteration the subset of dual equations are those that are associated with the occupied cells. Rather than writing out the entire set of dual inequalities and selecting the appropriate subset of dual equations, the modified distribution algorithm can be used. This algorithm permits one to solve for the dual variables u_i and v_j directly from the transportation table.

Example: Show that the Δ_{ij} values in the modified distribution algorithm are equal to the $c_{ij} - z_{ij}$ values of the primal simplex solution to the transportation problem.

The fact that Δ_{ij} and $c_{ij} - z_{ij}$ are equivalent measures of the change in the objective function from reallocating one unit to cell (i, j) can be shown by referring to the problem described in Table 8.20. The primal formulation of this problem was given on p. 303. One of the five equations in the primal formulation is redundant and is, therefore, omitted from the simplex tableau. In this example we omit the final equation.

The simplex tableau that corresponds to the northwest corner solution is given below. This tableau was obtained by solving the first four equations for the basic variables shown in the tableau. These variables correspond, of course, to the northwest corner solution to the problem.

	c_{ij}	5	4	4	5	3	6	
c_b	Basis	x_{11}	x_{12}	x_{13}	x_{21}	x_{22}	x_{23}	Solution
5	x_{11}	1	0	0	1	0	0	100
4	x_{12}	0	1	1	-1	0	0	100
3	x_{22}	0	0	-1	1	1	0	100
6	x_{23}	0	0	1	0	0	1	200
	z_{ij}	5	4	7	4	3	6	
	$c_{ij} - z_{ij}$	0	0	-3	1	0	0	

The Δ_{ij} values were found earlier to be $\Delta_{11} = 0$, $\Delta_{12} = 0$, $\Delta_{13} = -3$, $\Delta_{21} = 1$, $\Delta_{22} = 0$, and $\Delta_{23} = 0$. These are equivalent to the $c_{ij} - z_{ij}$ values given in the simplex tableau.

8.6 The Assignment Problem

The *assignment problem*, like the transportation problem, is a special case of the linear programming problem. The assignment problem occurs when n jobs must be assigned to n facilities on a one-to-one basis. The assignment is made with the objective of minimizing the overall cost of completing the jobs, or, alternatively, of maximizing the overall profit from the jobs.

To illustrate the assignment problem, consider the following example. A firm has five jobs that must be assigned to five work crews. Because of varying experience of the work crews, each work crew is not able to complete each job with the same effectiveness. The cost of each work crew to do each job is given by the cost matrix shown in Table 8.21. The objective is to assign the jobs to the work crews so as to minimize the total cost of completing all jobs.

Table 8.21

Job

	j $\diagdown$ *i*	*1*	*2*	*3*	*4*	*5*
	1	41	72	39	52	25
Work	2	22	29	49	65	81
Crew	3	27	39	60	51	40
	4	45	50	48	52	37
	5	29	40	45	26	30

8 6.1   ENUMERATING ALL POSSIBLE ASSIGNMENTS

One way to approach this problem would be to enumerate all possible assignments of work crew i to job j. Although this approach is possible for small problems, it rapidly becomes unmanageable as the size of the problem increases.

The number of possible assignments of n facilities to n jobs on a one-to-one basis is equal to $n(n - 1)(n - 2) \ldots 1$, or equivalently, by $n!$. Thus, the number of possible assignments in Table 8.21 is $5! = 120$. If only one additional work crew and job were added to the table, the possible number of assignments would increase to $6! = 720$. Quite obviously, enumerating all possible assignments is feasible for only very small problems. Consequently, it is necessary to develop an alternative solution technique.

8.6.2 LINEAR PROGRAMMING FORMULATION

One alternative to complete enumeration is to formulate the assignment problem as a linear programming problem. If c_{ij} is defined as the cost of assigning facility i to job j and x_{ij} is defined as the proportion of time that facility i is assigned to job j, the linear programming problem is

$$\text{Minimize:} \quad Z = \sum_{i=1}^{n} \sum_{j=1}^{n} c_{ij}x_{ij} \tag{8.6}$$

$$\text{Subject to:} \quad \sum_{i=1}^{n} x_{ij} = 1 \quad \text{for } j = 1, 2, \ldots, n \tag{8.7}$$

$$\sum_{j=1}^{n} x_{ij} = 1 \quad \text{for } i = 1, 2, \ldots, n \tag{8.8}$$

$$x_{ij} \geq 0$$

The first set of constraints is necessary to assure that each of the j jobs is assigned. The second set of constraints is required to make certain that exactly 100 percent of the time available to a facility is accounted for. For instance, the second set of constraints eliminates the possibility of assigning all jobs to the most efficient facility and no jobs to the other facilities. The final constraint eliminates the possibility of negative-valued variables.

An interesting result of the linear programming solution of the assignment problem is that the variables x_{ij} will have values of either zero or one. This occurs because of the fact that both the coefficients in the constraining equations and the right-hand-side values are equal to one. The implication of this result is that the optimal assignment always involves a one-to-one matching of facility to job. For instance, if $x_{12} = 1$, then facility 1 is assigned to job 2. On the other hand, if $x_{12} = 0$, then facility 1 is not assigned to job 2. Since the optimal solution can never include variables that have a fractional value, one facility will always be assigned to one and only one job. Conversely, one job will always be assigned to one and only one facility.

8.7 The Assignment Algorithm

The assignment problem can be solved by writing the problem as a linear programming problem and using the simplex algorithm. Fortunately, however, an algorithm that eliminates much of the computational burden required by the simplex algorithm has been developed for the assignment problem.

The algorithm is based on two facts. First, each facility must be assigned to one of the jobs. Second, the relative cost of assigning facility i to job j is not changed by the subtraction of a constant from either a column or a row of the cost matrix. To illustrate, consider the first row in Table 8.21. Since work crew 1 must be assigned to one of the five jobs, and since the relative costs of the assignment are not changed by the subtraction of a constant from each element in the row, we can subtract the minimum element in the row from all other elements in the row. Similarly, the relative costs of assigning work crew 2 to each of the five jobs is not changed by subtracting the minimum element in the second row from all other elements in the second row. The same principle is true for all rows of the assignment table. Table 8.22 gives the *reduced*

Table 8.22

Job

		1	2	3	4	5	Number Subtracted
	1	16	47	14	27	0	25
Work	2	0	7	27	43	59	22
Crew	3	0	12	33	24	13	27
	4	8	13	11	15	0	37
	5	3	14	19	0	4	26

cost matrix obtained by subtracting the minimum element in each row from all other elements in the row.

In some instances an optimal assignment is possible from the first reduced matrix. This occurs when an assignment can be made such that the total reduced cost of the assignment is zero. Since this is not the case in Table 8.22. we must further reduce the cost matrix.

A second reduced cost matrix can be obtained by subtracting the minimum element in each column from all other elements in the column. The rationale for this step is analogous to that used to obtain the first reduced matrix. Since each job must be assigned to one of the work crews, the relative cost of the assignment is not changed by the subtraction of a constant from all elements in the job column. The second reduced cost matrix, obtained by subtracting the minimum element in each column from all other elements in the column, is shown in Table 8.23.

An optimal assignment is possible from Table 8.23. The assignment is work crew 1 to job 5, 2 to 2, 3 to 1, 4 to 3, and 5 to 4. The cost of this assignment is found by summing the cost of the assignments in the original cost matrix, Table 8.21, and is $155.

It is not always possible to obtain an optimal assignment from the second

Table 8.23

Job

	j					
	i	1	2	3	4	5
	1	16	40	3	27	0
Work	2	0	0	16	43	59
Crew	3	0	5	22	24	13
	4	8	6	0	15	0
	5	3	7	8	0	4
	Number Subtracted	0	7	11	0	0

reduced cost matrix. To illustrate this case, consider the assignment problem given by Table 8.24. This problem is identical to the original problem with

Table 8.24

Job

	j					
	i	1	2	3	4	5
	1	41	72	39	52	25
Work	2	22	29	49	65	81
Crew	3	27	39	60	51	40
	4	45	50	48	26	37
	5	29	40	45	52	30

the exception that the final two elements in the fourth column have been interchanged. The reduced cost matrices are obtained in the manner described and are shown in Tables 8.25 and 8.26.

Table 8.25

Job

	j						Number Subtracted
	i	1	2	3	4	5	
	1	16	47	14	27	0	25
Work	2	0	7	27	43	59	22
Crew	3	0	12	33	24	13	27
	4	19	24	22	0	11	26
	5	0	11	16	23	1	29

Table 8.26

Job

i \ j	1	2	3	4	5
1	16	40	0	27	0
2	0	0	13	43	59
3	0	5	19	24	13
4	19	17	8	0	11
5	0	4	2	23	1
Number Subtracted	0	7	14	0	0

Work Crew labels rows 1–5 on the left.

An optimal assignment exists if the total reduced cost of the assignment is zero. Since the total reduced cost is not zero, an optimal assignment is not present. Consequently, the cost matrix must be reduced even further.

Table 8.26 can be reduced by use of a very simple technique. The technique involves drawing straight lines, either horizontally or vertically, that cover all zeroes in the reduced cost matrix. The zeros must be covered with as few lines as possible. If the minimum number of lines necessary to cover all zeros equals the number of assignments that must be made, an optimal solution has already been found. If the number of lines is less than the number of assignments, an additional reduction is necessary.

Four lines are required to cover the zeros in the reduced cost matrix of Table 8.26. Since five assignments must be made (i.e., the matrix has dimensions of 5 by 5), an additional reduction is necessary. The lines used to cover the zeros are shown in Table 8.27.

Table 8.27

Job

i \ j	1	2	3	4	5
1	16	40	0	27	0
2	0	0	13	43	59
3	0	5	19	24	13
4	19	17	8	0	11
5	0	4	2	23	1

Work Crew labels rows 1–5 on the left.

The reduction of Table 8.27 is made by first determining the smallest element not covered by a line. This number is subtracted from each uncovered element in the matrix and is added to those elements covered by two lines. Since the smallest uncovered number in Table 8.27 is 1, each uncovered number is reduced by 1 and each number covered by two lines is increased by 1. The resulting reduced cost matrix is given in Table 8.28.

Table 8.28

Job

	j / *i*	*1*	*2*	*3*	*4*	*5*
	1	17	40	0	27	0
	2	1	0	13	43	59
Work Crew	3	0	4	18	23	12
	4	20	17	8	0	11
	5	0	3	1	22	0

An assignment with a total reduced cost of zero is possible in Table 8.28. The assignment is work crew 1 to job 3, 2 to 2, 3 to 1, 4 to 4, and 5 to 5. The cost of this assignment is found from the original cost matrix, Table 8.24, and is $151.

The logic underlying the reduction of Table 8.27 to obtain Table 8.28 can be explained as follows. An optimal assignment is not possible in Table 8.27. Therefore, an additional reduction is necessary. This reduction is made by subtracting the smallest nonzero element from all elements in the matrix. Subtracting 1 from each number in the reduced cost matrix, Table 8.27, gives the cost matrix shown by Table 8.29.

Table 8.29

Job

	j / *i*	*1*	*2*	*3*	*4*	*5*
	1	15	39	−1	26	−1
	2	−1	−1	12	42	58
Work Crew	3	−1	4	18	23	12
	4	18	16	7	−1	10
	5	−1	3	1	22	0

Table 8.29 contains negative values. Since the objective is to obtain an assignment with reduced cost of zero, the negative numbers must be elimi-

nated. This can be done by adding a constant to the appropriate rows and/or columns. Adding 1 to each element in the first column gives the matrix shown in Table 8.30.

Table 8.30

Job

	i \ *j*	*1*	*2*	*3*	*4*	*5*
	1	16	39	−1	26	−1
Work	2	0	−1	12	42	58
Crew	3	0	4	18	23	12
	4	19	16	7	−1	10
	5	0	3	1	22	0

The negative numbers in the first, second, and fourth rows of Table 8.30 can be eliminated by adding 1 to each element in those rows. The resulting reduced cost matrix is identical to the matrix shown by Table 8.28.

The assignment algorithm can be summarized by a series of steps. These steps are:

1. Subtract the minimum element of each row in the assignment from all elements of the row. This gives the first reduced cost matrix.
2. Subtract the minimum element in each column of the first reduced matrix from all elements in the column. This gives the second reduced cost matrix.
3. Determine if an assignment with a total reduced cost of zero is possible from the second reduced matrix. If so, this assignment is optimal. If not, proceed to step 4.
4. Draw horizontal and vertical lines to cover all zeros. Use as small a number of lines as possible. Subtract the smallest element not covered by a line from all the elements not covered and add this element to all elements lying at the intersection of two lines.
5. Determine if an optimal assignment is possible. If not, repeat steps 4 and 5 until an optimal assignment is found.

The assignment algorithm, as described, applies only to minimization problems. In order to solve a maximization problem, the assignment matrix must be converted to a new matrix whose elements have reversed magnitudes.

The easiest way of reversing the magnitudes of the elements in the matrix is to subtract each element in the matrix from the largest element of the matrix. This has the effect of converting the maximization problem to a minimization problem. The algorithm can then be applied to the matrix with reversed magnitudes.

We have indicated that the assignment matrix must be square; i.e., the number of facilities must equal the number of jobs. There are many "real world" problems, however, in which the number of facilities is greater than the number of jobs or vice versa. These problems can be solved by adding a "dummy" row or column and applying the assignment algorithm to the modified square matrix. For instance, if there had been five jobs and six work crews in the problem discussed earlier in this section, we could have modified the problem to make the matrix square by adding a sixth column representing "idleness" (i.e., job 6). The c_{ij} entries in the column would be either zeros or some measure of the cost of the idle work crew. Conversely, had there been six jobs and five work crews, an additional row representing an imaginary work crew would have been necessary. In these cases, the optimal solution would include either an imaginary job (idle work) or an imaginary work crew.

Example: A management consulting firm has a backlog of four contracts. Work on these contracts must be started immediately. Three project leaders are available for assignment to the contracts. Because of varying work experience of the project leaders, the profit to the consulting firm will vary based on the assignment as shown below. The unassigned contract can be completed by subcontracting the work to an outside consultant. The profit on the subcontract is zero.

		Contract			
		1	*2*	*3*	*4*
	A	13	10	9	11
Project	B	15	17	13	20
Leader	C	6	8	11	7
	sub.	0	0	0	0

In order to determine the optimal assignment for this maximization problem, the magnitude of the profit matrix must be reversed. This is done by subtracting each element in the profit matrix from the largest element of the matrix. The resulting matrix is shown below.

Contract

		1	2	3	4
	A	7	10	11	9
Project	B	5	3	7	0
Leader	C	14	12	9	13
	sub.	20	20	20	20

The optimal solution is found by applying the assignment algorithm to the reversed magnitude matrix. The reader should verify that the optimal solution is project leader A to contract 1, B to 4, C to 3, and contract 2 to the outside consultant.

Example: In designing a production facility, it is important to locate the work centers so as to minimize the materials handling cost. In a specific example, three work centers are required to manufacture, assemble, and package a product. Four locations are available within the plant. The materials handling cost at each location for the work centers is given by the following cost matrix. Determine the location of work centers that minimizes total materials handling cost.

Location

		1	2	3	4
	Man.	18	15	16	13
Job	Ass'bly	16	11	X	15
	Pkg.	9	10	12	8

To apply the assignment algorithm, two modifications to the cost matrix are required. First, the matrix shows that assembly cannot be performed in location 3. Therefore, the cost of assembly at location 3 is represented by M. Second, a dummy job must be added to the matrix. The cost of this job at each location is zero. The modified cost matrix is given by the following table.

Location

		1	2	3	4
	Man.	18	15	16	13
Job	Ass'bly	16	11	M	15
	Pkg.	9	10	12	8
	Dummy	0	0	0	0

The optimal assignment is manufacturing at location 4, assembly at location 2, and packaging at location 1. Location 3 is not assigned to any of the three jobs; i.e., location 3 is assigned to the dummy row.

PROBLEMS

1. A manufacturer of inboard motor boats has three assembly plants where different models of the boats are made. The engines for the boats are purchased from a vendor and shipped to the assembly plants from the vendor's two manufacturing facilities. The cost of the engines, with the exception of shipping charges, is the same at each of the vendor's two manufacturing facilities. The supply of engines along with the number of boats scheduled for assembly during the scheduling period is shown below. The shipping costs from each engine plant to the three assembly plants are also shown. The manufacturer wishes to develop a shipping schedule that minimizes the total shipping cost while meeting the assembly requirements.

Engine Plant	Cost of Shipping 1	2	3	Engines Available
1	$40	$30	$20	500
2	$15	$25	$35	500
Boats scheduled for assembly	300	300	400	

(a) Establish the northwest corner solution.
(b) Determine the optimal solution, using the stepping stone algorithm.

2. A farm implement company has manufacturing facilities in Tulsa, Phoenix, and Portland, Oregon. Each of these facilities has the capability to manufacture the four major products made by the company. However, because of differences in equipment and plant design, the cost of manufacturing the four products differs at the three plants. These costs are shown in the table below.

The capacity of the plants also differs. The capacity of each plant and the requirements for each product are shown below. Both capacity and requirements have been expressed in terms of standard units. (This means that capacity and requirements are directly comparable.)

Plant	Product A	B	C	D	Capacity (std. units)
Tulsa	$300	$200	$500	$200	75
Phoenix	200	100	200	400	120
Portland	200	300	400	300	105
Requirements (std. units)	65	60	80	95	300

(a) Establish the Northwest Corner solution.

(b) Determine the optimal solution using the stepping stone algorithm.

3. The Baxter Glass Company produces disposable glass containers that are purchased by five soft-drink bottlers. The bottles are sold at a fixed delivered price of $0.25 per case. Orders for the current scheduling period have been received from the five bottlers and are as follows: 30,000 cases from bottler 1; 30,000 cases from bottler 2; 100,000 cases from bottler 3; 50,000 cases from bottler 4; and 40,000 cases from bottler 5.

The bottles are made at three plants. The monthly production capacities of the plants are 75,000 cases at plant 1, 100,000 cases at plant 2, and 125,000 cases at plant 3. The direct costs of production are $0.10 at plant 1, $0.09 at plant 2, and $0.08 at plant 3. The transportation costs of shipping a case from a plant to a bottler are shown below.

Plant	Bottler 1	2	3	4	5
1	$0.04	$0.06	$0.10	$0.13	$0.13
2	0.07	0.04	0.09	0.10	0.11
3	0.09	0.08	0.09	0.08	0.12

(a) Formulate the problem as a balanced transportation problem.

(b) Use the minimum cell method to determine the initial solution.

(c) Determine the optimal solution, using the stepping stone algorithm.

4. The Econo Car Rental Company has a shortage of rental cars in certain cities and an oversupply of cars in other cities. The imbalances are shown in the following table.

City	Cars Short	Cars Excess
Albany	—	15
Boston	—	20
Chicago	28	—
Cleveland	—	15
Dallas	—	20
Detroit	—	30
Kansas City	13	—
Miami	15	—
Philadelphia	24	—

The costs of transporting a car from the cities with an excess to those cities with a shortage are shown below.

	Transportation Cost ($) per Car			
From To:	Chicago	Kansas City	Miami	Philadelphia
Albany	$140	$210	$240	$ 50
Boston	150	220	230	60
Cleveland	70	190	200	80
Dallas	160	90	180	130
Detroit	60	110	190	90

The company wants to correct the imbalances at the minimum cost.

(a) Formulate the problem as a balanced transportation problem.

(b) Use the minimum cell method to determine the initial solution.

(c) Determine the optimal solution, using the stepping stone algorithm.

5. The Superior Oil Company has three oil refineries. These refineries are currently operating at less than maximum capacity. Superior has the opportunity to sell certain oil products to a competing oil company. Since the competing firm has alternative sources of supply, the sale will not change Superior's current market position. The potential profit from from the sale of the products is shown in the following table.

	Product				
Plant	Gasoline	Kerosene	Diesel	Jet Fuel	Asphalt
1	$0.165	—	$0.140	$0.125	$0.128
2	0.140	$0.146	0.126	—	0.133
3	—	0.139	0.134	0.125	0.130

As indicated in the table, the profit on a product differs at the several plants. In addition, certain products are not available at all plants. This is indicated by the dashes. Superior has excess capacity of 70,000 barrels at plant 1, 90,000 barrels at plant 2, and 40,000 barrels at plant 3. They have been requested to supply part or all of the following products: 60,000 barrels of gasoline, 15,000 barrels of kerosene, 40,000 barrels of diesel fuel, 25,000 barrels of jet fuel, and 20,000 barrels of asphalt. Management wants to schedule production of these products at the three plants so as to maximize profits.

(a) Formulate the problem as a balanced transportation problem.

(b) Use the penalty method to obtain the initial solution and verify that the initial solution is optimal.

6. The Quapaw Company operates three rock quarries that are located in central Oklahoma. Because of the lack of suitable concrete rock in that portion of the state, this company will be the principal supplier of concrete rock for a new turnpike currently under construction.

The concrete rock must be quarried and delivered to six concrete ready-mix plants that are situated along the road site. The quantity of rock that must be delivered to each of these plants is shown below.

	Concrete Plant Site					
	1	2	3	4	5	6
Rock required (cu. yd.)	120,000	80,000	160,000	100,000	80,000	100,000

The cost of the delivered rock is dependent on both the cost of quarrying the rock and the cost of trucking the rock to the ready-mix plant sites. The cost of quarrying the rock is related to factors such as the size of the quarry, the amount of overburden that must be removed, the amount of washing required, etc. These costs, per cubic yard of rock, are $2.40 at quarry 1, $3.20 at quarry 2, and $2.85 at quarry 3. The trucking cost per cubic yard is given in the following table.

From	To:	1	2	3	4	5	6
1		$1.80	$1.60	$0.80	$1.30	$1.40	$2.40
2		1.20	0.90	0.70	1.20	1.30	1.50
3		1.70	1.60	1.30	1.00	1.00	1.50

Because of differences in both equipment and rock, the capacities of the three quarries differ. The capacity of quarry 1 is 300,000 cubic yards, the capacity of quarry 2 is 180,000 cubic yards, and the capacity of quarry 3 is 240,000 cubic yards. The objective is to minimize the cost of the delivered rock.

(a) Formulate the problem as a balanced transportation problem.

(b) Use the penalty method to obtain the initial solution.

(c) Determine the optimal solution, using the modified distribution algorithm.

7. The Graham Electron Company has received a contract from Transcontinental Airlines, Inc., to provide metal detection devices. These devices are used by Transcontinental in the screening of passengers for both domestic and international flights. The contract specifies delivery of the metal detection devices according to the following schedule.

	Quarter			
	First	Second	Third	Fourth
Number	20	30	40	40

Graham has sufficient production capacity to meet the above schedule. However, because of contracts with other customers, the number of metal detection devices that can be manufactured in the next four quarters varies. In addition, Graham expects the cost of manufacture to increase. Maximum production, the unit cost of production, and the unit storage cost are given in the following table.

Quarter	Production	Unit Cost of Production	Unit Storage Cost per Quarter
1	30	$14,000	$1000
2	45	16,000	1000
3	40	15,000	1000
4	25	17,000	1000

No storage costs are incurred for devices that are delivered in the same quarter as produced. The objective is to schedule production and delivery so as to minimize cost.

(a) Formulate the problem as a balanced transportation problem.

(b) Verify that the initial solution found by using the minimum cell method is optimal.

8. Lacy's Department Store must place orders for five styles of their very popular hand-carved wooden statues. Lacy's has been able to find only three manufacturers of these particular statues. The number of statues available from each manufacturer is shown below.

Manufacturer	No. of Statues
1	200
2	300
3	450

All statues, regardless of style, are sold at a retail price of $225. The cost of the statues varies, however, from one manufacturer to another. These costs are shown in the following table.

Manufacturer	Style				
	1	*2*	*3*	*4*	*5*
1	$200	$190	$140	$210	$160
2	150	200	130	190	160
3	180	150	180	200	250

The potential sales of the statues exceeds the total supply. The potential sales are: style 1, 150; style 2, 200; style 3, 350; style 4, 200; style 5, 300. Since the price of all five styles of statues is the same, Lacy's objective is to minimize cost.

(a) Formulate the problem as a balanced transportation problem.

(b) Determine the initial solution, using the penalty method.

(c) Find the optimal solution, using the modified distribution algorithm.

9. A large oil company has oil reserves in four different locations. This oil must be transported from these locations to five oil refineries. The transportation costs, supplies, and requirements are shown in the following balanced transportation table.

	Transportation Cost (per barrel) Refinery					Maximum Yearly Supply (thousands of barrels)
Reserve	1	2	3	4	5	
1	$6.50	$2.50	$5.00	$3.00	$5.00	1100
2	4.00	0.50	3.00	3.00	3.50	920
3	0.50	6.00	2.00	3.50	3.50	620
4	5.50	8.00	3.50	5.00	1.00	1020
Yearly demand (thousands of barrels)	840	400	1000	600	820	3660

Use the modified distribution algorithm to determine the shipments from the reserves to the refineries that minimize total shipping cost.

10. Five girls in a secretarial pool must be assigned to five different jobs. From past records, the time that each girl takes to do each job is known. These times, in hours, are shown in the following table.

Girl	Job 1	2	3	4	5
1	3	7	8	4	6
2	4	8	7	5	6
3	5	7	9	5	7
4	4	6	8	6	8
5	6	9	10	7	9

Assuming that each girl can be assigned to only one job, determine the optimal assignment.

11. A television repair shop employs six technicians. These technicians must be assigned to five different repair jobs. Because of their different specialties and levels of skill, the time required for each technician to complete each job varies. The times required are shown in the following table. All entries are in minutes.

Technician	Job 1	2	3	4	5
1	180	160	240	300	150
2	190	150	250	320	170
3	150	170	230	340	140
4	170	200	210	310	190
5	220	190	260	330	160
6	190	180	300	310	210

The objective is to minimize the total repair time. The unassigned technician will remain idle. Formulate the problem as an assignment problem and determine the optimal solution.

12. A major aerospace company has received a government contract for the production of an advanced fighter/bomber. As is usually the case, a large number of the systems required for the aircraft must be obtained from subcontractors. To illustrate, the avionics have been divided into six different packages and bids have been requested for each package. Eight major avionics manufacturers have submitted bids on the different packages. These bids are shown in the following table. A dash indicates that the avionics manufacturer declined to bid on an individual package. All dollar figures are in millions of dollars.

Manufacturer	1	2	Avionics Package 3	4	5	6
1	$4.2	$6.8	$1.4	$3.6	$9.4	$6.2
2	4.4	—	1.6	3.5	9.5	6.4
3	4.3	7.0	1.5	3.5	9.6	6.1
4	5.0	7.0	1.4	3.7	10.0	6.0
5	5.0	6.9	1.6	3.6	9.5	5.9
6	3.9	6.7	1.7	—	9.6	6.1
7	4.1	6.8	1.5	3.5	9.4	6.1
8	4.2	6.9	1.7	3.5	—	6.0

The scheduling is such that only one avionics package can be subcontracted to any single manufacturer. The aerospace company's objective is, of course, to minimize the cost of procuring the entire avionics system.
(a) Formulate the problem as an assignment problem.
(b) Determine the optimal solution.

13. KECT, the local educational television station, must schedule their weekend television programming. In addition to KECT, there are three network affiliate stations in the local viewing area. The programming for these three network affiliates has already been announced.

On the basis of past experience, KECT realizes that they can attract only a small portion of the total viewing public. Nevertheless, their objective is to maximize their total program exposure.

The station has eight, one-hour programs that have been selected for the weekend prime-time viewing hours (i.e., 7:00-11:00 p.m. on Saturdays and Sundays). The number of viewers they can expect depends on the popularity of the network shows scheduled during these hours as well as the popularity of KECT's shows. Management's estimates of the number of viewers each show can attract during the prime-time viewing hours are given in the following table. All entries in the table are in thousands of viewers.

Day & Time		1	2	3	Show 4	5	6	7	8
Sat.	7:00– 8:00	18	12	25	16	30	5	15	8
	8:00– 9:00	14	14	30	18	32	6	18	6
	9:00–10:00	12	15	22	18	36	9	17	4
	10:00–11:00	10	20	16	14	40	13	16	4
Sun.	7:00– 8:00	20	11	20	15	28	7	14	7
	8:00– 9:00	16	12	26	17	34	11	17	5
	9:00–10:00	14	14	20	25	40	10	16	3
	10:00–11:00	12	16	14	14	30	6	13	2

As an illustration of the table, management expects show 1 to have 18,000 viewers if scheduled on Saturdays from 7:00–8:00 p.m. The number of viewers would drop to 14,000, however, if the show were scheduled from 8:00 to 9:00 p.m. on the same day.

(a) Formulate the problem as an assignment problem.

(b) Determine the optimal solution.

14. Nash, Smith, and Co., Certified Public Accountants, provides auditing and tax services for a large number of firms in Northern California. Mr. Nash, the managing senior partner of the firm, must assign jobs to the staff (junior) and senior members of the firm.

The accounting firm currently has six staff accountants and three senior accountants available for assignments. The major differences in the two classifications is in experience, the staff accountants normally having less than three years' experience and the senior accountants three to six years' experience.

The firm currently has a backlog of five auditing jobs and three tax jobs that must be assigned. Mr. Nash has estimated the time required by each accountant to complete each job. These entries are shown in the following table in units of days. A dash indicates that the accountant is not qualified for the job.

Staff Acc't†	Audit					Tax		
	1	2	3	4	5	1	2	3
Mr. Hill	10	—	5	21	13	—	—	—
Mr. Nutter	12	20	7	18	15	—	15	—
Mr. Hamilton	10	18	6	17	16	18	13	18
Mr. Pryor	8	—	8	19	15	—	19	22
Mr. Pratt	12	22	6	19	13	16	14	20
Mr. Wagner	14	21	7	20	16	—	16	19
Senior Acc't								
Mr. Redman	7	15	5	16	12	10	12	15
Mr. Lochen	8	17	5	15	11	12	14	17
Mr. Savich	8	14	5	15	11	13	14	17

† Entries are in 8-hour days.

The billing rate is $15 per hour for staff accountants and $20 per hour for senior accountants. The objective is to minimize the total billings to the customers.

(a) Formulate the problem as an assignment problem.
(b) Determine the optimal solution.

SUGGESTED REFERENCES

The references for this chapter are listed in Chapter 5.

Chapter 9

Integer
Programming

One of the assumptions of the linear programming problem is that the variables are continuous, i.e., that a variable can assume fractional as well as integer values. This assumption is unrealistic for certain types of problems. Consequently, we must introduce an algorithm for solving *integer linear programming* problems. The integer linear programming problem has the same form as the linear programming problem, the exception being that the values of the variables are constrained to be *discrete* or *integer* nonnegative numbers.

There are many ingenious uses of integer programming. Some of these are illustrated by examples in the following section. Unfortunately, the solution algorithms currently available for solving integer programming problems usually require much more computational effort than the simplex algorithm. For this reason, only the most elementary integer programming problems can be solved without the aid of an electronic computer. Moreover, the current state of the art is such that even with a computer a solution cannot always be guaranteed for certain large problems.

Several solution algorithms are available for solving integer programming problems. These include the cutting plane algorithm, the implicit enumeration algorithm, and the branch and bound algorithm.† Although each of these

† These algorithms are discussed in the very readable text by Donald Plane and Claude McMillan, *Discrete Optimization*, Prentice-Hall, Inc., 1971.

algorithms is important in integer programming, an explanation of all three algorithms is beyond the scope of an introductory text. We shall instead concentrate on one of the more widely used algorithms, the branch and bound algorithm. This algorithm is discussed in Sec. 9.2.

In spite of the difficulties of solving certain large-scale integer programming problems, the integer programming model is one of the most useful of the many operations research techniques. The interested reader is urged to study this model further.

9.1 Applications of Integer Programming

A number of types of problems can be formulated as integer programming problems. A sampling of these problems is provided in this section.

THE KNAPSACK PROBLEM. Our first example of an integer programming problem is the "knapsack" problem. A typical example of the knapsack problem involves a salesman who stocks a counter display in a retail store. The counter display is of limited area, and the salesman wishes to include the most profitable items in the display. There are eight items to choose from. These are shown in Table 9.1, along with the expected profit contribution from each item and the areas required by the item.

Table 9.1

Item Number	Name	Profit	Area (sq ft)
1	Movie camera	$40	4
2	Projector	50	7
3	Portable radio	6	1
4	Portable TV	60	8
5	Binoculars	15	1
6	Telescope	30	2
7	Cassette stereo	20	1
8	Opera glasses	10	1

There are several constraints that must be taken into account by the salesman. First, the display must take up no more than 15 square feet. Second the movie camera cannot be displayed without the projector and vice versa. Third, at least one item of optical equipment must be included. Fourth, the cassette stereo cannot be included without the portable TV. The portable TV can be included, however, without the stereo.

Since this is a display counter, the salesman would not include more than one of each item in the display. The variables are, therefore, $x_j = 0$ or 1, for

$j = 1, 2, 3, \ldots, 8$, where $x_j = 0$ represents the jth item excluded from the display and $x_j = 1$ represents the item included in the display.†

The objective is to maximize the profit contribution. The objective function is

$$\text{Maximize:} \quad Z = 40x_1 + 50x_2 + 6x_3 + 60x_4$$
$$+ 15x_5 + 30x_6 + 20x_7 + 10x_8$$

The area constraint is

$$4x_1 + 7x_2 + 1x_3 + 8x_4 + 1x_5 + 2x_6 + 1x_7 + 1x_8 \leq 15$$

The constraint that the movie camera cannot be displayed without the projector and vice versa is

$$x_1 - x_2 = 0$$

The fact that at least one item of optical equipment must be included is described by the constraint

$$x_5 + x_6 + x_8 \geq 1$$

Since the cassette stereo cannot be included without the portable TV, we must include the constraint

$$x_4 \geq x_7 \quad \text{or} \quad x_4 - x_7 \geq 0$$

A final constraint is that the variables have values of 0 or 1. Since the integer programming algorithms, like the simplex algorithm, limits the solution to nonnegative valued variables, this constraint is included by

$$x_j \leq 1 \quad \text{and} \quad \text{integer}$$

and

$$x_j \geq 0 \quad \text{for } j = 1, 2, 3, \ldots, 8$$

CAPITAL EQUIPMENT PROBLEM. A second example of integer programming involves the selection of capital equipment. In this example we assume that a firm has a number of alternative choices for plant modernization and expansion. The alternative choices, together with the expenditures required for the projects and the net present value of the profits expected from the investments, are shown in Table 9.2. ‡

The total expenditures that can be undertaken by the firm during the current planning horizons are limited to $30 million. Management believes that additional plant facilities are required; consequently, either project 1, 3, or 4 must be included. However, no more than one of these three projects can be undertaken. Project 2, numerical control, is available only with the new plant.

---

† Because the variables x_j can have values of only zero or one, this integer programming problem is referred to as a *zero-one* programming problem.

‡ Net present value is the criterion most often used for investments.

Table 9.2

Project Number	Description	Expenditures Required (millions)	Net Present Value of Profits (millions)
1	Build new plant	$13	$8
2	Numerical control for new plant	3	2
3	Modernize existing plant	8	3
4	Expand current plant	5	2
5	Computerize accounting system	7	4
6	Expand product line	11	6

The objective of the firm is to maximize the net present value of profits. If we let x_j represent the jth project, the objective function is

Maximize: $Z = 8x_1 + 2x_2 + 3x_3 + 2x_4 + 4x_5 + 6x_6$

The budget constraint is

$$13x_1 + 3x_2 + 8x_3 + 5x_4 + 7x_5 + 11x_6 \leq 30$$

The constraint requiring additional plant facilities is

$$x_1 + x_3 + x_4 = 1$$

Numerical control is available only with the new plant; therefore,

$$x_1 - x_2 \geq 0$$

Quite obviously, no more than one of any individual project would be undertaken.† Therefore,

$$x_j \leq 1 \quad \text{and} \quad \text{integer}$$

and

$$x_j \geq 0 \quad \text{for } j = 1, 2, 3, \ldots, 6$$

PLANT LOCATION—SETUP COST. One of the significant applications of integer programming involves the incorporation of setup costs in the selection of plant locations, product development, advertising programs, etc. As an illustration, consider the problem of a power company in selecting from among alternative locations for power plants. The company has forecast power requirements for the western United States. From this forecast, it is clear that additional power generating plants will be required. From a survey of available sites, m sites have been selected as possible locations for power plants. The area served by the power company has been divided into n geographical regions, each representing a source of demand.

The potential supply of power from each of the m possible locations is largely dependent on the geographical and environmental characteristics of the location. The potential supply of power, in thousands of kilowatt hours,

---

† This is another example of a zero-one integer programming problem.

is represented by S_i, for each of the $i = 1, 2, 3, \ldots, m$ sites. The yearly demand for power in each of the n geographical regions is forecast to be D_j, for $j = 1, 2, 3, \ldots, n$.

Three separate types of costs must be considered in selecting the sites. These are construction cost, power generation and transmission cost, and power line cost. The fixed cost of constructing a new plant or enlarging an existing plant at location i, expressed as a yearly charge, is given by F_i. The power generation and transmission cost from location i to region j is dependent on the cost of generating power at location i and the distance from location i to region j. This cost, per thousand kilowatt hours, is represented by C_{ij}. The cost of constructing and maintaining power lines from location i to region j, again expressed as a yearly charge, is represented by L_{ij}.

The amount of power transmitted from location i to region j is given by x_{ij}. The objective of the power company is to minimize the overall yearly cost of supplying the required power.

To formulate this problem as an integer programming problem, we must define two additional variables. These variables are

$$y_i = \begin{cases} 1, & \text{if location } i \text{ is selected} \\ 0, & \text{otherwise} \end{cases}$$

and

$$z_{ij} = \begin{cases} 1, & \text{if power lines from location } i \text{ to region } j \\ & \text{are required} \\ 0, & \text{otherwise} \end{cases}$$

for all i and j.

The objective function involves minimizing the fixed cost, power generation and transportation cost, and the cost of constructing and maintaining power lines. The objective, therefore, is to

$$\text{Minimize:} \quad C = \sum_{i=1}^{m} F_i y_i + \sum_{i=1}^{m} \sum_{i=1}^{n} C_{ij} x_{ij} + \sum_{i=1}^{m} \sum_{i=1}^{n} L_{ij} z_{ij}$$

The purpose of y_i and z_{ij} is to include the fixed costs and power line costs for facilities required in the problem.

The amount of power transmitted from location i to region j must not exceed the total supply available at location i. This constraint is expressed as

$$\sum_{j=1}^{n} x_{ij} \leq S_i y_i \qquad \text{for } i = 1, 2, 3, \ldots, m$$

If $y_i = 1$ in the constraint, we are assured that the total power transmitted from location i does not exceed the supply available from that location. If $y_i = 0$, location i is not selected as a power site, and the constraint guarantees that power is not transmitted from that location.

The problem also requires that demand at each of the j regions be satisfied. This is included by the constraint

$$\sum_{i=1}^{m} x_{ij} \geq D_j \qquad \text{for } j = 1, 2, 3, \ldots, m$$

It is necessary to insure that the cost of constructing and maintaining the necessary power lines is included in the objective function. This can be done by adding a constraint that forces z_{ij} to equal 1 if power lines are needed to transmit power from location i to region j. Without such a constraint, the minimum value of the objective function would occur for all z_{ij} equal 0.

This constraint is

$$x_{ij} \leq U z_{ij} \qquad \text{for all } i \text{ and } j$$

The constant U is an arbitrarily large number. Any value of U is acceptable as long as U equals or exceeds the largest possible value of x_{ij}.

To illustrate the constraint, assume that power lines are required from location 1 to region 2. If this is the case, $z_{12} = 1$ and the cost of constructing and maintaining the power lines, L_{12}, is included in the objective function. Since U equals or exceeds the largest possible value of x_{ij}, the constraint $x_{12} \leq U$ is satisfied.

To illustrate the opposite case, assume that the optimal solution does not include transmission of power from location 1 to region 2. In this case, $z_{12} = 0$ and the cost of constructing and maintaining power lines from location 1 to region 2, $L_{12}z_{12} = 0$, does not add to the value of the objective function. Moreover, since there are no power lines between location 1 and region 2, x_{12} must equal zero. This is guaranteed by the constraint $x_{12} \leq U z_{12}$, or, alternatively, $x_{12} \leq 0$.

The permissible values of y_i and z_{ij} are 0 and 1. Consequently, we must include the constraints

$$y_i, z_{ij} \leq 1 \text{ and integer} \qquad \text{for all } i \text{ and } j$$

Nonnegative constraints are also required. Thus,

$$x_{ij}, y_i, z_{ij} \geq 0 \qquad \text{for all } i \text{ and } j$$

The solution to this problem, obtained by using an integer programming algorithm, will give the values of x_{ij}, y_i, and z_{ij} that lead to minimum construction, operation, and maintenance costs.†

---

† In this problem the variable x_{ij} need not have integer values. Problems of this type are termed *mixed-integer* programming problems.

9.2 The Branch and Bound Algorithm

The branch and bound algorithm is based on the fact that the value of the objective function for the optimal integer solution to a linear programming maximization problem can never be greater than the optimal noninteger solution. Conversely, the value of the objective function for the optimal integer solution to a linear programming minimization problem can never be less than the optimal noninteger solution. The optimal solution found by ignoring the integer requirements, therefore, provides an upper or lower bound on the value of the objective function for the integer solution. This important relationship between integer and noninteger solutions is illustrated by Fig. 9.1. The optimal noninteger solution to the linear programming

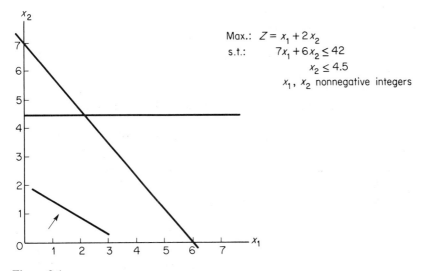

Max.: $Z = x_1 + 2x_2$
s.t.: $7x_1 + 6x_2 \leq 42$
             $x_2 \leq 4.5$
          x_1, x_2 nonnegative integers

Figure 9.1

problem is $x_1 = 2\frac{1}{7}$, $x_2 = 4\frac{1}{2}$ and $Z = 11\frac{1}{7}$. The objective function for the optimal integer solution to the linear programming problem can, therefore, not exceed $Z = 11\frac{1}{7}$.

The optimal integer solution to the linear programming problem in Fig. 9.1 can be found by the addition of constraints. The constraints are added to force the noninteger variables to integers. This process is termed *branching*. The resulting linear programming problems are referred to as *descendants* of the original problem.

Branching begins by selecting one variable with a noninteger value x_j from the optimal noninteger solution to the original problem. Two branches are required for the noninteger variable x_j. One branch represents all descendant problems in which $x_j \leq y_j$, where y_j is the largest integer that is less than the optimal noninteger solution value of x_j. The other branch represents the descendants in which $x_j \geq y_j + 1$. As stated above, these constraints are included in all subsequent descendant problems.

To illustrate what is meant by branching, consider the variables with noninteger values in the optimal solution to the linear programming problem in Fig. 9.1. The value of one noninteger variable is $x_1 = 2\frac{1}{7}$. The largest integer that is less than the optimal noninteger solution value of x_1 is $y_1 = 2$. Two branches are formed, based on the value of this variable. One includes all descendants for which $x_1 \leq 2$, the other the descendants for $x_1 \geq 3$. The branches are shown in Fig. 9.2 as nodes 2 and 3. Note that all branches stemming from node 2 include the constraint $x_1 \leq 2$. Similarly, all branches from node 3 include the constraint $x_1 \geq 3$.

The solution to the linear programming problem at node 2 is $x_1 = 2$, $x_2 = 4\frac{1}{2}$, and $Z_2 = 11$. Since the solution value of x_2 is noninteger, branches from node 2 are required. These are shown in Fig. 9.2 as nodes 4 and 5. As before, the purpose of these branches is to exclude the possibility of noninteger values of the variables, i.e., to integerize the noninteger variable. Therefore, one branch adds the constraint $x_2 \leq 4$; the other, the constraint $x_2 \geq 5$.

The solution at node 4 is $x_1 = 2$, $x_2 = 4$, and $Z_4 = 10$. This solution was found by resolving the original linear programming problem with the branch constraints. Since the solution is integer, no additional descendants from node 4 are required.

The branch from node 2 to node 5 added the constraint $x_2 \geq 5$. This constraint is inconsistent with the original constraint $x_2 \leq 4.5$. Consequently, no feasible solution exists at node 5, and the branching terminates at this node.

The branch from node 1 to node 3 added the constraint $x_1 \geq 3$. The solution to the linear programming problem at node 3 is $x_1 = 3$, $x_2 = 3\frac{1}{2}$, and $Z_3 = 10$. Even though the solution value of x_2 is noninteger, no additional descendants from node 3 are necessary. As explained earlier, the value of the objective function for the integer solution will never exceed the value of the objective function for the noninteger solution. The value of the objective function for the noninteger solution at node 3 is $Z_3 = 10$. The value of the objective function at any nodes descending from node 3 can, therefore, not exceed $Z_3 = 10$. Since we have found an integer solution at node 4 with an objective function value of $Z_4 = 10$ and since the value of the objective function for the descendants from node 3 cannot exceed $Z_4 = 10$, descendants from node 3 are unnecessary. The branch and bound algorithm thus terminates with the optimal integer solution $x_1 = 2$, $x_2 = 4$, and $Z = 10$.

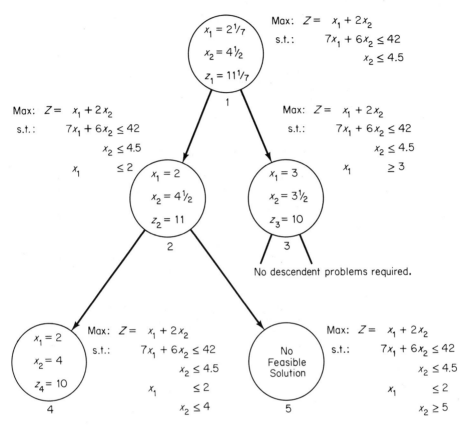

Figure 9.2

A flow diagram for the branch and bound algorithm is shown in Fig. 9.3. In studying this diagram and the accompanying example, it is important to remember that the algorithm is based on the fact that an integer solution at any descendant node can never be better than the noninteger solution at the preceding node. This one fact eliminates much of the branching that would otherwise be required to obtain integer solutions at all descendant nodes.

The following example illustrates the flow diagram given by Fig. 9.3.

Example: Solve the following integer programming problem.

$$\text{Maximize:} \quad Z = 2x_1 + 3x_2$$
$$\text{Subject to:} \quad 5x_1 + 7x_2 \leq 35$$
$$4x_1 + 9x_2 \leq 36$$
$$x_1, x_2 \text{ nonnegative integers}$$

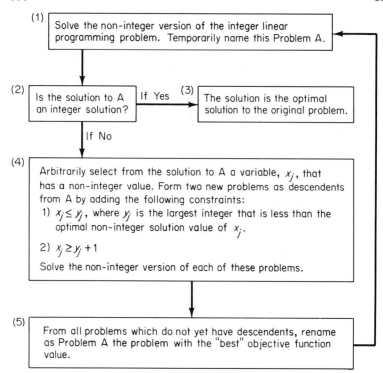

(1) Solve the non-integer version of the integer linear programming problem. Temporarily name this Problem A.

(2) Is the solution to A an integer solution?   If Yes   (3) The solution is the optimal solution to the original problem.

If No

(4) Arbitrarily select from the solution to A a variable, x_j, that has a non-integer value. Form two new problems as descendents from A by adding the following constraints:

1) $x_j \leq y_j$, where y_j is the largest integer that is less than the optimal non-integer solution value of x_j.

2) $x_j \geq y_j + 1$

Solve the non-integer version of each of these problems.

(5) From all problems which do not yet have descendents, rename as Problem A the problem with the "best" objective function value.

Figure 9.3

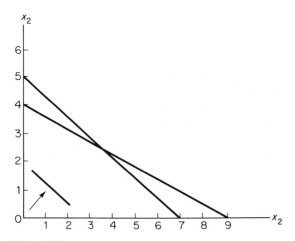

Figure 9.4

The graphical formulation of the noninteger linear programming problem is shown in Fig. 9.4. The optimal noninteger solution is $x_1 = 3\frac{12}{17}$, $x_2 = 2\frac{6}{17}$, and $Z = 14\frac{8}{17}$. This problem temporarily becomes Problem A and is shown at node 1 in Fig. 9.5.

The solution to the original linear programming problem is noninteger. We therefore create two descendants of Problem A as described in Box 5 of the

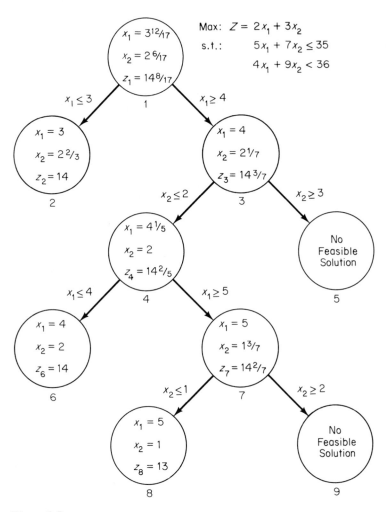

Figure 9.5

flow diagram in Fig. 9.3. These descendants are represented by nodes 2 and 3 in Fig. 9.5.

From the two descendants, we rename as Problem A the problem with the best objective function. Since the value of the objective function at node 3 exceeds that at node 2, the problem at node 3 is renamed as Problem A. As shown in the flow diagram, we now return to Box 2.

The solution to the new Problem A is noninteger. Constraints are added for the noninteger variable, as described in Box 4. These descendants are shown by nodes 4 and 5 in Fig. 9.5.

Following the instructions in Box 5 of the flow diagram, we rename the descendant with the best objective function value as Problem A. The candidates are the problems at nodes 2, 4, and 5. The problem at node 5 has no feasible solution and thus cannot be selected. Comparing the solutions at nodes 2 and 4, we see that the value of the objective function at node 4 is the largest, and the problem at node 4 is thus renamed as Problem A.

Since the solution at node 4 is noninteger, descendants of this problem must be created. These are shown by nodes 6 and 7 in Fig. 9.5. The solution at node 6 is integer. It does not, however, have the best objective function value. The problem at node 7 is, therefore, renamed as Problem A. The descendants of the problem at node 7 are shown at nodes 8 and 9.

From all problems without descendants in Fig. 9.5, the problem with the best objective function value is renamed as Problem A. The problems at nodes 2, 5, 6, 8, and 9 have no descendants. The problems at nodes 5 and 9 have no feasible solutions and are, therefore, eliminated from consideration. Of the remaining problems, those at nodes 2 and 6 have the largest objective function value. Since the problem at node 6 has an integer solution, it is renamed as Problem A. Returning to Box 3 of the flow diagram, we see that this solution is optimal. The solution is $x_1 = 4$, $x_2 = 2$, and $Z = 14$.

PROBLEMS

1. A company plans to introduce six new products during the next five years. Using the subscript i ($i = 1, 2, \ldots, 6$) to represent the product, the subscript j ($j = 1, 2, \ldots, 5$) to represent the year, and the variable $x_{ij} = 0$ or 1 to represent not introducing ($x_{ij} = 0$) or introducing ($x_{ij} = 1$) product i in year j, show how each of the following stipulations can be described by a linear integer constraint. Consider each of the following parts separately.
   (a) Product 1 must be introduced no later than year 3.
   (b) No more than two products can be introduced in any single year.

(c) Product 2 cannot be introduced before product 1. They can, however, be introduced during the same year.

(d) Products 4 and 5 must be introduced the same year.

(e) At least one of the products must be introduced in year 1.

2. The Unified Power and Light Company must add additional electrical power plants in order to meet the anticipated increase in demand for electrical power. Five types of electrical power plants are available. The relevant information on each type of plant is given in the following table.

Plant Type	Annual Capacity (millions of megawatt hr)	Peak Capacity (megawatts)	Initial Cost (millions of $)	T.D.I. & O.C.† (millions of $)
Nuclear	4.7	600	$ 70	$ 90
Coal	3.1	500	45	80
Hydroelectric	6.8	1100	180	130
Hydroelectric	5.5	750	120	100
Hydroelectric	2.5	300	60	60

† Total discounted investment and operating cost.

The combination of plants selected must have peak capacity of not less than 3800 megawatts and annual capacity of not less than 22 million megawatt-hours. Because of political and environmental considerations, the electrical system cannot include more than two nuclear generators. In addition, there is only one location suitable for a large hydroelectric plant (i.e., a plant with annual capacity of over 5 million megawatt-hours). There are, however, numerous sites for the smaller hydroelectric plants. The authorized budget limitation for the expansion program is $450 million. The objective is to minimize the total discounted investment and operating cost. Formulate the problem as an integer programming problem.

3. Cherokee Airlines, Inc., is a small charter airline that provides charter air service and aircraft rentals to local individuals and companies. The company has the problem of updating its current fleet of aircraft.

Cherokee uses three types of aircraft in its charter and rental business. These are (1) single engine, retractables (S.E.R.), (2) light twins (L.T.), and (3) medium twins (M.T.). Revenue from the single-engine and light twin engine aircraft comes mainly from rentals to businessmen/pilots. The medium twins are used primarily for charters.

The president of Cherokee has investigated the current production aircraft and has decided that the aircraft shown in the following table best suit the requirements of the company.

Manufacturer	Model	Type	Expected Rental (hr per yr)	Expected Profit ($ per hr)	Purchase Price
Beech	Bonanza	S.E.	500	$10	$ 60,000
Cessna	Centurion	S.E.	500	9	55,000
Piper	Comanche	S.E.	520	8	50,000
Beech	Baron	L.T.	400	14	110,000
Piper	Aztec	L.T.	390	12	100,000
Beech	Queenaire	M.T.	350	40	250,000
Cessna	401	M.T.	340	36	200,000
Piper	Navaho	M.T.	360	30	160,000

The objective is to select the aircraft mix that maximizes the total profit. At least one single-engine aircraft must be included in the new fleet. No more than two medium twin-engine aircraft should be included. Because of insurance requirements, no more than one model of light twin can be included (e.g., two Barons or two Aztecs could be included, but not one of each). The maximum expenditure for the total fleet is limited to $1,000,000. Formulate the problem as an integer programming problem.

4. Save-King Grocery Stores, Inc., is a rapidly growing chain of "discount" supermarkets. Save-King currently has stores in the eastern and New England states. However, their long-range plans include expanding into certain western states.

   One of the important factors in the planned expansion is the location of their western warehouse facilities. Two factors must be considered in locating the warehouses: (1) the transportation cost from the warehouses to the supermarkets; and (2) the fixed cost of building and operating a warehouse. Quite obviously, there is a trade-off between these two costs. A relatively small number of warehouses have the effect of reducing the fixed costs of building and operating the warehouses. However, the transportation costs under this plan are relatively large. Conversely, a large number of warehouses means an increase in the fixed and operating cost but a reduction in the transportation costs. The objective is to minimize the total distribution system cost.

   Based on detailed studies of the expansion program, the following information is available. There are m possible warehouse sites that must supply n supermarkets. The maximum yearly capacity of each of the m possible warehouses is known and is represented by S_i for $i = 1, 2, \ldots, m$, where S_i is in units of square feet of warehouse space. Similarly, the yearly requirements of each supermarket for warehouse space is also known and is represented by D_j, for $j = 1, 2, \ldots, n$, where D_j is in units of square feet of warehouse space. The transportation cost from warehouse i to supermarket j is c_{ij}, where c_{ij} is in units of dollars per square foot. (Assume that loading, unloading, and handling costs are included in c_{ij}.) The total cost of building and operating a warehouse, expressed on a yearly basis, is given by F_i for $i = 1, 2, \ldots, m$. Formulate the problem as a mixed integer programming problem. (*Hint:* Let x_{ij} represent the amount of warehouse i area allocated to supermarket j and let $y_i = 0, 1$ represent nonselection or selection of a warehouse site.)

5. The Baxter Glass Company, referred to in Problem 3, Chapter 8, has the opportunity to purchase an additional bottling plant. This plant has the capacity to produce 100,000 cases during the scheduling period. The direct cost of production is $0.08 per case. The transportation costs from the new plant to each of Baxter's five customers are given in the following table.

Bottler				
1	*2*	*3*	*4*	*5*
$0.08	$0.09	$0.07	$0.11	$0.09

Baxter estimates of the fixed cost of operating the three current plants and the new plant are $12,000 at Plant 1, $10,000 at Plant 2, $13,000 at Plant 3, and $14,000 at the new plant. By closing a plant, the fixed cost of the plant can be avoided. Formulate the problem as an integer programming problem.

6. Use the branch and bound algorithm to solve the following integer programming problems.

(a)   Maximize: $Z = 3x_1 + 4x_2$
          Subject to: $3x_1 + 2x_2 \leq 8$
                          $x_1 + 4x_2 \leq 10$
                              $x_1, x_2 \geq 0$ and integer

(b)   Maximize: $Z = x_1 + x_2$
          Subject to: $3x_1 + 2x_2 \leq 24$
                          $2x_1 + 3x_2 \leq 28$
                        $-x_1 + 2x_2 \leq 10$
                              $x_1, x_2 \geq 0$ and integer

(c)   Maximize: $Z = 2x_1 + 3x_2$
          Subject to: $2x_1 + x_2 \leq 22$
                          $x_1 + 3x_2 \leq 25$
                              $x_1, x_2 \geq 0$ and integer

(d)   Minimize: $Z = 4x_1 + 3x_2$
          Subject to: $x_1 \geq 4$
                               $x_2 \geq 2$
                  $3x_1 + 2x_2 \geq 20$
                             $x_1, x_2 \geq 0$ and integer

SUGGESTED REFERENCES

ABADIE, J., *Integer and Nonlinear Programming* (New York, N.Y.: American Elsevier Publishing Co., Inc., 1970).

HADLEY, G., Nonlinear and Dynamic Programming (Reading, Mass.: Addison-Wesley Publishing Company, Inc., 1964).

HILLIER, FREDERICK and GERALD LIEBERMAN, *Introduction to Operations Research* (San Francisco, Ca.: Holden-Day, Inc., 1967).

PLANE, DONALD and CLAUDE MCMILLAN, *Discrete Optimization* (Englewood Cliffs, N.J.: Prentice-Hall, Inc., 1971).

WAGNER, HARVEY M., *Principles of Operations Research* (Englewood Cliffs, N.J.: Prentice-Hall, Inc., 1969).

Selected Answers

to

Odd-Numbered Questions

CHAPTER 1

1. No. The individuals that will vote in an upcoming election are not known with certainty.
3. Yes. It is not necessary to have a listing of all family units with incomes of less than $6000 to describe this group as a set.
5. $\{a, b, c\}$, $\{a, b\}$, $\{a, c\}$, $\{b, c\}$, $\{a\}$, $\{b\}$, $\{c\}$, $\{\ \ \}$
7. 64
9. (a) Q', (b) ϕ, (c) U
11. 60
13. (a) 20, (b) 280
17. 12
19. 630

CHAPTER 2

1. (a) 0, (c) 2, (e) -0.5, (g) -3
3. (a) $f(x) = 3000 + \frac{2}{3}x$,   (c) $f(x) = 3500 + 10x$
   (e) $p = 11.50 - 0.001q$ (g) $y = 10,000 + 500t$
5. $A \cap B = \{4\}$, $A \cup B = \{-5, -4, -3, -2, -1, 0, 1, 2, 3, 4, 5\}$
7. Domain: $x = 0, 3, 4, 5$; Range: $f(x) = 6, 7, 8, 10$
9. (a) $2x + 3y \leq 80$ for $x, y \geq 0$
   (c) $20,000x + 60,000y + 80,000z \leq 1,200,000$ for x, y, z nonnegative integers
11. $f(x) = -45 + 9x$ for $5 \leq x \leq 12$; Range: $0 \leq f(x) \leq 63$; Breakeven: 5 hours
13. $f(x) = -10,000,000 + 6000x - 0.5x^2$
15. $f(x) = -16 + 5.5x - 0.25x^2$
17.
$$f(x) = \begin{cases} 100 & \text{for } 0 \leq x \leq 1000 \\ 125 - 0.025x & \text{for } 1000 < x \leq 3000 \\ 20 + 0.01x & \text{for } 3000 < x \leq 4000 \end{cases}$$
19. $x_A = 6$, $x_B = 3$
21. An infinite number of solutions.
23. (a) Consistent, unique solution $x = 7$, $y = 5$
    (c) Inconsistent

CHAPTER 3

1. (a) $A - B = (-4, -2, 7)$,
   (c) The product of two row vectors is not defined.
3. (a) $A + B = \begin{pmatrix} 14 \\ 2 \\ 8 \end{pmatrix}$,   (c) $3A - 2B = \begin{pmatrix} 2 \\ 16 \\ -6 \end{pmatrix}$
5. $A(B + C) = AB + AC = 192$
7. (a) $A + B = \begin{pmatrix} 5 & 0 \\ 5 & 12 \\ -4 & 9 \end{pmatrix}$ (c) $3A - B = \begin{pmatrix} 7 & 20 \\ 3 & 12 \\ 9 & 7 \end{pmatrix}$
9. $A(B + C) = AB + AC = \begin{pmatrix} 72 & 84 \\ -12 & -60 \\ 60 & 96 \end{pmatrix}$

11. $VW = \begin{pmatrix} 10 & 16 & 6 \\ -10 & -16 & -6 \\ 20 & 32 & 12 \end{pmatrix}$

13. $AI = IA = \begin{pmatrix} 6 & 3 & 8 \\ 4 & 2 & 5 \\ 3 & 1 & 7 \end{pmatrix}$

15. (a) $x_1 = 4$, $x_2 = 6$, (c) $x_1 = 3$, $x_2 = 5$, $x_3 = 4$

17. (a) $A^{-1} = \begin{pmatrix} 3.5 & -1.5 \\ -2 & 1 \end{pmatrix}$, (c) $A^{-1} = \begin{pmatrix} 2 & -3 & 1 \\ 1 & 3 & -2 \\ -1 & -1 & 1 \end{pmatrix}$

19. $(I - A)^{-1}D = \begin{pmatrix} 124.32 \\ 143.30 \\ 159.76 \\ 159.19 \end{pmatrix}$

CHAPTER 4

1. (a) $A^t = \begin{pmatrix} 3 & 4 \\ 1 & 3 \end{pmatrix}$, (c) $A^t = \begin{pmatrix} 2 & 3 \\ 1 & -2 \\ -3 & 4 \end{pmatrix}$

3. $AB = \begin{pmatrix} 30 & 54 \\ 45 & 78 \\ 31 & 83 \\ 40 & 87 \end{pmatrix}$

5. minor $a_{11} = 9$, minor $a_{12} = -17$, minor $a_{13} = 29$
   minor $a_{21} = -21$, minor $a_{22} = -34$, minor $a_{23} = 6$
   minor $a_{31} = 26$, minor $a_{32} = 0$, minor $a_{33} = -39$

7. $|A| = 221$

9. (a) $|A| = 16$, (b) $|A| = 191$

11. adj $A = (\text{cof } A)^t = \begin{pmatrix} 9 & 21 & 26 \\ 17 & -34 & 0 \\ 29 & -6 & -39 \end{pmatrix}$

13. (a) $A^{-1} = \begin{pmatrix} \frac{1}{2} & \frac{1}{4} & -1 \\ 0 & -\frac{1}{2} & 1 \\ -\frac{1}{2} & -\frac{1}{4} & 2 \end{pmatrix}$,

(c) $A^{-1} = -\frac{1}{64}\begin{pmatrix} 16 & -16 & 0 \\ -12 & -4 & 16 \\ -6 & 14 & -24 \end{pmatrix}$

15. (a) $x_1 = 6$, $x_2 = -4$, $x_3 = 2$ (c) $x_1 = 40$, $x_2 = 42$, $x_3 = -3$

17. (a) $r = 2$, (c) $r = 2$, (e) $r = 1$

CHAPTER 5

1. (a) $x_1 = 6$, $x_2 = 7$, $Z = 72$
   (c) $x_1 = 12.27$, $x_2 = 3.83$, $C = 88.16$
3. $x_1 = 4.4$, $x_2 = 5.5$, Max $Z = 9.9$, Min $-Z = -9.9$
5. $x_1 = 461.5$, $x_2 = 615.4$, Profit $= \$16,923$
7. $x_1 = 200$, $x_2 = 600$, Profit $= \$520$
9. $x_1 = 5$ oz., $x_2 = 1$ oz., Cost $= \$0.19$
11. Let x_j = number of units manufactured, for $j = 1, 2, 3$

    Maximize: $Z = 0.15x_1 + 0.12x_2 + 0.09x_3$

    Subject to:
    $$0.03x_1 + 0.02x_2 + 0.03x_3 \leq 1000$$
    $$0.11x_1 + 0.14x_2 + 0.20x_3 \leq 4000$$
    $$0.30x_1 + 0.20x_2 + 0.26x_3 \leq 9000$$
    $$0.08x_1 + 0.07x_2 + 0.08x_3 \leq 3000$$
    $$x_1 \geq 9000, x_2 \geq 9000, x_3 \geq 6000$$

13. Let x_{ijk} represent the amount of grain i in feed j sold at price k, for $i = 1$ (Wheat) and 2 (Barley), $j = 1$ (Fertilex) and 2 (Multiplex), and $k = 1$ (regular price) and 2 (discount price).

    Maximize: $Z = 1.10x_{111} + 0.94x_{112} + 1.23x_{211} + 1.07x_{212}$
    $$+ 0.86x_{121} + 0.69x_{122} + 0.99x_{221} + 0.82x_{222}$$

    Subject to:
    $$x_{111} + x_{112} + x_{121} + x_{122} \leq 1000$$
    $$x_{211} + x_{212} + x_{221} + x_{222} \leq 1200$$
    $$x_{111} + x_{211} \leq 99$$
    $$x_{121} + x_{221} \leq 99$$
    $$x_{111} \geq 2x_{211}$$
    $$x_{112} \geq 2x_{212}$$
    $$x_{221} \geq 2x_{121}$$
    $$x_{222} \geq 2x_{122}$$
    $$x_{ijk} \geq 0$$

15. Let x_{ij} represent the amount of blending component i in brand j for $i = 1, 2$ and $j = 1, 2, 3$.

    Maximize: $Z = 0.23x_{11} + 0.28x_{12} + 0.33x_{13} + 0.13x_{21} + 0.18x_{22} + 0.23x_{23}$

    Subject to:
    $$x_{11} + x_{12} + x_{13} \leq 4000$$
    $$x_{21} + x_{22} + x_{23} \leq 2000$$
    $$10x_{11} + 50x_{21} \geq 20(x_{11} + x_{21})$$
    $$10x_{12} + 50x_{22} \geq 30(x_{12} + x_{22})$$
    $$10x_{13} + 50x_{23} \geq 40(x_{13} + x_{23})$$
    $$x_{ij} \geq 0$$

17. Let x_{ij} represent job i scheduled during period j, for $i = 1, 2$ and $j = 1, 2$.
    Assumptions: 1) Jobs scheduled during the day can be completed during the day or night.
    2) Jobs scheduled during the night are constrained by the night capacity and the overflow of day jobs.

Maximize: $Z = 275(x_{11} + x_{12}) + 125(x_{21} + x_{22}) + 225(x_{31} + x_{32})$

Subject to:
$$1200(x_{11} + x_{12}) + 1400(x_{21} + x_{22}) + 800(x_{31} + x_{32}) \leq 13,400$$
$$20(x_{11} + x_{12}) + 15(x_{21} + x_{22}) + 35(x_{31} + x_{32}) \leq 400$$
$$100(x_{11} + x_{12}) + 60(x_{21} + x_{22}) + 80(x_{31} + x_{32}) \leq 1050$$
$$1200x_{12} + 1400x_{22} + 800x_{32} \leq 9200$$
$$20x_{12} + 15x_{22} + 35x_{32} \leq 250$$
$$100x_{12} + 60x_{22} + 80x_{32} \leq 650$$
$$x_{ij} \geq 0$$

19. Let x_j represent the number of advertisements in media j, for $j = 1, 2, 3$.

Media	Effectiveness Coefficient
1	$0.80(0.4) + 0.70(0.2) + 0.15(0.4) = 0.52$
2	$0.70(0.4) + 0.80(0.2) + 0.20(0.4) = 0.52$
3	$0.20(0.4) + 0.60(0.2) + 0.40(0.4) = 0.36$

Maximize: $Z = 0.52(600,000)x_1 + 0.52(800,000)x_2 + 0.36(300,000)x_3$

Subject to:
$$600x_1 + 800x_2 + 450x_3 \leq 20,000$$
$$x_1 \leq 12, x_2 \leq 24, x_3 \leq 12$$
$$x_1 \geq 3, x_2 \geq 6, x_3 \geq 2$$

CHAPTER 6

1. (a) Multiple optimal solutions: $x_1 = 8, x_2 = 0, x_3 = 0, x_4 = 28, Z = 24$;
   and $x_1 = 5.45, x_2 = 3.82, x_3 = 0, x_4 = 0, Z = 24$
   (c) $x_1 = 2.8, x_2 = 4.4, x_3 = 0, x_4 = 0, x_5 = 0, Z = 54.8$

3. (a) $x_1 = 13.75, x_2 = 15, x_3 = 20, x_4 = 0, x_5 = 0, x_6 = 0, Z = 525$

5. (a) $x_1 = 20, x_2 = 6, x_3 = 96, x_4 = 0, Z = 152$

7. (a) No feasible solution,   (b) Unbounded solution,
   (c) No feasible solution,   (d) Unbounded solution

9. 533.3 type 1200 lamps, 133.3 type 1201 lamps, profit = \$893.3. The marginal value of assembly labor is \$8.67 and the marginal value of wiring labor is \$1.67.

11. 61.5 Toots, 149.25 Wheets, 0 Honks, Profit \$109.76.

13. 99 pounds of Fertilex at regular price, 301 pounds of Fertilex at discount price, 99 pounds of Multiplex at regular price, 1701 pounds of Multiplex at discount price. Total profit is $2116.67.

15. Total production is 5000 quarts of Brand S and 1000 quarts of Brand SSS. Profit is $1280.

17. 1.05 payroll jobs scheduled during the day, 1.45 payroll jobs scheduled during the night, 3.68 inventory jobs scheduled during the day, 6.32 inventory jobs scheduled during the night. Total profit is $2937.50.

CHAPTER 7

1. (a)   Maximize: $\quad P = 234w_1 + 214w_2$
                Subject to: $\quad\quad\quad 13w_1 + 16w_2 \leq 2$
$$18w_1 + 14w_2 \leq 3$$
$$w_1, w_2 \geq 0$$

(c)   Maximize: $\quad P = 42w_1 + 60w_2 + 63w_3$
                Subject to: $\quad\quad 12w_1 + 5w_2 + 3w_3 \leq 2$
$$7w_1 + 12w_2 + 21w_3 \leq 1$$
$$w_1, w_2, w_3 \geq 0$$

3. (a) $x_1 = 20$, $x_2 = 6$, $x_3 = 0$, $x_4 = 96$, $C = 1520$
      $w_1 = 14.5$, $w_2 = 0$, $w_3 = 18.2$, $w_4 = 0$, $w_5 = 0$, $P = 1520$

5. (a) $x_1 = 0$, $x_2 = 25.33$, $x_3 = 0$, $x_4 = 20.67$, $C = 152$

7. $x_1 = 0$, $x_2 = 8$, $x_3 = 9$, $x_4 = 5$, $C = \$786$

9. $x_1 = 4.8$, $x_2 = 12.6$, $x_3 = 0$, $x_4 = 4.8$, $x_5 = 0$, $C = \$0.444$

CHAPTER 8

1. (a)

	A_1	A_2	A_3	Available
E_1	300	200		500
E_2		100	400	500
Scheduled	300	300	400	1000

(b)

	A_1	A_2	A_3	Available
E_1		100	400	500
E_2	300	200		500
Scheduled	300	300	400	1000

The total shipping cost is $20,500.

3. (a)

	1	2	3	4	5	6	Capacity
1	0.14	0.16	0.20	0.23	0.23	0	75
2	0.16	0.13	0.18	0.19	0.20	0	100
3	0.17	0.16	0.17	0.16	0.20	0	125
Req'd.	30	30	100	40	40	50	300

(b)

	1	2	3	4	5	6	Capacity
1	30		5		40		75
2		30	70				100
3			25	50		50	125
Req'd.	30	30	100	50	40	50	300

(c)

	1	2	3	4	5	6	Capacity
1	30					45	75
2		30	25		40	5	100
3			75	50			125
Req'd.	30	30	100	50	40	50	300

The total cost is $41,350.

5. (a)

Plant	Gas.	Kero.	Diesel	Jet	Asph.	Slack	Capacity
1	0.165	$-M$	0.140	0.125	0.138	0	70,000
2	0.140	0.146	0.126	$-M$	0.133	0	90,000
3	$-M$	0.139	0.134	0.134	0.130	0	40,000
Req'd.	60,000	15,000	40,000	25,000	20,000	40,000	200,000

(b)

Plant	Gas.	Kero.	Diesel	Jet	Asph.	Slack	Capacity
1	60,000		10,000				70,000
2		15,000	15,000		20,000	40,000	90,000
3			15,000	25,000			40,000
Req'd.	60,000	15,000	40,000	25,000	20,000	40,000	200,000

The initial solution is optimal. Profit is $24,400.

9.

	1	2	3	4	5	Supply
1		240	260	600		1100
2	220	160	540			920
3	620					620
4			200		820	1020
Req'd.	840	400	1000	600	820	3660

The total transportation cost is $6,310,000.

11.

	1	2	3	4	5	6
1	20	0	20	⓪	0	10
2	40	⓪	40	30	30	20
3	⓪	20	20	50	0	20
4	20	50	0	20	50	20
5	50	20	30	20	0	0
6	20	10	70	0	50	⓪

An optimal assignment is shown by the circled elements. The total time is 970 minutes.

13. (a) Since the problem involves maximizing the number of viewers, it is necessary to reverse the magnitudes of the entries. This reversed magnitude tableau is

	1	2	3	4	5	6	7	8
1	22	28	15	24	10	35	25	32
2	26	26	10	22	8	34	22	34
3	28	25	18	22	4	21	23	36
4	30	20	24	26	0	27	24	36
5	20	29	20	25	12	33	26	33
6	24	28	14	23	16	29	23	35
7	26	26	20	15	0	30	24	37
8	28	24	26	26	10	34	27	38

(b)

	1	2	3	4	5	6	7	8
1	3	2	3	0	1	8	1	⓪
2	9	2	⓪	0	1	9	0	4
3	15	5	12	4	1	⓪	5	10
4	20	3	21	11	⓪	9	9	13
5	⓪	2	7	0	2	5	1	0
6	6	3	3	0	8	3	⓪	4
7	16	9	17	⓪	0	12	9	4
8	11	⓪	16	4	3	9	5	7

CHAPTER 9

1. (a) $x_{11} + x_{12} + x_{13} = 1$

(c) $x_{11} \geq x_{12}$

$x_{11} + x_{12} \geq x_{21} + x_{22}$

$$x_{11} + x_{12} + x_{13} \geq x_{21} + x_{22} + x_{23}$$
$$x_{11} + x_{12} + x_{13} + x_{14} \geq x_{21} + x_{22} + x_{23} + x_{24}$$
$$x_{11} + x_{12} + x_{13} + x_{14} + x_{15} \geq x_{21} + x_{22} + x_{23} + x_{24} + x_{25}$$

(e) $x_{11} + x_{21} + x_{31} + x_{41} + x_{51} = 1$

3. Let: x_j for $j = 1, 2, \ldots, 8$ represent the number of each type of aircraft to be purchased for the fleet.

$$q = \begin{cases} 1 & \text{if Barons are purchased} \\ 0 & \text{if Aztecs are purchased} \end{cases}$$

Maximize: $\quad Z = 5000x_1 + 4500x_2 + 4160x_3 + 5600x_4$
$$+ 4680x_5 + 14{,}000x_6 + 12{,}240x_7 + 10{,}800x_8$$

Subject to: $\quad\quad\quad\quad x_1 + x_2 + x_3 \geq 1$
$$x_6 + x_7 + x_8 \leq 2$$
$$x_4 \leq Mq$$
$$x_5 \leq M(1 - q)$$
$$60{,}000x_1 + 55{,}000x_2 + 50{,}000x_3 + 110{,}000x_4 + 100{,}000x_5$$
$$+ 250{,}000x_6 + 200{,}000x_7 + 160{,}000x_8 \leq 1{,}000{,}000$$
$$z \leq 1$$
$$x_j, z \geq 0 \text{ and integer}$$

5. Let: x_{ij} represent the number of cases from plant i to bottler j, for $i = 1, 2, 3, 4$ and $j = 1, 2, 3, 4, 5$.

c_{ij} = direct cost of production at plant i plus the transportation cost from plant i to bottler j

$$y_i = \begin{cases} 1 & \text{if plant } i \text{ remains open} \\ 0 & \text{if plant } i \text{ is closed} \end{cases}$$

Minimize: $\quad Z = \sum_{i=1}^{4} \sum_{j=1}^{5} c_{ij}x_{ij} + 12{,}000y_1 + 10{,}000y_2$
$$+ 13{,}000y_3 + 14{,}000y_4$$

Subject to: $\quad \sum_{i=1}^{4} x_{i1} = 30{,}000; \quad \sum_{i=1}^{4} x_{i2} = 30{,}000; \quad \sum_{i=1}^{4} x_{i3} = 100{,}000;$

$$\sum_{i=1}^{4} x_{i4} = 50{,}000; \quad \sum_{i=1}^{4} x_{i5} = 40{,}000 \text{ (Demand Constraints)}$$

$$\sum_{j=1}^{5} x_{1j} \leq 75{,}000y_1; \quad \sum_{j=1}^{5} x_{2j} \leq 100{,}000y_2;$$

$$\sum_{j=1}^{5} x_{3j} \leq 125{,}000y_3; \quad \sum_{j=1}^{5} x_{4j} \leq 100{,}000y_4 \text{ (Capacity}$$
$$\text{Constraints)}$$

$$y_i \leq 1 \text{ and integer}$$
$$y_i, x_{ij} \geq 0$$

Note: x_{ij} would theoretically be restricted to integer values. Rounding of noninteger values of x_{ij} would, however, be acceptable in this type of problem.

6. (a) $x_1 = 1, x_2 = 2, z = 11$

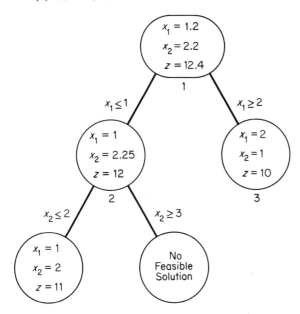

(b) $x_1 = 4, x_2 = 6, z = 10$

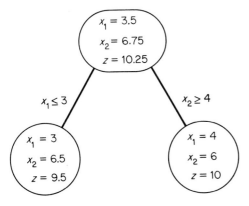

Index

353